Data Mining Applications for Empowering Knowledge Societies

Hakikur Rahman
Sustainable Development Networking Foundation (SDNF), Bangladesh

Information Science REFERENCE

INFORMATION SCIENCE REFERENCE

Hershey · New York

Director of Editorial Content:	Kristin Klinger
Managing Development Editor:	Kristin M. Roth
Assistant Managing Development Editor:	Jessica Thompson
Assistant Development Editor:	Deborah Yahnke
Senior Managing Editor:	Jennifer Neidig
Managing Editor:	Jamie Snavely
Assistant Managing Editor:	Carole Coulson
Copy Editor:	Erin Meyer
Typesetter:	Sean Woznicki
Cover Design:	Lisa Tosheff
Printed at:	Yurchak Printing Inc.

Published in the United States of America by
Information Science Reference (an imprint of IGI Global)
701 E. Chocolate Avenue, Suite 200
Hershey PA 17033
Tel: 717-533-8845
Fax: 717-533-8661
E-mail: cust@igi-global.com
Web site: http://www.igi-global.com

and in the United Kingdom by
Information Science Reference (an imprint of IGI Global)
3 Henrietta Street
Covent Garden
London WC2E 8LU
Tel: 44 20 7240 0856
Fax: 44 20 7379 0609
Web site: http://www.eurospanbookstore.com

Library of Congress Cataloging-in-Publication Data

Data mining applications for empowering knowledge societies / Hakikur Rahman, editor.
p. cm.

Summary: "This book presents an overview on the main issues of data mining, including its classification, regression, clustering, and ethical issues"--Provided by publisher.

Includes bibliographical references and index.

ISBN 978-1-59904-657-0 (hardcover) -- ISBN 978-1-59904-659-4 (ebook)
1. Data mining. 2. Knowledge management. I. Rahman, Hakikur, 1957-
QA76.9.D343D38226 2009
005.74--dc22

2008008466

British Cataloguing in Publication Data
A Cataloguing in Publication record for this book is available from the British Library.

All work contributed to this book set is original material. The views expressed in this book are those of the authors, but not necessarily of the publisher.

If a library purchased a print copy of this publication, please go to http://www.igi-global.com/agreement for information on activating the library's complimentary electronic access to this publication.

Table of Contents

Foreword xi
Preface xii
Acknowledgment xxii

Section I
Education and Research

Chapter I
Introduction to Data Mining Techniques via Multiple Criteria Optimization
Approaches and Applications 1
Yong Shi, University of the Chinese Academy of Sciences, China and University of Nebraska at Omaha, USA
Yi Peng, University of Nebraska at Omaha, USA
Gang Kou, University of Nebraska at Omaha, USA
Zhengxin Chen, University of Nebraska at Omaha, USA

Chapter II
Making Decisions with Data: Using Computational Intelligence Within a
Business Environment 26
Kevin Swingler, University of Stirling, Scotland
David Cairns, University of Stirling, Scotland

Chapter III
Data Mining Association Rules for Making Knowledgeable Decisions 43
A.V. Senthil Kumar, CMS College of Science and Commerce, India
R. S. D. Wahidabanu, Govt. College of Engineering, India

Section II
Tools, Techniques, Methods

Chapter IV
Image Mining: Detecting Deforestation Patterns Through Satellites 55
Marcelino Pereira dos Santos Silva, Rio Grande do Norte State University, Brazil
Gilberto Câmara, National Institute for Space Research, Brazil
Maria Isabel Sobral Escada, National Institute for Space Research, Brazil

Chapter V
Machine Learning and Web Mining: Methods and Applications in Societal Benefit Areas 76
Georgios Lappas, Technological Educational Institution of Western Macedonia, Kastoria Campus, Greece

Chapter VI
The Importance of Data Within Contemporary CRM 96
Diana Luck, London Metropolitan University, UK

Chapter VII
Mining Allocating Patterns in Investment Portfolios 110
Yanbo J. Wang, University of Liverpool, UK
Xinwei Zheng, University of Durham, UK
Frans Coenen, University of Liverpool, UK

Chapter VIII
Application of Data Mining Algorithms for Measuring Performance Impact of Social Development Activities 136
Hakikur Rahman, Sustainable Development Networking Foundation (SDNF), Bangladesh

Section III
Applications of Data Mining

Chapter IX
Prospects and Scopes of Data Mining Applications in Society Development Activities 162
Hakikur Rahman, Sustainable Development Networking Foundation, Bangladesh

Chapter X
Business Data Warehouse: The Case of Wal-Mart 189
Indranil Bose, The University of Hong Kong, Hong Kong
Lam Albert Kar Chun, The University of Hong Kong, Hong Kong
Leung Vivien Wai Yue, The University of Hong Kong, Hong Kong
Li Hoi Wan Ines, The University of Hong Kong, Hong Kong
Wong Oi Ling Helen, The University of Hong Kong, Hong Kong

Chapter XI
Medical Applications of Nanotechnology in the Research Literature 199
Ronald N. Kostoff, Office of Naval Research, USA
Raymond G. Koytcheff, Office of Naval Research, USA
Clifford G.Y. Lau, Institute for Defense Analyses, USA

Chapter XII
Early Warning System for SMEs as a Financial Risk Detector 221
Ali Serhan Koyuncugil, Capital Markets Board of Turkey, Turkey
Nermin Ozgulbas, Baskent University, Turkey

Chapter XIII
What Role is "Business Intelligence" Playing in Developing Countries?
A Picture of Brazilian Companies 241
Maira Petrini, Fundação Getulio Vargas, Brazil
Marlei Pozzebon, HEC Montreal, Canada

Chapter XIV
Building an Environmental GIS Knowledge Infrastructure 262
Inya Nlenanya, Center for Transportation Research and Education, Iowa State University, USA

Chapter XV
The Application of Data Mining for Drought Monitoring and Prediction 280
Tsegaye Tadesse, National Drought Mitigation Center, University of Nebraska, USA
Brian Wardlow, National Drought Mitigation Center, University of Nebraska, USA
Michael J. Hayes, National Drought Mitigation Center, University of Nebraska, USA

Compilation of References 292

About the Contributors 325

Index 330

Detailed Table of Contents

Foreword xi
Preface xii
Acknowledgment xxii

Section I
Education and Research

Chapter I
Introduction to Data Mining Techniques via Multiple Criteria Optimization
Approaches and Applications 1
Yong Shi, University of the Chinese Academy of Sciences, China
and University of Nebraska at Omaha, USA
Yi Peng, University of Nebraska at Omaha, USA
Gang Kou, University of Nebraska at Omaha, USA
Zhengxin Chen, University of Nebraska at Omaha, USA

This chapter presents an overview of a series of multiple criteria optimization-based data mining methods that utilize multiple criteria programming to solve various data mining problems and outlines some research challenges. At the same time, this chapter points out to several research opportunities for the data mining community.

Chapter II
Making Decisions with Data: Using Computational Intelligence Within a
Business Environment 26
Kevin Swingler, University of Stirling, Scotland
David Cairns, University of Stirling, Scotland

This chapter identifies important barriers to the successful application of computational intelligence techniques in a commercial environment and suggests a number of ways in which they may be overcome. It further identifies a few key conceptual, cultural, and technical barriers and describes different ways in which they affect business users and computational intelligence practitioners. This chapter aims to provide knowledgeable insight for its readers through outcome of a successful computational intelligence project.

Chapter III
Data Mining Association Rules for Making Knowledgeable Decisions 43
A. V. Senthil Kumar, CMS College of Science and Commerce, India
R. S. D. Wahidabanu, Govt. College of Engineering, India

This chapter describes two popular data mining techniques that are being used to explore frequent large itemsets in the database. The first one is called closed directed graph approach where the algorithm scans the database once making a count on possible 2-itemsets from which only the 2-itemsets with a minimum support are used to form the closed directed graph and explores possible frequent large itemsets in the database. In the second one, dynamic hashing algorithm where large 3-itemsets are generated at an earlier stage that reduces the size of the transaction database after trimming and thereby cost of later iterations will be reduced. However, this chapter envisages that these techniques may help researchers not only to understand about generating frequent large itemsets, but also finding association rules among transactions within relational databases, and make knowledgeable decisions.

Section II
Tools, Techniques, Methods

Chapter IV
Image Mining: Detecting Deforestation Patterns Through Satellites 55
Marcelino Pereira dos Santos Silva, Rio Grande do Norte State University, Brazil
Gilberto Câmara, National Institute for Space Research, Brazil
Maria Isabel Sobral Escada, National Institute for Space Research, Brazil

This chapter presents with relevant definitions on remote sensing and image mining domain, by referring to related work in this field and demonstrates the importance of appropriate tools and techniques to analyze satellite images and extract knowledge from this kind of data. A case study, the Amazonia with deforestation problem is being discussed, and effort has been made to develop strategy to deal with challenges involving Earth observation resources. The purpose is to present new approaches and research directions on remote sensing image mining, and demonstrates how to increase the analysis potential of such huge strategic data for the benefit of the researchers.

Chapter V
Machine Learning and Web Mining: Methods and Applications in Societal Benefit Areas 76
Georgios Lappas, Technological Educational Institution of Western Macedonia, Kastoria Campus, Greece

This chapter reviews contemporary researches on machine learning and Web mining methods that are related to areas of social benefit. It further demonstrates that machine learning and web mining methods may provide intelligent Web services of social interest. The chapter also discusses about the growing interest of researchers in recent days for using advanced computational methods, such as machine learning and Web mining, for better services to the public.

Chapter VI
The Importance of Data Within Contemporary CRM 96
Diana Luck, London Metropolitan University, UK

This chapter search for the importance of customer relationship management (CRM) in the product development and service elements as well as organizational structure and strategies, where data takes as the pivotal dimension around which the concept of CRM revolves in contemporary terms. Subsequently it has tried to demonstrate how these processes are associated with data management, namely: data collection, data collation, data storage and data mining, and are becoming essential components of CRM in both theoretical and practical aspects.

Chapter VII
Mining Allocating Patterns in Investment Portfolios 110
Yanbo J. Wang, University of Liverpool, UK
Xinwei Zheng, University of Durham, UK
Frans Coenen, University of Liverpool, UK

This chapter has introduced the concept of "one-sum" weighted association rules (WARs) and named such WARs as allocating patterns (ALPs). Here, an algorithm is being proposed to extract hidden and interesting ALPs from data. The chapter further points out that ALPs can be applied in portfolio management, and modeling a collection of investment portfolios as a one-sum weighted transaction-database, ALPs can be applied to guide future investment activities.

Chapter VIII
Application of Data Mining Algorithms for Measuring Performance Impact
of Social Development Activities 136
Hakikur Rahman, Sustainable Development Networking Foundation (SDNF), Bangladesh

This chapter focuses to data mining applications and their utilizations in devising performance-measuring tools for social development activities. It has provided justifications to include data mining algorithm for establishing specifically derived monitoring and evaluation tools that may be used for various social development applications. Specifically, this chapter gave in-depth analytical observations for establishing knowledge centers with a range of approaches and put forward a few research issues and challenges to transform the contemporary human society into a knowledge society.

Section III
Applications of Data Mining

Chapter IX
Prospects and Scopes of Data Mining Applications in Society Development Activities 162
Hakikur Rahman, Sustainable Development Networking Foundation, Bangladesh

Chapter IX focuses on a few areas of social development processes and put forwards hints on application of data mining tools, through which decision-making would be easier. Subsequently, it has put forward potential areas of society development initiatives, where data mining applications can be incorporated. The focus area may vary from basic social services, like education, health care, general commodities, tourism, and ecosystem management to advanced uses, like database tomography.

Chapter X
Business Data Warehouse: The Case of Wal-Mart 189
Indranil Bose, The University of Hong Kong, Hong Kong
Lam Albert Kar Chun, The University of Hong Kong, Hong Kong
Leung Vivien Wai Yue, The University of Hong Kong, Hong Kong
Li Hoi Wan Ines, The University of Hong Kong, Hong Kong
Wong Oi Ling Helen, The University of Hong Kong, Hong Kong

This chapter highlights on business data warehouse and discusses about the retailing giant Wal-Mart. Here, the planning and implementation of the Wal-Mart data warehouse is being described and its integration with the operational systems is being discussed. This chapter has also highlighted some of the problems that have been encountered during the development process of the data warehouse, and provided some future recommendations about Wal-Mart data warehouse.

Chapter XI
Medical Applications of Nanotechnology in the Research Literature 199
Ronald N. Kostoff, Office of Naval Research, USA
Raymond G. Koytcheff, Office of Naval Research, USA
Clifford G.Y. Lau, Institute for Defense Analyses, USA

Chapter XI examines medical applications literatures that are associated with nanoscience and nanotechnology research. For this research, authors have retrieved about 65000 nanotechnology records in 2005 from the Science Citation Index/ Social Science Citation Index (SCI/SSCI) using a comprehensive 300+ term query, and in this chapter they intend to facilitate the nanotechnology transition process by identifying the significant application areas. Specifically, it has identified the main nanotechnology health applications from today's vantage point, as well as the related science and infrastructure. The medical applications were ascertained through a fuzzy clustering process, and metrics were generated using text mining to extract technical intelligence for specific medical applications/ applications groups.

Chapter XII
Early Warning System for SMEs as a Financial Risk Detector 221
Ali Serhan Koyuncugil, Capital Markets Board of Turkey, Turkey
Nermin Ozgulbas, Baskent University, Turkey

This chapter introduces an early warning system for SMEs (SEWS) as a financial risk detector that is based on data mining. During the development of an early warning system, it compiled a system in which qualitative and quantitative data about the requirements of enterprises are taken into consideration. Moreover, an easy to understand, easy to interpret and easy to apply utilitarian model is targeted by discovering the implicit relationships between the data and the identification of effect level of every factor related to the system. This chapter eventually shows the way of empowering knowledge society from SME's point of view by designing an early warning system based on data mining.

Chapter XIII
What Role is "Business Intelligence" Playing in Developing Countries?
A Picture of Brazilian Companies 241
Maira Petrini, Fundação Getulio Vargas, Brazil
Marlei Pozzebon, HEC Montreal, Canada

Chapter XIII focuses at various business intelligence (BI) projects in developing countries, and specifically highlights on Brazilian BI projects. Within a broad enquiry about the role of BI playing in developing countries, two specific research questions were explored in this chapter. The first one tried to determine whether the approaches, models or frameworks are tailored for particularities and the contextually situated business strategy of each company, or if they are "standard" and imported from "developed" contexts. The second one tried to analyze what type of information is being considered for incorporation by BI systems; whether they are formal or informal in nature; whether they are gathered from internal or external sources; whether there is a trend that favors some areas, like finance or marketing, over others, or if there is a concern with maintaining multiple perspectives; who in the firms is using BI systems, and so forth.

Chapter XIV
Building an Environmental GIS Knowledge Infrastructure 262
Inya Nlenanya, Center for Transportation Research and Education,
Iowa State University, USA

In Chapter XIV, the author proposes a simple and accessible conceptual geographical information system (GIS) based knowledge discovery interface that can be used as a decision making tool. The chapter also addresses some issues that might make this knowledge infrastructure stimulate sustainable development, especially emphasizing sub-Saharan African region.

Chapter XV
The Application of Data Mining for Drought Monitoring and Prediction 280
Tsegaye Tadesse, National Drought Mitigation Center, University of Nebraska, USA
Brian Wardlow, National Drought Mitigation Center, University of Nebraska, USA
Michael J. Hayes, National Drought Mitigation Center, University of Nebraska, USA

Chapter XV discusses about the application of data mining to develop drought monitoring utilities, which enable monitoring and prediction of drought's impact on vegetation conditions. The chapter also summarizes current research using data mining approaches to build up various types of drought monitoring tools and explains how they are being integrated with decision support systems, specifically focusing drought monitoring and prediction in the United States.

Compilation of References 292

About the Contributors 325

Index 330

Foreword

Advances in information technology and data collection methods have led to the availability of larger data sets in government and commercial enterprises, and in a wide variety of scientific and engineering disciplines. Consequently, researchers and practitioners have an unprecedented opportunity to analyze this data in much more analytic ways and extract intelligent and useful information from it.

The traditional approach to data analysis for decision making has been shifted to merge business and scientific expertise with statistical modeling techniques in order to develop experimentally verified solutions for explicit problems. In recent years, a number of trends have emerged that have started to challenge this traditional approach. One trend is the increasing accessibility of large volumes of high-dimensional data, occupying database tables with many millions of rows and many thousands of columns. Another trend is the increasing dynamic demand for rapidly building and deploying data-driven analytics. A third trend is the increasing necessity to present analysis results to end-users in a form that can be readily understood and assimilated so that end-users can gain the insights they need to improve the decisions they make.

Data mining tools sweep through databases and identify previously hidden patterns in one step. An example of pattern discovery is the analysis of retail sales data to identify seemingly unrelated products that are often purchased together. Other pattern discovery problems include detecting fraudulent credit card transactions and identifying anomalous data that could represent data entry keying errors. Data mining algorithms embody techniques that have existed for at least 10 years, but have only recently been implemented as mature, reliable, understandable tools that consistently outperform older statistical methods.

This book has specifically focused on applying data mining techniques to design, develop, and evaluate social advancement processes that have been applied in several developing economies. This book provides a overview on the main issues of data mining (including its classification, regression, clustering, association rules, trend detection, feature selection, intelligent search, data cleaning, privacy and security issues, etc.) and knowledge enhancing processes as well as a wide spectrum of data mining applications such as computational natural science, e-commerce, environmental study, financial market study, network monitoring, social service analysis, and so forth.

This book will be highly acceptable to researchers, academics and practitioners, including GOs and NGOs for further research and study, especially who would be working in the aspect of monitoring and evaluation of projects; follow-up activities on development projects, and be an invaluable scholarly content for development practitioners.

Dr. Abdul Matin Patwari
Vice Chancellor, The University of Asia Pacific
Dhaka, Bangladesh.

Preface

Data mining may be characterized as the process of extracting intelligent information from large amounts of raw data, and day-by-day becoming a pervasive technology in activities as diverse as using historical data to predict the success of a awareness raising campaign by looking into pattern sequence formations, or a promotional operation by looking into pattern sequence transformations, or a monitoring tool by looking into pattern sequence repetitions, or a analysis tool by looking into pattern sequence formations.

Theories and concepts on data mining recently added to the arena of database and researches in this aspect do not go beyond more than a decade. Very minor research and development activities have been observed in the 1990's, along the immense prospect of information and communication technologies (ICTs). Organized and coordinated researches on data mining started in 2001, with the advent of various workshops, seminars, promotional campaigns, and funded researches. International conferences on data mining organized by Institute of Electrical and Electronics Engineers, Inc. (since 2001), Wessex Institute of Technology (since 1999), Society for Industrial and Applied Mathematics (since 2001), Institute of Computer Vision and applied Computer Sciences (since 1999), and World Academy of Science are among the leaders in creating awareness on advanced research activities on data mining and its effective applications. Furthermore, these events reveal that the theme of research has been shifting from fundamental data mining to information engineering and/or information management along these years.

Data mining is a promising and relatively new area of research and development, which can provide important advantages to the users. It can yield substantial knowledge from data primarily gathered through a wide range of applications. Various institutions have derived considerable benefits from its application and many other industries and disciplines are now applying the methodology in increasing effect for their benefit.

Subsequently, collective efforts in machine learning, artificial intelligence, statistics, and database communities have been reinforcing technologies of knowledge discovery in databases to extract valuable information from massive amounts of data in support of intelligent decision making. Data mining aims to develop algorithms for extracting new patterns from the facts recorded in a database, and up till now, data mining tools adopted techniques from statistics, network modeling and visualization to classify data and identify patterns. Ultimately, knowledge recovery aims to enable an information system to transform information to knowledge through hypothesis, testing and theory formation. It sets new challenges for database technology: new concepts and methods are needed for basic operations, query languages, and query processing strategies (Witten & Frank, 2005; Yuan, Buttenfield, Gehagen & Miller, 2004).

However, data mining does not provide any straightforward analysis, nor does it necessarily equate with machine learning, especially in a situation of relatively larger databases. Furthermore, an exhaustive statistical analysis is not possible, though many data mining methods contain a degree of nondeterminism to enable them to scale massive datasets.

At the same time, successful applications of data mining are not common, despite the vast literature now accumulating on the subject. The reason is that, although it is relatively straightforward to find pattern or structure in data, but establishing its relevance and explaining its cause are both very difficult tasks. In addition, much of what that has been discovered so far may well be known to the expert. Therefore, addressing these problematic issues requires the synthesis of underlying theory from the databases, statistics, algorithms, machine learning, and visualization (Giudici, 2003; Hastie, Tibshirani & Friedman, 2001; Yuan, Buttenfield, Gehagen & Miller, 2004).

Along these perspectives, to enable practitioners in improving their researches and participate actively in solving practical problems related to data explosion, optimum searching, qualitative content management, improved decision making, and intelligent data mining a complete guide is the need of the hour. A book featuring all these aspects can fill an extremely demanding knowledge gap in the contemporary world.

Furthermore, data mining is not an independently existed research subject anymore. To understand its essential insights, and effective implementations one must open the knowledge periphery in multidimensional aspects. Therefore, in this era of information revolution data mining should be treated as a cross-cutting and cross-sectoral feature. At the same time, data mining is becoming an interdisciplinary field of research driven by a variety of multidimensional applications. On one hand it entails techniques for machine learning, pattern recognition, statistics, algorithm, database, linguistic, and visualization. On the other hand, one finds applications to understand human behavior, such as that of the end user of an enterprise. It also helps entrepreneurs to perceive the type of transactions involved, including those needed to evaluate risks or detect scams.

The reality of data explosion in multidimensional databases is a surprising and widely misunderstood phenomenon. For those about to use an OLAP (online analytical processing) product, it is critically important to understand what data explosion is, what causes it, and how it can be avoided, because the consequences of ignoring data explosion can be very costly, and, in most cases, result in project failure (Applix, 2003), while enterprise data requirements grow at 50-100% a year, creating a constant storage infrastructure management challenge (Intransa, 2005).

Concurrently, the database community draws much of its motivation from the vast digital datasets now available online and the computational problems involved in analyzing them. Almost without exception, current databases and database management systems are designed without to knowledge or content, so the access methods and query languages they provide are often inefficient or unsuitable for mining tasks. The functionality of some existing methods can be approximated either by sampling the data or reexpressing the data in a simpler form. However, algorithms attempt to encapsulate all the important structure contained in the original data, so that information loss is minimal and mining algorithms can function more efficiently. Therefore, sampling strategies must try to avoid bias, which is difficult if the target and its explanation are unknown.

These are related to the core technology aspects of data mining. Apart from the intricate technology context, the applications of data mining methods lag in the development context. Lack of data has been found to inhibit the ability of organizations to fully assist clients, and lack of knowledge made the government vulnerable to the influence of outsiders who did have access to data from countries overseas. Furthermore, disparity in data collection demands a coordinated data archiving and data sharing, as it is extremely crucial for developing countries.

The technique of data mining enables governments, enterprises, and private organizations to carry out mass surveillance and personalized profiling, in most cases without any controls or right of access to examine this data. However, to raise the human capacity and establish effective knowledge systems from the applications of data mining, the main focus should be on sustainable use of resources and the associated systems under specific context (ecological, climatic, social and economic conditions) of developing countries. Research activities should also focus on sustainable management of vulnerable resources and apply integrated management techniques, with a view to support the implementation of the provisions related to research and sustainable use of existing resources (EC, 2005).

To obtain advantages of data mining applications, the scientific issues and aspects of archiving scientific and technology data can include the discipline specific needs and practices of scientific communities as well as interdisciplinary assessments and methods. In this context, data archiving can be seen primarily as a program of practices and procedures that support the collection, long-term preservation, and low cost access to, and dissemination of scientific and technology data. The tasks of the data archiving include: digitizing data, gathering digitized data into archive collections, describing the collected data to support long term preservation, decreasing the risks of losing data, and providing easy ways to make the data accessible. Hence, data archiving and the associated data centers need to be part of the day-to-day practice of science. This is particularly important now that much new data is collected and generated digitally, and regularly (Codata, 2002; Mohammadian, 2004).

So far, data mining has existed in the form of discrete technologies. Recently, its integration into many other formats of ICTs has become attractive as various organizations possessing huge databases began to realize the potential of information hidden there (Hernandez, Göhring & Hopmann, 2004). Thereby, the Internet can be a tremendous tool for the collection and exchange of information, best practices, success cases and vast quantities of data. But it is also becoming increasingly congested and its popular use raises issues about authentication and evaluation of information and data. Interoperability is another issue, which provides significant challenges. The growing number and volume of data sources, together with the high-speed connectivity of the Internet and the increasing number and complexity of data sources, are making interoperability and data integration an important research and industry focus. Moreover, incompatibilities between data formats, software systems, methodologies and analytical models are creating barriers to easy flow and creation of data, information and knowledge (Carty, 2002). All these demand, not only technology revolution, but also tremendous uplift of human capacity as a whole.

Therefore, the challenge of human development taking into account the social and economic background while protecting the environment confronts decision makers like national governments, local communities and development organizations. A question arises, as how can new technology for information and communication be applied to fulfill this task (Hernandez, Göhring & Hopmann, 2004)? This book gives a review of data mining and decision support techniques and their requirement to achieve sustainable outcomes. It looks into authenticated global approaches on data mining and shows its capabilities as an effective instrument on the base of its application as real projects in the developing countries. The applications are on development of algorithms, computer security, open and distance learning, online analytical processing, scientific modeling, simple warehousing, and social and economic development process.

Applying data mining techniques in various aspects of social development processes could thereby empower the society with proper knowledge, and would produce economic products by raising their economic capabilities.

On the other hand, coupled to linguistic techniques data mining has produced a new field of text mining. This has considerably increased the applications of data mining to extract ideas and sentiment from a wide range of sources, and opened up new possibilities for data mining that can act as a bridge between the technology and physical sciences and those related to social sciences. Furthermore, data mining today is recognized as an important tool to analyze and understand the information collected by governments, businesses and scientific centers. In the context of novel data, text, and Web-mining application areas are emerging fast and these developments call for new perspectives and approaches in the form of inclusive researches.

Similarly, info-miners in the distance learning community are using one or more info-mining tools. They offer a high quality open and distance learning (ODL) information retrieval and search services.

Thus, ICT based info-mining services will likely be producing huge digital libraries such as e-books, journals, reports and databases on DVD and similar high-density information storage media. Most of these off-line formats are PC-accessible, and can store considerably more information per unit than a CD-ROM (COL, 2003). Hence, knowledge enhancement processes can be significantly improved through proper use of data mining techniques.

Thus, data mining techniques are gradually becoming essential components of corporate intelligence systems and are progressively evolving into a pervasive technology within activities that range from the utilization of historical data to predicting the success of an awareness campaign, or a promotional operation in search of succession patterns used as monitoring tools, or in the analysis of genome chains or formation of knowledge banks. In reality, data mining is becoming an interdisciplinary field driven by various multidimensional applications. On one hand it involves schemes for machine learning, pattern recognition, statistics, algorithm, database, linguistic, and visualization. On the other hand, one finds its applications to understand human behavior, or to understand the type of transactions involved, or to evaluate risks or detect frauds in an enterprise. Data mining can yield substantial knowledge from raw data that are primarily gathered for a wide range of applications. Various institutions have derived significant benefits from its application, and many other industries and disciplines are now applying the modus operandi in increasing effect for their overall management development.

This book tries to examine the meaning and role of data mining in terms of social development initiatives and its outcomes in developing economies in terms of upholding knowledge dimensions. At the same time, it gives an in-depth look into the critical management of information in developed countries with a similar point of view. Furthermore, this book provides an overview on the main issues of data mining (including its classification, regression, clustering, association rules, trend detection, feature selection, intelligent search, data cleaning, privacy and security issues, etc.) and knowledge enhancing processes as well as a wide spectrum of data mining applications such as computational natural science, e-commerce, environmental study, business intelligence, network monitoring, social service analysis, and so forth to empower the knowledge society.

WHERE THE BOOK STANDS

In the global context, a combination of continual technological innovation and increasing competitiveness makes the management of information a huge challenge and requires decision-making processes built on reliable and opportune information, gathered from available internal and external sources. Although the volume of acquired information is immensely increasing, this does not mean that people are able to derive appropriate value from it (Maira & Marlei, 2003). This deserves authenticated investigation on information archival strategies and demands years of continuous investments in order to put in place a technological platform that supports all development processes and strengthens the efficiency of the operational structure. Most organizations are supposed to have reached at a certain level where the implementation of IT solutions for strategic levels becomes achievable and essential. This context explains the emergence of the domain generally known as "intelligent data mining", seen as an answer to the current demands in terms of data/information for decision-making with the intensive utilization of information technology.

The objective of the book is to examine the meaning and role of data mining in a particular context (i.e., in terms of development initiatives and its outcomes), especially in developing countries and transitional economies. If the management of information is a challenge even to enterprises in developed countries, what can be said about organizations struggling in unstable contexts such as developing ones? The book has tried to focus on data mining application in developed countries' context, too.

With the unprecedented rate at which data is being collected today in almost all fields of human endeavor, there is an emerging demand to extract useful information from it for economic and scientific benefit of the society. Intelligent data mining enables the community to take advantages out of the gathered data and information by taking intelligent decisions. This increases the knowledge content of each member of the community, if it can be applied to practical usage areas. Eventually, a knowledge base is being created and a knowledge-based society will be established.

However, data mining involves the process of automatic discovery of patterns, sequences, transformations, associations, and anomalies in massive databases, and is a enormously interdisciplinary field representing the confluence of several disciplines, including database systems, data warehousing, machine learning, statistics, algorithms, data visualization, and high-performance computing (LCPS, 2001; UN, 2004). A book of this nature, encompassing such omnipotent subject area has been missing in the contemporary global market, intends to fill in this knowledge gap.

In this context, this book provides an overview on the main issues of data mining (including its classification, regression, clustering, association rules, trend detection, feature selection, intelligent search, data cleaning, privacy and security issues, and etc.) and knowledge enhancing processes as well as a wide spectrum of data mining applications such as computational natural science, e-commerce, environmental study, financial market study, machine learning, Web mining, nanotechnology, e-tourism, and social service analysis.

Apart from providing insight into the advanced context of data mining, this book has emphasized on:

- Development and availability of shared data, metadata, and products commonly required across diverse societal benefit areas
- Promoting research efforts that are necessary for the development of tools required in all societal benefit areas
- Encouraging and facilitating the transition from research to operations of appropriate systems and techniques
- Facilitating partnerships between operational groups and research groups
- Developing recommended priorities for new or augmented efforts in human capacity building
- Contributing to, access, and retrieve data from global data systems and networks
- Encouraging the adoption of existing and new standards to support broader data and information usability
- Data management approaches that encompass a broad perspective on the observation of data life cycle, from input through processing, archiving, and dissemination, including reprocessing, analysis and visualization of large volumes and diverse types of data
- Facilitating recording and storage of data in clearly defined formats, with metadata and quality indications to enable search, retrieval, and archiving as easily accessible data sets
- Facilitating user involvement and conducting outreach at global, regional, national and local levels
- Complete and open exchange of data, metadata, and products within relevant agencies and national policies and legislations

ORGANIZATION OF CHAPTERS

Altogether this book has fifteen chapters and they are divided into three sections: Education and Research; Tools, Techniques, Methods; and Applications of Data Mining. Section I has three chapters, and they discuss policy and decision-making approaches of data mining for sociodevelopment aspects in technical and semitechnical contexts. Section II is comprised of five chapters and they illustrate tools, techniques, and methods of data mining applications for various human development processes and scientific research. The third section has seven chapters and those chapters show various case studies, practical applications and research activities on data mining applications that are being used in the social development processes for empowering the knowledge societies.

Chapter I provides an overview of a series of multiple criteria optimization-based data mining methods that utilize multiple criteria programming (MCP) to solve various data mining problems. Authors state that data mining is being established on the basis of many disciplines, such as machine learning, databases, statistics, computer science, and operation research and each field comprehends data mining from its own perspectives by making distinct contributions. They further state that due to the difficulty of accessing the accuracy of hidden data and increasing the predicting rate in a complex large-scale database, researchers and practitioners have always desired to seek new or alternative data mining techniques. Therefore, this chapter outlines a few research challenges and opportunities at the end.

Chapter II identifies some important barriers to the successful application of computational intelligence (CI) techniques in a commercial environment and suggests various ways in which they may be overcome. It states that CI offers new opportunities to a business that wishes to improve the efficiency of their operations. In this context, this chapter further identifies a few key conceptual, cultural, and technical barriers and describes different ways in which they affect the business users and the CI practitioners. This chapter aims to provide knowledgeable insight for its readers through outcome of a successful computational intelligence project and expects that by enabling both parties to understand each other's perspectives, the true potential of CI may be realized.

Chapter III describes two data mining techniques that are used to explore frequent large itemsets in the database. In the first technique called closed directed graph approach. The algorithm scans the database once making a count on 2-itemsets possible from which only the 2-itemsets with a minimum support are used to form the closed directed graph and explores frequent large itemsets in the database. In the second technique, dynamic hashing algorithm large 3-itemsets are generated at an earlier stage that reduces the size of the transaction database after trimming and thereby cost of later iterations will be reduced. Furthermore, this chapter predicts that the techniques may help researchers not only to understand about generating frequent large itemsets, but also finding association rules among transactions within relational databases, and make knowledgeable decisions.

It is observed that daily, different satellites capture data of distinct contexts, and among which images are processed and stored by many institutions. In **Chapter IV** authors present relevant definitions on remote sensing and image mining domain, by referring to related work in this field and indicating about the importance of appropriate tools and techniques to analyze satellite images and extract knowledge from this kind of data. As a case study, the Amazonia deforestation problem is being discussed; as well INPE's effort to develop and spread technology to deal with challenges involving Earth observation resources. The purpose is to present relevant technologies, new approaches and research directions on remote sensing image mining, and demonstrating how to increase the analysis potential of such huge strategic data for the benefit of the researchers.

Chapter V reviews contemporary research on machine learning and Web mining methods that are related to areas of social benefit. It demonstrates that machine learning and Web mining methods may provide intelligent Web services of social interest. The chapter also reveals a growing interest for using advanced computational methods, such as machine learning and Web mining, for better services to the public, as most research identified in the literature has been conducted during recent years. The chapter tries to assist researchers and academics from different disciplines to understand how Web mining and machine learning methods are applied to Web data. Furthermore, it aims to provide the latest developments on research in this field that is related to societal benefit areas.

In recent times, customer relationship management (CRM) can be related to sales, marketing and even services automation. Additionally, the concept of CRM is increasingly associated with cost savings and streamline processes as well as with the engendering, nurturing and tracking of relationships with customers. **Chapter VI** seeks to illustrate how, although the product and service elements as well as organizational structure and strategies are central to CRM, data is the pivotal dimension around which the concept revolves in contemporary terms, and subsequently tried to demonstrate how these processes are associated with data management, namely: data collection, data collation, data storage and data mining, which are becoming essential components of CRM in both theoretical and practical aspects.

In **Chapter VII**, authors have introduced the concept of "one-sum" weighted association rules (WARs) and named such WARs as allocating patterns (ALPs). An algorithm is also being proposed to extract hidden and interesting ALPs from data. The chapter further point out that ALPs can be applied in portfolio management. Modeling a collection of investment portfolios as a one-sum weighted transaction-database that contains hidden ALPs can do this, and eventually those ALPs, mined from the given portfolio-data, can be applied to guide future investment activities.

Chapter VIII is focused to data mining applications and their utilizations in formulating performance-measuring tools for social development activities. In this context, this chapter provides justifications to include data mining algorithm to establish specifically derived monitoring and evaluation tools for various social development applications. In particular, this chapter gave in-depth analytical observations to establish knowledge centers with a range of approaches and finally it put forward a few research issues and challenges to transform the contemporary human society into a knowledge society.

Chapter IX highlightes a few areas of development aspects and hints application of data mining tools, through which decision-making would be easier. Subsequently, this chapter has put forward potential areas of society development initiatives, where data mining applications can be introduced. The focus area may vary from basic education, health care, general commodities, tourism, and ecosystem management to advanced uses, like database tomography. This chapter also provides some future challenges and recommendations in terms of using data mining applications for empowering knowledge society.

Chapter X focuses on business data warehouse and discusses the retailing giant, Wal-Mart. In this chapter, the planning and implementation of the Wal-Mart data warehouse is being described and its integration with the operational systems is discussed. It also highlighted some of the problems that have been encountered during the development process of the data warehouse, including providing some future recommendations.

In **Chapter XI** medical applications literature associated with nanoscience and nanotechnology research was examined. Authors retrieved about 65,000 nanotechnology records in 2005 from the Science Citation Index/ Social Science Citation Index (SCI/SSCI) using a comprehensive 300+ term query. This chapter intends to facilitate the nanotechnology transition process by identifying the significant application areas. It also identified the main nanotechnology health applications from today's vantage point, as well as the related science and infrastructure. The medical applications were identified through a fuzzy clustering process, and metrics were generated using text mining to extract technical intelligence for specific medical applications/ applications groups.

Chapter XII introduces an early warning system for SMEs (SEWS) as a financial risk detector that is based on data mining. Through a study this chapter composes a system in which qualitative and quantitative data about the requirements of enterprises are taken into consideration, during the development of an early warning system. Moreover, during the formation of this system; an easy to understand, easy to interpret and easy to apply utilitarian model is targeted by discovering the implicit relationships between the data and the identification of effect level of every factor related to the system. This chapter also shows the way of empowering knowledge society from SME's point of view by designing an early warning system based on data mining. Using this system, SME managers could easily reach financial management, risk management knowledge without any prior knowledge and expertise.

Chapter XIII looks at various business intelligence (BI) projects in developing countries, and specifically focuses on Brazilian BI projects. Authors poised this question that, if the management of IT is a challenge for companies in developed countries, what can be said about organizations struggling in unstable contexts such as those often prevailing in developing countries. Within this broad enquiry about the role of BI playing in developing countries, two specific research questions are explored in this chapter. The purpose of the first question is to determine whether those approaches, models, or frameworks are tailored for particularities and the contextually situated business strategy of each company, or if they are "standard" and imported from "developed" contexts. The purpose of the second one is to analyze: what type of information is being considered for incorporation by BI systems; whether they are formal or informal in nature; whether they are gathered from internal or external sources; whether there is a trend that favors some areas, like finance or marketing, over others, or if there is a concern with maintaining multiple perspectives; who in the firms is using BI systems, and so forth.

Technologies such as geographic information systems (GIS) enable geo-spatial information to be gathered, modified, integrated, and mapped easily and cost effectively. However, these technologies generate both opportunities and challenges for achieving wider and more effective use of geo-spatial information in stimulating and sustaining sustainable development through elegant policy making. In **Chapter XIV**, the author proposes a simple and accessible conceptual knowledge discovery interface that can be used as a tool. Moreover, the chapter addresses some issues that might make this knowledge infrastructure stimulate sustainable development, especially emphasizing sub-Saharan African region.

Finally, **Chapter XV** discusses the application of data mining to develop drought monitoring tools that enable monitoring and prediction of drought's impact on vegetation conditions. The chapter also summarizes current research using data mining approaches (e.g., association rules and decision-tree methods) to develop various types of drought monitoring tools and briefly explains how they are being integrated with decision support systems. This chapter also introduces how data mining can be used to enhance drought monitoring and prediction in the United States, and at the same time, assist others to understand how similar tools might be developed in other parts of the world.

CONCLUSION

Data mining is becoming an essential tool in science, engineering, industrial processes, healthcare, and medicine. The datasets in these fields are large, complex, and often noisy. However, extracting knowledge from raw datasets requires the use of sophisticated, high-performance and principled analysis techniques and algorithms, based on sound statistical foundations. In turn, these techniques require powerful visualization technologies; implementations that must be carefully tuned for enhanced performance; software systems that are usable by scientists, engineers, and physicians as well as researchers.

Data mining, as stated earlier, is denoted as *the extraction of hidden predictive information from large databases*, and it is a powerful new technology with great potential to help enterprises focus on the most important information in their data warehouses. Data mining tools predict future trends and behaviors, allowing entrepreneurs to make proactive, knowledge-driven decisions. The automated, prospective analyses offered by data mining move beyond the analyses of past events provided by retrospective constituents typical of decision support systems. Data mining tools can answer business questions that traditionally were too time consuming to resolve. They scour databases for hidden patterns, finding predictive information that experts may miss because it lies outside their expectations.

In effect, data mining techniques are the result of a long process of research and product development. This evolution began when business data was first stored on computers, continued with improvements in data access, and more recently, generated technologies that allow users to navigate through their data in real time. Thus, data mining takes this evolutionary progression beyond retrospective data access and navigation to prospective and proactive information delivery. Furthermore, data mining algorithms allow researchers to device unique decision-making tools from emancipated data varying in nature. Foremost, applying data mining techniques extremely valuable utilities can be devised that could raise the knowledge content at each tier of society segments.

However, in terms of accumulated literature and research contexts, not many publications are available in the field of data mining applications in social development phenomenon, especially in the form of a book. By taking this as a baseline, compiled literature seems to be extremely valuable in the context of utilizing data mining and other information techniques for the improvement of skills development, knowledge management, and societal benefits. Similarly, Internet search engines do not fetch sufficient bibliographies in the field of data mining for development perspective. Due to the high demand from researchers' in the aspect of ICTD, a book of this format stands to be unique. Moreover, utilization of new ICTs in the form of data mining deserves appropriate intervention for their diffusion at local, national, regional, and global levels.

It is assumed that numerous individuals, academics, researchers, engineers, professionals from government and nongovernment security and development organizations will be interested in this increasingly important topic for carrying out implementation strategies towards their national development. This book will assist its readers to understand the key practical and research issues related to applying data mining in development data analysis, cyber acclamations, digital deftness, contemporary CRM, investment portfolios, early warning system in SMEs, business intelligence, and intrinsic nature in the context of society uplift as a whole and the use of data and information for empowering knowledge societies.

Most books of data mining deal with mere technology aspects, despite the diversified nature of its various applications along many tiers of human endeavor. However, there are a few activities in recent years that are producing high quality proceedings, but it is felt that compilation of contents of this nature from advanced research outcomes that have been carried out globally may produce a demanding book among the researchers.

REFERENCES

Applix (2003). *OLAP data scalability: Ignore the OLAP data explosion at great cost.* A White Paper. Westborough, MA: Applix, Inc.

Carty, A. J. (2002, September 29). Scientific and technical data: Extending the frontiers of research. In *Proceedings of the Opening Address at the 18th International CODATA Conference*, Montreal, Quebec.

Codata (2002, May 21-22). In *Proceedings of the Workshop on Archiving Scientific and Technical Data, Committee on Data for Science and Technology* (CODATA), Pretoria, South Africa.

COL (2003). *Find information faster: COL's "Info-mining" tools*. Vancouver, BC: Clippings, Commonwealth of Learning.

EC (2005). *Integrating and strengthening the European Research Area*, 2005 Work Programme (SP1-10). European Commission.

Hernandez, V., Göhring, W., & Hopmann, C. (2004, Nov. 30-Dec. 3). Sustainable decision support for environmental problems in developing countries: Applying multicriteria spatial analysis on the Nicaragua Development Gateway niDG. In *Proceedings of the Workshop on Binding EU-Latin American IST Research Initiatives for Enhancing Future Co-Operation*. Santo Domingo, Costa Rica.

Giudici, P. (2003). *Applied data mining: Statistical methods for business and industry*. John Wiley.

Hastie, T., Tibshirani, R., & Friedman, J. (2001) (Eds.). *The elements of statistical learning: Data mining, inference, and prediction*. Springer Verlag.

Intransa (2005). *Managing storage growth with an affordable and flexible IP SAN: A highly cost-effective storage solution that leverages existing IT resources*. San Jose, CA: Intransa, Inc.

LCPS (2001, September 11-12). Draft workshop report. In *Proceedings of the International Consultative Workshop, The Digital Initiative for Development Agency (DID), The Lebanese Center for Policy Studies (LCPS)*, Beirut.

Maira, P. & Marlei, P. (2003, June 16-21). The value of "business intelligence" in the context of developing countries. In *Proceedings of the 11th European Conference on Information Systems, ECIS 2003*, Naples, Italy. Retrieved April 6, 2008, http://is2.lse.ac.uk/asp/aspecis/20030119.pdf

Mohammadian, M. (2004). *Intelligent agents for data mining and information retrieval*. Hershey, PA: Idea Group Publishing.

UN (2004, June 16). *Draft Sao Paulo Consensus*, UNCTAD XI Multi-Stakeholder Partnerships, United Nations Conference on Trade and Development, TD/L.380/Add.1, Sao Paulo.

Witten, I. H. & Frank, E. (2005). *Data mining: Practical machine learning tools and techniques* (2nd ed). Morgan Kaufmann.

Yuan, M., Buttenfield, B., Gehagen, M. & Miller, H. (2004). Geospatial data mining and knowledge discovery. In R. B. McMaster & E. L. Usery (Eds.), *A research agenda for geographic information science* (pp. 365-388). Boca Raton, FL: CRC Press.

Acknowledgment

The editor would like to acknowledge the assistance from all involved in the entire accretion of manuscripts, painstaking review process, and methodical revision of the book, without whose support the project could not have been satisfactorily completed. I am indebted to all the authors who provided their relentless and generous supports, but reviewers who were most helpful and provided comprehensive, thorough and creative comments are: Ali Serhan Koyuncugil, Georgios Lappas, and Paul Henman. Thanks go to my close friends at UNDP, and colleagues at SDNF and ICMS for their wholehearted encouragements during the entire process.

Special thanks also go to the dedicated publishing team at IGI Global. Particularly to Kristin Roth, Jessica Thompson, and Jennifer Neidig for their continuous suggestions, supports and feedbacks via e-mail for keeping the project on schedule, and to Mehdi Khosrow-Pour and Jan Travers for their enduring professional supports. Finally, I would like to thank all my family members for their love and support throughout this period.

Hakikur Rahman, Editor
SDNF, Bangladesh
September 2007

Section I
Education and Research

Chapter I
Introduction to Data Mining Techniques via Multiple Criteria Optimization Approaches and Applications

Yong Shi
University of the Chinese Academy of Sciences, China and University of Nebraska at Omaha, USA

Yi Peng
University of Nebraska at Omaha, USA

Gang Kou
University of Nebraska at Omaha, USA

Zhengxin Chen
University of Nebraska at Omaha, USA

ABSTRACT

This chapter provides an overview of a series of multiple criteria optimization-based data mining methods, which utilize multiple criteria programming (MCP) to solve data mining problems, and outlines some research challenges and opportunities for the data mining community. To achieve these goals, this chapter first introduces the basic notions and mathematical formulations for multiple criteria optimization-based classification models, including the multiple criteria linear programming model, multiple criteria quadratic programming model, and multiple criteria fuzzy linear programming model. Then it presents the real-life applications of these models in credit card scoring management, HIV-1 associated dementia (HAD) neuronal damage and dropout, and network intrusion detection. Finally, the chapter discusses research challenges and opportunities.

INTRODUCTION

Data mining has become a powerful information technology tool in today's competitive business world. As the sizes and varieties of electronic datasets grow, the interest in data mining is increasing rapidly. Data mining is established on the basis of many disciplines, such as machine learning, databases, statistics, computer science, and operations research. Each field comprehends data mining from its own perspective and makes its distinct contributions. It is this multidisciplinary nature that brings vitality to data mining. One of the application roots of data mining can be regarded as statistical data analysis in the pharmaceutical industry. Nowadays the financial industry, including commercial banks, has benefited from the use of data mining. In addition to statistics, decision trees, neural networks, rough sets, fuzzy sets, and vector support machines have gradually become popular data mining methods over the last 10 years. Due to the difficulty of accessing the accuracy of hidden data and increasing the predicting rate in a complex large-scale database, researchers and practitioners have always desired to seek new or alternative data mining techniques. This is a key motivation for the proposed multiple criteria optimization-based data mining methods.

The objective of this chapter is to provide an overview of a series of multiple criteria optimization-based methods, which utilize the multiple criteria programming (MCP) to solve classification problems. In addition to giving an overview, this chapter lists some data mining research challenges and opportunities for the data mining community. To achieve these goals, the next section introduces the basic notions and mathematical formulations for three multiple criteria optimization-based classification models: the multiple criteria linear programming model, multiple criteria quadratic programming model, and multiple criteria fuzzy linear programming model. The third section presents some real-life applications of these models, including credit card scoring management, classifications on HIV-1 associated dementia (HAD) neuronal damage and dropout, and network intrusion detection. The chapter then outlines research challenges and opportunities, and the conclusion is presented.

MULTIPLE CRITERIA OPTIMIZATION-BASED CLASSIFICATION MODELS

This section explores solving classification problems, one of the major areas of data mining, through the use of multiple criteria mathematical programming-based methods (Shi, Wise, Luo, & Lin, 2001; Shi, Peng, Kou, & Chen, 2005). Such methods have shown its strong applicability in solving a variety of classification problems (e.g., Kou et al., 2005; Zheng et al., 2004).

Classification

Although the definition of classification in data mining varies, the basic idea of classification can be generally described as to "predicate the most likely state of a categorical variable (the class) given the values of other variables" (Bradley, Fayyad, & Mangasarian, 1999, p. 6). Classification is a two-step process. The first step constructs a predictive model based on training dataset. The second step applies the predictive model constructed from the first step to testing dataset. If the classification accuracy of testing dataset is acceptable, the model can be used to predicate unknown data (Han & Kamber, 2000; Olson & Shi, 2005).

Using the multiple criteria programming, the classification task can be defined as follows: *for a given set of variables in the database, the boundaries between the classes are represented by scalars in the constraint availabilities.* Then, the standards of classification are measured by minimizing the total overlapping of data and maximizing the distances of every data to its class boundary simultaneously. Through the algorithms of MCP, an "optimal" solution of variables (so-called classifier) for the data observations is determined for the separation of the given classes. Finally, the resulting classifier can be used to predict the unknown data for discovering the hidden patterns of data as possible knowledge. Note that MCP differs from the known support vector machine (SVM) (e.g., Mangasarian, 2000; Vapnik, 2000). While the former uses multiple measurements to separate each data from different classes, the latter searches the minority of the data (support vectors) to represent the majority in classifying the data. However, both can be generally regarded as in the same category of optimization approaches to data mining.

In the following, we first discuss a generalized multi-criteria programming model formulation, and then explore several variations of the model.

A Generalized Multiple Criteria Programming Model Formulation

This section introduces a generalized multi-criteria programming method for classification. Simply speaking, this method is to classify observations into distinct groups based on two criteria for data separation. The following models represent this concept mathematically:

Given an r-dimensional attribute vector $a=(a_1,...a_r)$, let $A_i=(A_{i1},...,A_{ir})\in R^r$ be one of the sample records of these attributes, where $i=1,...,n$; n represents the total number of records in the dataset. Suppose two groups G_1 and G_2 are predefined. A boundary scalar b can be selected to separate these two groups. A vector $X=(x_1,...,X_r)^T\in R^r$ can be identified to establish the following linear inequations (Fisher, 1936; Shi et al., 2001):

- $A_i X < b, \forall A_i \in G_1$
- $A_i X \geq b, \forall A_i \in G_2$

To formulate the criteria and complete constraints for data separation, some variables need to be introduced. In the classification problem, $A_i X$ is the score for the ith data record. Let α_i be the overlapping of two-group boundary for record A_i (external measurement) and β_i be the distance of record A_i from its adjusted boundary (internal measurement). The overlapping α_i means the distance of record A_i to the boundary b if A_i is misclassified into another group. For instance, in Figure 1 the "black dot" located to the right of the boundary b belongs to G_1, but it was misclassified by the boundary b to G_2. Thus, the distance between b and the "dot" equals α_i. Adjusted boundary is defined as b-α^* or b+α^*, while α^* represents the maximum of overlapping (Freed & Glover, 1981, 1986). Then, a mathematical function $f(\alpha)$ can be used to describe the relation of all overlapping α_i, while another mathematical function $g(\beta)$ represents the aggregation of all distances β_i. The final classification accuracies depend on simultaneously minimizing $f(\alpha)$ and maximizing $g(\beta)$. Thus, a generalized bi-criteria programming method for classification can be formulated as:

(Generalized Model) *Minimize* $f(\alpha)$ and *Maximize* $g(\beta)$

Subject to:

$$A_i X - \alpha_i + \beta_i - b = 0, \forall A_i \in G_1,$$
$$A_i X + \alpha_i - \beta_i - b = 0, \forall A_i \in G_2,$$

where A_i, $i = 1, \ldots, n$ are given, X and b are unrestricted, and $\alpha = (\alpha_1,...\alpha_n)^T$, $\beta=(\beta_1,...\beta_n)^T$; α_i, $\beta_i \geq 0$, $i = 1, \ldots, n$.

All variables and their relationships are represented in Figure 1. There are two groups in Figure 1: "black dots" indicate G_1 data objects, and "stars" indicate G_2 data objects. There is one misclassified data object from each group if the boundary scalar b is used to classify these two groups, whereas adjusted boundaries b-α^* and b+α^* separate two groups without misclassification.

Figure 1. Two-group classification model

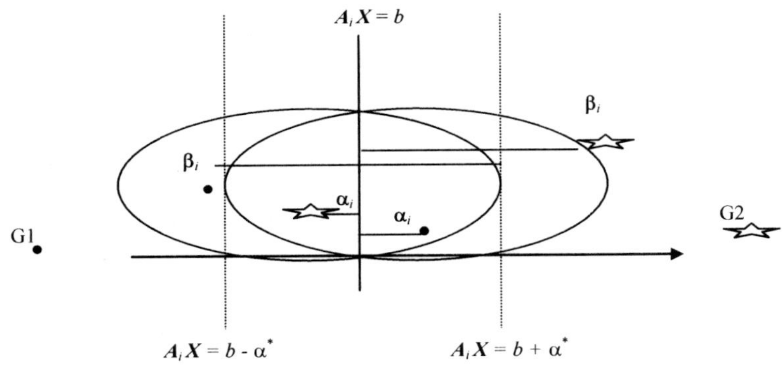

Based on the above generalized model, the following subsection formulates a multiple criteria linear programming (MCLP) model and a multiple criteria quadratic programming (MCQP) model.

Multiple Criteria Linear and Quadratic Programming Model Formulation

Different forms of $f(\alpha)$ and $g(\beta)$ in the generalized model will affect the classification criteria. Commonly $f(\alpha)$ (or $g(\beta)$) can be component-wise and non-increasing (or non-decreasing) functions. For example, in order to utilize the computational power of some existing mathematical programming software packages, a sub-model can be set up by using the norm to represent $f(\alpha)$ and $g(\beta)$. This means that we can assume $f(\alpha) = ||\alpha||_p$ and $g(\beta) = ||\beta||_q$. To transform the bi-criteria problems of the generalized model into a single-criterion problem, we use weights $w_\alpha > 0$ and $w_\beta > 0$ for $||\alpha||_p$ and $||\beta||_q$, respectively. The values of w_α and w_β can be pre-defined in the process of identifying the optimal solution. Thus, the generalized model is converted into a single criterion mathematical programming model as:

Model 1: *Minimize* $w_\alpha\ ||\alpha||_p - w_\beta\ ||\beta||_q$

Subject to:

$$A_iX - \alpha_i + \beta_i - b = 0,\ \forall\ A_i \in G_1,$$
$$A_iX + \alpha_i - \beta_i - b = 0,\ \forall A_i \in G_2,$$

where A_i, $i = 1, \ldots, n$ are given, X and b are unrestricted, and $\alpha = (\alpha_1, \ldots, \alpha_n)^T$, $\beta = (\beta_1, \ldots \beta_n)^T$; $\alpha_i, \beta_i \geq 0$, $i = 1, \ldots, n$.

Based on Model 1, mathematical programming models with any norm can be theoretically defined. This study is interested in formulating a linear and a quadratic programming model. Let $p = q = 1$, then $||\alpha||_1 = \sum_{i=1}^{n} \alpha_i$ and $||\beta||_1 = \sum_{i=1}^{n} \beta_i$. Let $p = q = 2$, then $||\alpha||_2 = \sum_{i=1}^{n} \alpha_i^2$ and $||\beta||_2 = \sum_{i=1}^{n} \beta_i^2$.

The objective function in Model 1 can now be an MCLP model or MCQP model.

Model 2: MCLP

$$\text{Minimize } w_a \sum_{i=1}^{n} \alpha_i - w_\beta \sum_{i=1}^{n} \beta_i$$

Subject to:

$$A_iX - \alpha_i + \beta_i + b = 0,\ \forall A_i \in G_1,$$
$$A_iX + \alpha_i - \beta_i - b = 0,\ \forall A_i \in G_2,$$

where A_i, $i = 1, \ldots, n$ are given, X and b are unrestricted, and $\alpha=(\alpha_1,\ldots\alpha_n)^T$, $\beta = (\beta_1,\ldots\beta_n)^T$; α_i, $\beta_i \geq 0$, $i = 1, \ldots, n$.

Model 3: MCQP

$$\text{Minimize } w_a \sum_{i=1}^{n} \alpha_i^2 - w_\beta \sum_{i=1}^{n} \beta_i^2$$

Subject to:

$$A_i X - \alpha_i + \beta_i - b = 0, \ \forall A_i \in G_1,$$
$$A_i X + \alpha_i - \beta_i - b = 0, \ \forall A_i \in G_2,$$

where A_i, $i = 1, \ldots, n$ are given, X and b are unrestricted, and $\alpha = (\alpha_1,\ldots,\alpha_n)^T$, $\beta = (\beta_1,\ldots\beta_n)^T$; α_i, $\beta_i \geq 0$, $i = 1, \ldots, n$.

Remark 1

There are some issues related to MCLP and MCQP that can be briefly addressed here:

1. In the process of finding an optimal solution for MCLP problem, if some β_i is too large with given $w_\alpha > 0$ and $w_\beta > 0$ and all α_i relatively small, the problem may have an unbounded solution. In the real applications, the data with large β_i can be detected as "outlier" or "noisy" in the data preprocessing, which should be removed before classification.
2. Note that although variables X and b are unrestricted in the above models, $X=0$ is an "insignificant case" in terms of data separation, and therefore it should be ignored in the process of solving the problem. For $b = 0$, however, may result a solution for the data separation depending on the data structure. From experimental studies, a pre-defined value of b can quickly lead to an optimal solution if the user fully understands the data structure.
3. Some variations of the generalized model, such as MCQP, are NP-hard problems. Developing algorithms directly to solve these models can be a challenge. Although in application we can utilize some existing commercial software, the theoretical-related problem will be addressed in later in this chapter.

Multiple Criteria Fuzzy Linear Programming Model Formulation

It has been recognized that in many decision-making problems, instead of finding the existing "optimal solution" (a goal value), decision makers often approach a "satisfying solution" between upper and lower aspiration levels that can be represented by the upper and lower bounds of acceptability for objective payoffs, respectively (Charnes & Cooper, 1961; Lee, 1972; Shi & Yu, 1989; Yu, 1985). This idea, which has an important and pervasive impact on human decision making (Lindsay & Norman 1972), is called the decision makers' goal-seeking concept. Zimmermann (1978) employed it as the basis of his pioneering work on FLP. When FLP is adopted to classify the 'good' and 'bad' data, a fuzzy (satisfying) solution is used to meet a threshold for the accuracy rate of classifications, although the fuzzy solution is a near optimal solution.

According to Zimmermann (1978), in formulating an FLP problem, the objectives (*Minimize* $\Sigma_i\alpha_i$ and *Maximize* $\Sigma_i\beta_i$) and constraints ($A_i X = b + \alpha_i - \beta_i$, $A_i \in G$; $A_i X = b - \alpha_i + \beta_i$, $A_i \in B$) of the generalized model are redefined as fuzzy sets F and X with corresponding membership functions $\mu_F(x)$ and $\mu_X(x)$ respectively. In this case the fuzzy decision set D is defined as $D = F \cup X$, and the membership function is defined as $\mu_D(x) = \{\mu_F(x), \mu_X(x)\}$. In a maximal problem, x_1 is a "better" decision than x_2 if $\mu_D(x_1) \geq \mu_D(x_2)$. Thus, it can be considered appropriately to select x^* such that $\max_x \mu_D(x) = \max_x \min\{\mu_F(x), \mu_X(x)\} = \min\{\mu_F(x^*), \mu_X(x^*)\}$ is the maximized solution.

Let y_{1L} be *Minimize* $\Sigma_i\alpha_i$ and y_{2U} be *Maximize* $\Sigma_i\beta_i$, then one can assume that the value of *Maximize* $\Sigma_i\alpha_i$ to be y_{1U} and that of *Minimize* $\Sigma_i\beta_i$ to be y_{2L}. If the "upper bound" y_{1U} and the "lower bound" y_{2L} do not exist for the formulations, they can be estimated. Let $F_1\{x: y_{1L} \leq \Sigma_i\alpha_i \leq y_{1U}\}$ and $F_2\{x: y_{2L} \leq \Sigma_i\beta_i \leq y_{2U}\}$ and their membership functions can be expressed respectively by:

$$\mu_{F_1}(x) = \begin{cases} 1, & if\ \Sigma_i \alpha_i \geq y_{1U} \\ \dfrac{\Sigma_i \alpha_i - y_{1L}}{y_{1U} - y_{1L}}, & if\ y_{1L} < \Sigma_i \alpha_i < y_{1U} \\ 0, & if\ \Sigma_i \alpha_i \leq y_{1L} \end{cases}$$

and

$$\mu_{F_2}(x) = \begin{cases} 1, & if\ \Sigma_i \beta_i \geq y_{2U} \\ \dfrac{\Sigma_i \beta_i - y_{2L}}{y_{2U} - y_{2L}}, & if\ y_{2L} < \Sigma_i \beta_i < y_{2U} \\ 0, & if\ \Sigma_i \beta_i \leq y_{2L} \end{cases}$$

Then the fuzzy set of the objective functions is $F = F_1 \cap F_2$, and its membership function is $\mu_F(x) = \min\{\mu_{F_1}(x), \mu_{F_2}(x)\}$. Using the crisp constraint set $X = \{x: A_i X = b + \alpha_i - \beta_i, A_i \in G; A_i X = b - \alpha_i + \beta_i, A_i \in B\}$, the fuzzy set of the decision problem is $D = F_1 \cap F_2 \cap X$, and its membership function is $\mu_D(x) = \mu_{F_1 \cap F_2 \cap X}(x)$.

Zimmermann (1978) has shown that the "optimal solution" of $\max_x \mu_D(x) = \max_x \min\{\mu_{F_1}(x), \mu_{F_2}(x), \mu_X(x)\}$ is an efficient solution of a variation of the generalized model when $f(\alpha) = \Sigma_i\alpha_i$ and $g(\beta) = \Sigma_i\beta_i$. Then, this problem is equivalent to the following linear program (He, Liu, Shi, Xu, & Yan, 2004):

Model 4: FLP

Maximize ξ

Subject to:

$$\xi \leq \frac{\Sigma_i \alpha_i - y_{1L}}{y_{1U} - y_{1L}}$$

$$\xi \leq \frac{\Sigma_i \beta_i - y_{2L}}{y_{2U} - y_{2L}}$$

$$A_i X = b + \alpha_i - \beta_i,\ A_i \in G,$$
$$A_i X = b - \alpha_i + \beta_i,\ A_i \in B,$$

where A_i, y_{1L}, y_{1U}, y_{2L} and y_{2U} are known, X and b are unrestricted, and $\alpha_i, \beta_i, \xi \geq 0$.

Note that Model 4 will produce a value of ξ with $1 > \xi \geq 0$. To avoid the trivial solution, one can set up $\xi > \varepsilon \geq 0$, for a given ε. Therefore, seeking *Maximum* ξ in the FLP approach becomes the standard of determining the classifications between 'good' and 'bad' records in the database. A graphical illustration of this approach can be seen from Figure 2; any point of hyper plane $0 < \xi < 1$ over the shadow area represents the possible determination of classifications by the FLP method. Whenever Model 4 has been trained to meet the given threshold τ, it is said that the better classifier has been identified.

A procedure of using the FLP method for data classifications can be captured by the flowchart of Figure 2. Note that although the boundary of two classes b is the unrestricted variable in Model 4, it can be presumed by the analyst according to the structure of a particular database. First, choosing a proper value of b can speed up solving Model 4. Second, given a threshold τ, the best data separation can be selected from a number of results determined by different b values. Therefore, the parameter b plays a key role in this chapter to achieve and guarantee the desired accuracy rate τ. For this reason, the FLP classification method uses b as an important control parameter as shown in Figure 2.

REAL-LIFE APPLICATIONS USING MULTIPLE CRITERIA OPTIMIZATION APPROACHES

The models of multiple criteria optimization data mining in this chapter have been applied in credit

Figure 2. A flowchart of the fuzzy linear programming classification method

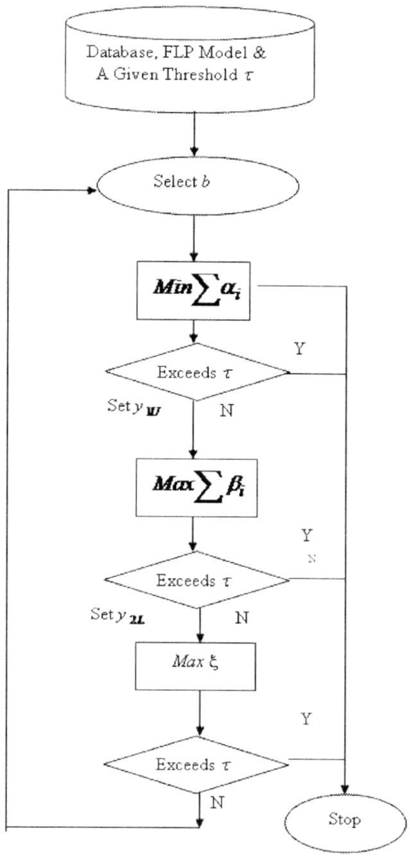

card portfolio management (He et al., 2004; Kou, Liu, Peng, Shi, Wise, & Xu, 2003; Peng, Kou, Chen, & Shi, 2004; Shi et al., 2001; Shi, Peng, Xu, & Tang, 2002; Shi et al., 2005), HIV-1-mediated neural dendritic and synaptic damage treatment (Zheng et al., 2004), network intrusion detection (Kou et al., 2004a; Kou, Peng, Chen, Shi, & Chen. 2004b), and firms bankruptcy analyses (Kwak, Shi, Eldridge, & Kou, 2006). These approaches are also being applied in other ongoing real-life data mining projects, such as anti-gene and antibody analyses, petroleum drilling and exploration, fraud management, and financial risk evaluation. In order to let the reader understand the usefulness of the models, the key experiences in some applications are reported as below.

Credit Card Portfolio Management

The goal of credit card accounts classification is to produce a "blacklist" of the credit cardholders; this list can help creditors to take proactive steps to minimize charge-off loss. In this study, credit card accounts are classified into two groups: 'good' or 'bad'. From the technical point of view, we need first construct a number of classifiers and then choose one that can find more bad records. The research procedure consists of five steps. The first step is *data cleaning*. Within this step, missing data cells and outliers are removed from the dataset. The second step is *data transformation*. The dataset is transformed in accord with the format requirements of MCLP software (Kou & Shi, 2002) and LINGO 8.0, which is a software tool for solving nonlinear programming problems (LINDO Systems Inc.). The third step is *datasets selection*. The training dataset and the testing dataset are selected according to a heuristic process. The fourth step is *model formulation and classification*. The two-group MCLP and MCQP models are applied to the training dataset to obtain optimal solutions. The solutions are then applied to the testing dataset within which class labels are removed for validation. Based on these scores, each record is predicted as either bad (bankrupt account) or good (current account). By comparing the predicted labels with original labels of records, the classification accuracies of multiple-criteria models can be determined. If the classification accuracy is acceptable by data analysts, this solution will be applied to future unknown credit card records or applications to make predictions. Otherwise, data analysts can modify the boundary and attributes values to get another set of optimal solutions. The fifth step is *results' presentation*. The acceptable classification results are summarized in tables or figures and presented to end users.

Credit Card Dataset

The credit card dataset used in this chapter is provided by a major U.S. bank. It contains 5,000 records and 102 variables (38 original variables and 64 derived variables). The data were collected from June 1995 to December 1995, and the cardholders were from 28 states of the United States. Each record has a class label to indicate its credit status: either 'good' or 'bad'. 'Bad' indicates a bankruptcy credit card account and 'good' indicates a good status account. Among these 5,000 records, 815 are bankruptcy accounts and 4,185 are good status accounts. The 38 original variables can be divided into four categories: balance, purchase, payment, and cash advance. The 64 derived variables are created from the original 38 variables to reinforce the comprehension of cardholders' behaviors, such as times over-limit in last two years, calculated interest rate, cash as percentage of balance, purchase as percentage to balance, payment as percentage to balance, and purchase as percentage to payment. For the purpose of credit card classification, the 64 derived variables were chosen to compute the model since they provide more precise information about credit cardholders' behaviors.

Experimental Results of MCLP

Inspired by the k-fold cross-validation method in classification, this study proposed a heuristic process for training and testing dataset selections. Standard *k*-fold cross-validation is not used because the majority-vote ensemble method used later on in this chapter may need hundreds of voters. If standard k-fold cross-validation was employed, *k* should be equal to hundreds. The following paragraph describes the heuristic process.

First, the bankruptcy dataset (815 records) is divided into 100 intervals (each interval has eight records). Within each interval, seven records are randomly selected. The number of seven is determined according to empirical results of *k*-fold cross-validation. Thus 700 'bad' records are obtained. Second, the good-status dataset (4,185 records) is divided into 100 intervals (each interval has 41 records). Within each interval, seven records are randomly selected. Thus the total of 700 'good' records is obtained. Third, the 700 bankruptcy and 700 current records are combined to form a training dataset. Finally, the remaining 115 bankruptcy and 3,485 current accounts become the testing dataset. According to this procedure, the total possible combinations of this selection equals $(C_8^7 \times C_{41}^7)^{100}$. Thus, the possibility of getting identical training or testing datasets is approximately zero. The across-the-board thresholds of 65% and 70% are set for the 'bad' and 'good' class, respectively. The values of thresholds are determined from previous experience. The classification results whose predictive accuracies are below these thresholds will be filtered out.

The whole research procedure can be summarized using the following algorithm:

Algorithm 1

Input: The data set $A = \{A_1, A_2, A_3, \ldots, A_n\}$, boundary b

Output: The optimal solution, $X^* = (x_1^*, x_2^*, x_3^*, \ldots, x_{64}^*)$, the classification score $MCLP_i$

Step 1: Generate the Training set and the Testing set from the credit card data set.

Step 2: Apply the two-group MCLP model to compute the optimal solution $X^* = (x_1^*, x_2^*, \ldots, x_{64}^*)$ as the best weights of all 64 variables with given values of control parameters (b, α^*, β^*) in Training set.

Step 3: The classification score $MCLP_i = A_i X^*$ against of each observation in the Training set is calculated against the boundary b to check the performance measures of the classification.

Step 4: If the classification result of Step 3 is acceptable (i.e., the found performance measure is larger or equal to the given threshold), go to the next step. Otherwise, arbitrarily choose different values of control parameters (b, α^*, β^*) and go to Step 1.
Step 5: Use $X^* = (x_1^*, x_2^*, \ldots, x_{64}^*)$ to calculate the MCLP scores for all A_i in the Testing set and conduct the performance analysis. If it produces a satisfying classification result, go to the next step. Otherwise, go back to Step 1 to reformulate the Training Set and Testing Set.
Step 6: Repeat the whole process until a preset number (e.g., 999) of different X^* are generated for the future ensemble method.
End.

Using Algorithm 1 to the credit card dataset, classification results were obtained and summarized. Due to the space limitation, only a part (10 out of the total 500 cross-validation results) of the results is summarized in Table 1 (Peng et al., 2004). The columns "Bad" and "Good" refer to the number of records that were correctly classified as "bad" and "good," respectively. The column "Accuracy" was calculated using correctly classified records divided by the total records in that class. For instance, 80.43% accuracy of Dataset 1 for bad record in the training dataset was calculated using 563 divided by 700 and means that 80.43% of bad records were correctly classified. The average predictive accuracies for bad and good groups in the training dataset are 79.79% and 78.97%, and the average predictive accuracies for bad and good groups in the testing dataset are 68% and 74.39%. The results demonstrated that a good separation of bankruptcy and good status credit card accounts is observed with this method.

Improvement of MCLP Experimental Results with Ensemble Method

In credit card bankruptcy predictions, even a small percentage of increase in the classification accuracy can save creditors millions of dollars. Thus it is necessary to investigate possible techniques that can improve MCLP classification results. The technique studied in this experiment is majority-vote ensemble. An ensemble consists of two fundamental elements: a set of trained classifiers and an aggregation mechanism that organizes these classifiers into the output ensemble. The aggregation mechanism can be an average or a

Table 1. MCLP credit card accounts classification

Cross Validation	Training Set (700 Bad +700 Good)				Testing Set (115 Bad +3485 Good)			
	Bad	Accuracy	Good	Accuracy	Bad	Accuracy	Good	Accuracy
DataSet 1	563	80.43%	557	79.57%	78	67.83%	2575	73.89%
DataSet 2	546	78.00%	546	78.00%	75	65.22%	2653	76.13%
DataSet 3	564	80.57%	560	80.00%	75	65.22%	2550	73.17%
DataSet 4	553	79.00%	553	79.00%	78	67.83%	2651	76.07%
DataSet 5	548	78.29%	540	77.14%	78	67.83%	2630	75.47%
DataSet 6	567	81.00%	561	80.14%	79	68.70%	2576	73.92%
DataSet 7	556	79.43%	548	78.29%	77	66.96%	2557	73.37%
DataSet 8	562	80.29%	552	78.86%	79	68.70%	2557	73.37%
DataSet 9	566	80.86%	557	79.57%	83	72.17%	2588	74.26%
DataSet 10	560	80.00%	554	79.14%	80	69.57%	2589	74.29%

majority vote (Zenobi & Cunningham, 2002). Weingessel, Dimitriadou, and Hornik (2003) have reviewed a series of ensemble-related publications (Dietterich, 2000; Lam, 2000; Parhami, 1994; Bauer & Kohavi, 1999; Kuncheva, 2000). Previous research has shown that an ensemble can help to increase classification accuracy and stability (Opitz & Maclin, 1999). A part of MCLP's optimal solutions was selected to form ensembles. Each solution will have one vote for each credit card record, and final classification result is determined by the majority votes. Algorithm 2 describes the ensemble process:

Algorithm 2

Input: The data set $A = \{A_1, A_2, A_3, \ldots, A_n\}$, boundary b, a certain number of solutions, $X^* = (x_1^*, x_2^*, x_3^*, \ldots, x_{64}^*)$
Output: The classification score $MCLP_i$ and the prediction P_i
Step 1: A committee of certain odd number of classifiers X^* is formed.
Step 2: The classification score $MCLP_i = A_i X^*$ against each observation is calculated against the boundary b by every member of the committee. The performance measures of the classification will be decided by majorities of the committee. If more than half of the committee members agreed in the classification, then the prediction P_i for this observation is successful, otherwise the prediction is failed.
Step 3: The accuracy for each group will be computed by the percentage of successful classification in all observations.
End.

The results of applying Algorithm 2 are summarized in Table 2 (Peng et al., 2004). The average predictive accuracies for bad and good groups in the training dataset are 80.8% and 80.6%, and the average predictive accuracies for bad and good groups in the testing dataset are 72.17% and 76.4%. Compared with previous results, ensemble technique improves the classification accuracies. Especially for bad records classification in the testing set, the average accuracy increased 4.17%. Since bankruptcy accounts are the major cause of creditors' loss, predictive accuracy for bad records is considered to be more important than for good records.

Experimental Results of MCQP

Based on the MCQP model and the research procedure described in previous sections, similar experiments were conducted to get MCQP results. LINGO 8.0 was used to compute the optimal solutions. The whole research procedure for MCQP is summarized in Algorithm 3:

Table 2. MCLP credit card accounts classification with ensemble

Ensemble Results	Training Set (700 Bad data+700 Good data)				Testing Set (115 Bad data+3485 Good data)			
No. of Voters	Bad	Accuracy	Good	Accuracy	Bad	Accuracy	Good	Accuracy
9	563	80.43%	561	80.14%	81	70.43%	2605	74.75%
99	565	80.71%	563	80.43%	83	72.17%	2665	76.47%
199	565	80.71%	566	80.86%	83	72.17%	2656	76.21%
299	568	81.14%	564	80.57%	84	73.04%	2697	77.39%
399	567	81.00%	567	81.00%	84	73.04%	2689	77.16%

Algorithm 3

Input: The data set $A = \{A_1, A_2, A_3,\ldots, A_n\}$, boundary b
Output: The optimal solution, $X^* = (x_1^*\ x_2^*, x_3^*, \ldots, x_{64}^*)$, the classification score $MCQP_i$
Step 1: Generate the Training set and Testing set from the credit card data set.
Step 2: Apply the two-group MCQP model to compute the compromise solution $X^* = (x_1^*, x_2^*, \ldots, x_{64}^*)$ as the best weights of all 64 variables with given values of control parameters (b, α^*, β^*) using LINGO 8.0 software.
Step 3: The classification score $MCQP_i = A_i X^*$ against each observation is calculated against the boundary b to check the performance measures of the classification.
Step 4: If the classification result of Step 3 is acceptable (i.e., the found performance measure is larger or equal to the given threshold), go to the next step. Otherwise, choose different values of control parameters (b, α^*, β^*) and go to Step 1.
Step 5: Use $X^* = (x_1^*, x_2^*,\ldots, x_{64}^*)$ to calculate the MCQP scores for all A_i in the test set and conduct the performance analysis. If it produces a satisfying classification result, go to the next step. Otherwise, go back to Step 1 to reformulate the Training Set and Testing Set.
Step 6: Repeat the whole process until a preset number of different X^* are generated.
End.

A part (10 out of the total 38 results) of the results is summarized in Table 3.

The average predictive accuracies for bad and good groups in the training dataset are 86.61% and 73.29%, and the average predictive accuracies for bad and good groups in the testing dataset are 81.22% and 68.25%. Compared with MCLP, MCQP has lower predictive accuracies for good records. Nevertheless, bad group classification accuracies of the testing set using MCQP increased from 68% to 81.22%, which is a remarkable improvement.

Improvement of MCQP with Ensemble Method

Similar to the MCLP experiment, the majority-vote ensemble discussed previously was applied

Table 3. MCQP credit card accounts classification

Cross Validation	Training Set (700 Bad data+700 Good data)				Testing Set (115 Bad data+3485 Good data)			
	Bad	Accuracy	Good	Accuracy	Bad	Accuracy	Good	Accuracy
DataSet 1	602	86.00%	541	77.29%	96	83.48%	2383	68.38%
DataSet 2	614	87.71%	496	70.86%	93	80.87%	2473	70.96%
DataSet 3	604	86.29%	530	75.71%	95	82.61%	2388	68.52%
DataSet 4	616	88.00%	528	75.43%	95	82.61%	2408	69.10%
DataSet 5	604	86.29%	547	78.14%	90	78.26%	2427	69.64%
DataSet 6	614	87.71%	502	71.71%	94	81.74%	2328	66.80%
DataSet 7	610	87.14%	514	73.43%	95	82.61%	2380	68.29%
DataSet 8	582	83.14%	482	68.86%	93	80.87%	2354	67.55%
DataSet 9	614	87.71%	479	68.43%	90	78.26%	2295	65.85%
DataSet 10	603	86.14%	511	73.00%	93	80.87%	2348	67.37%

to MCQP to examine whether it can make an improvement. The results are represented in Table 4. The average predictive accuracies for bad and good groups in the training dataset are 89.18% and 74.68%, and the average predictive accuracies for bad and good groups in the testing dataset are 85.61% and 68.67%. Compared with previous MCQP results, majority-vote ensemble improves the total classification accuracies. Especially for bad records in testing set, the average accuracy increased 4.39%.

Experimental Results of Fuzzy Linear Programming

Applying the fuzzy linear programming model discussed earlier in this chapter to the same credit card dataset, we obtained some FLP classification results. These results are compared with the decision tree, MCLP, and neural networks (see Tables 5 and 6). The software of decision tree is the commercial version called C5.0 (C5.0 2004), while software for both neural network and MCLP were developed at the Data Mining Lab, University of Nebraska at Omaha, USA (Kou & Shi, 2002).

Note that in both Table 5 and Table 6, the columns T_g and T_b respectively represent the number of good and bad accounts identified by a method, while the rows of good and bad represent the actual numbers of the accounts.

Classifications on HIV-1 Mediated Neural Dendritic and Synaptic Damage Using MCLP

The ability to identify neuronal damage in the dendritic arbor during HIV-1-associated dementia (HAD) is crucial for designing specific therapies for the treatment of HAD. A two-class model of multiple criteria linear programming (MCLP) was proposed to classify such HIV-1 mediated neuronal dendritic and synaptic damages. Given certain classes, including treatments with brain-derived neurotrophic factor (BDNF), glutamate, gp120, or non-treatment controls from our in vitro experimental systems, we used the two-class MCLP model to determine the data patterns between classes in order to gain insight about neuronal dendritic and synaptic damages under different treatments (Zheng et al., 2004). This knowledge can be applied to the design and study of specific therapies for the prevention or reversal of neuronal damage associated with HAD.

Table 4. MCQP credit card accounts classification with ensemble

Ensemble Results	Training Set (700 Bad data+700 Good data)				Testing Set (115 Bad data+3485 Good data)			
No. of Voters	Bad	Accuracy	Good	Accuracy	Bad	Accuracy	Good	Accuracy
3	612	87.43%	533	76.14%	98	85.22%	2406	69.04%
5	619	88.43%	525	75.00%	95	82.61%	2422	69.50%
7	620	88.57%	525	75.00%	97	84.35%	2412	69.21%
9	624	89.14%	524	74.86%	100	86.96%	2398	68.81%
11	625	89.29%	525	75.00%	99	86.09%	2389	68.55%
13	629	89.86%	517	73.86%	100	86.96%	2374	68.12%
15	629	89.86%	516	73.71%	98	85.22%	2372	68.06%
17	632	90.29%	520	74.29%	99	86.09%	2379	68.26%
19	628	89.71%	520	74.29%	100	86.96%	2387	68.49%

Database

The data produced by laboratory experimentation and image analysis was organized into a database composed of four classes (G1-G4), each of which has nine attributes. The four classes are defined as the following:

- **G1:** Treatment with the neurotrophin BDNF (brain-derived neurotrophic factor, 0.5 ng/ml, 5 ng/ml, 10 ng/mL, and 50 ng/ml), this factor promotes neuronal cell survival and has been shown to enrich neuronal cell cultures (Lopez et al., 2001; Shibata et al., 2003).
- **G2:** Non-treatment, neuronal cells are kept in their normal media used for culturing (Neurobasal media with B27, which is a neuronal cell culture maintenance supplement from Gibco, with glutamine and penicillin-streptomycin).
- **G3:** Treatment with glutamate (10, 100, and 1,000 □M). At low concentrations, glutamate acts as a neurotransmitter in the brain. However, at high concentrations, it has been shown to be a neurotoxin by over-stimulating NMDA receptors. This factor has been shown to be upregulated in HIV-1-infected macrophages (Jiang et al., 2001) and thereby linked to neuronal damage by HIV-1 infected macrophages.
- **G4:** Treatment with gp120 (1 nanoM), an HIV-1 envelope protein. This protein could interact with receptors on neurons and interfere with cell signaling leading to neuronal damage, or it could also indirectly induce neuronal injury through the production of other neurotoxins (Hesselgesser et al., 1998; Kaul, Garden, & Lipton, 2001; Zheng et al., 1999).

The nine attributes are defined as:

- x1 = The number of neurites

Table 5. Learning comparisons on balanced 280 records

Decision Tree	T_g	T_b	Total
Good	***138***	2	140
Bad	13	**127**	140
Total	151	129	**280**
Neural Network	T_g	T_b	Total
Good	**116**	24	140
Bad	14	**126**	140
Total	130	150	**280**
MCLP	T_g	T_b	Total
Good	**134**	6	140
Bad	7	***133***	140
Total	141	139	**280**
FLP	T_g	T_b	Total
Good	**127**	13	140
Bad	13	**127**	140
Total	140	140	**280**

Table 6. Comparisons on prediction of 5,000 records

Decision Tree	T_g	T_b	Total
Good	**2180**	2005	4185
Bad	141	**674**	815
Total	2321	2679	**5000**
Neural Network	T_g	T_b	Total
Good	**2814**	1371	4185
Bad	176	**639**	815
Total	2990	2010	**5000**
MCLP	T_g	T_b	Total
Good	***3160***	1025	4185
Bad	484	**331**	815
Total	3644	1356	**5000**
FLP	T_g	T_b	Total
Good	**2498**	1687	4185
Bad	113	***702***	815
Total	2611	2389	**5000**

- x2 = The number of arbors
- x3 = The number of branch nodes
- x4 = The average length of arbors
- x5 = The ratio of neurite to arbor
- x6 = The area of cell bodies
- x7 = The maximum length of the arbors
- x8 = The culture time (during this time, the neuron grows normally and BDNF, glutamate, or gp120 have not been added to affect growth)
- x9 = The treatment time (during this time, the neuron was growing under the effects of BDNF, glutamate, or gp120)

The database used in this chapter contained 2,112 observations. Among them, 101 are on G1, 1,001 are on G2, 229 are on G3, and 781 are on G4.

Comparing with the traditional mathematical tools in classification, such as neural networks, decision tree, and statistics, the two-class MCLP approach is simple and direct, free of the statistical assumptions, and flexible by allowing decision makers to play an active part in the analysis (Shi, 2001).

Results of Empirical Study Using MCLP

By using the two-class model for the classifications on {G1, G2, G3, and G4}, there are six possible pairings: G1 vs. G2; G1 vs. G3; G1 vs. G4; G2 vs. G3; G2 vs. G4; and G3 vs. G4. In the cases of G1 vs. G3 and G1 vs. G4, we see these combinations would be treated as redundancies, therefore they are not considered in the pairing groups. G1 through G3 or G4 is a continuum. G1 represents an enrichment of neuronal cultures, G2 is basal or maintenance of neuronal culture, and G3/G4 are both damage of neuronal cultures. There would never be a jump between G1 to G3/G4 without traveling through G2. So, we used the following four two-class pairs: G1 vs. G2; G2 vs. G3; G2 vs. G4; and G3 vs. G4. The meanings of these two-class pairs are:

- G1 vs. G2 shows that BDNF should enrich the neuronal cell cultures and increase neuronal network complexity—that is, more dendrites and arbors, more length to dendrites, and so forth.
- G2 vs. G3 indicates that glutamate should damage neurons and lead to a decrease in dendrite and arbor number including dendrite length.
- G2 vs. G4 should show that gp120 causes neuronal damage leading to a decrease in dendrite and arbor number and dendrite length.
- G3 vs. G4 provides information on the possible difference between glutamate toxicity and gp120-induced neurotoxicity.

Given a threshold of training process that can be any performance measure, we have carried out the following steps:

Algorithm 4

Step 1: For each class pair, we used the Linux code of the two-class model to compute the compromise solution $X^* = (x_1^*,..., x_9^*)$ as the best weights of all nine neuronal variables with given values of control parameters (b, α^*, β^*).

Step 2: The classification score $MCLP_i = A_i X^*$ against of each observation has been calculated against the boundary b to check the performance measures of the classification.

Step 3: If the classification result of Step 2 is acceptable (i.e., the given performance measure is larger or equal to the given threshold), go to Step 4. Otherwise, choose different values of control parameters (b, α^*, β^*) and go to Step 1.

Step 4: For each class pair, use $X^* = (x_1^*,..., x_9^*)$ to calculate the MCLP scores for all A_i in the test set and conduct the performance analysis.

According to the nature of this research, we define the following terms, which have been widely used in the performance analysis as:

TP (True Positive) = the number of records in the first class that has been classified correctly
FP (False Positive) = the number of records in the second class that has been classified into the first class
TN (True Negative) = the number of records in the second class that has been classified correctly
FN (False Negative) = the number of records in the first class that has been classified into the second class

Then we have four different performance measures:

$$\text{Sensitivity} = \frac{TP}{TP+FN}$$

$$\text{Positive Predictivity} = \frac{TP}{TP+FP}$$

$$\text{False-Positive Rate} = \frac{FP}{TN+FP}$$

$$\text{Negative Predictivity} = \frac{TN}{FN+TN}$$

The "positive" represents the first-class label while the "negative" represents the second-class label in the same class pair. For example, in the class pair {G1 vs. G2}, the record of G1 is "positive" while that of G2 is "negative." Among the above four measures, more attention is paid to sensitivity or false-positive rates because both measure the correctness of classification on class-pair data analyses. Note that in a given a class pair, the sensitivity represents the corrected rate of the first class, and one minus the false positive rate is the corrected rate of the second class by the above measure definitions.

Considering the limited data availability in this pilot study, we set the across-the-board threshold of 55% for sensitivity [or 55% of (1- false positive rate)] to select the experimental results from training and test processes. All 20 of the training and test sets, over the four class pairs, have been computed using the above procedure. The results against the threshold are summarized in Tables 7 to 10. As seen in these tables, the sensitivities for the comparison of all four pairs are higher than 55%, indicating that good separation among individual pairs is observed with this method. The results are then analyzed in terms of both positive predictivity and negative predictivity for the prediction power of the MCLP method on neuron injuries. In Table 7, G1 is the number of observations predefined as BDNF treatment, G2 is the number of observations predefined as non-treatment, N1 means the number of obser-

Table 7. Classification results with G1 vs. G2

Training	N1	N2	Sensitivity	Positive Predictivity	False Positive Rate	Negative Predictivity
G1	55 (TP)	34 (FN)	61.80%	61.80%	38.20%	61.80%
G2	34 (FP)	55 (TN)				
Test	N1	N2	Sensitivity	Positive Predictivity	False Positive Rate	Negative Predictivity
G1	11 (TP)	9 (FN)	55.00%	3.78%	30.70%	98.60%
G2	280 (FP)	632 (TN)				

Table 8. Classification results with G2 vs. G3

Training	N2	N3	Sensitivity	Positive Predictivity	False Positive Rate	Negative Predictivity
G2	126 (TP)	57 (FN)	68.85%	68.48%	31.69%	68.68%
G3	58 (FP)	125 (TN)				
Test	N2	N3	Sensitivity	Positive Predictivity	False Positive Rate	Negative Predictivity
G2	594 (TP)	224 (FN)	72.62%	99.32%	8.70%	15.79%
G3	4 (FP)	42 (TN)				

Table 9. Classification results with G2 vs. G4

Training	N2	N4	Sensitivity	Positive Predictivity	False Positive Rate	Negative Predictivity
G2	419(TP)	206 (FN)	67.04%	65.88%	34.72%	66.45%
G4	217 (FP)	408 (TN)				
Test	N2	N4	Sensitivity	Positive Predictivity	False Positive Rate	Negative Predictivity
G2	216 (TP)	160 (FN)	57.45%	80.90%	32.90%	39.39%
G4	51 (FP)	104 (TN)				

Table 10. Classification results with G3 vs. G4

Training	N3	N4	Sensitivity	Positive Predictivity	False Positive Rate	Negative Predictivity
G3	120(TP)	40 (FN)	57.45%	80.90%	24.38%	75.16%
G4	39 (FP)	121 (TN)				
Test	N3	N4	Sensitivity	Positive Predictivity	False Positive Rate	Negative Predictivity
G3	50 (TP)	19 (FN)	72.46%	16.78%	40.00%	95.14%
G4	248 (FP)	372 (TN)				

vations classified as BDNF treatment, and N2 is the number of observations classified as non-treatment. The meanings of other pairs in Tables 8 to 10 can be similarly explained. In Table 7 for {G1 vs. G2}, both positive predictivity and negative predictivity are the same (61.80%) in the training set. However, the negative predictivity of the test set (98.60%) is much higher than that of the positive predictivity (3.78%). The prediction of G1 in the training set is better than that of the test set, while the prediction of G2 in test outperforms that of training. This is due to the small size of G1. In Table 3 for {G2 vs. G3}, the positive predictivity (68.48%) is almost equal to the negative predictivity (68.68%) of the training set. The positive predictivity (99.32%) is much higher than the negative predictivity (15.79%) of the test set. As a result, the prediction of G2 in the test set is better than in the training set, but the prediction of G3 in the training set is better than in the test set.

The case of Table 9 for {G2 vs. G4} is similar to that of Table 8 for {G2 vs. G3}. We see that the separation of G2 in test (80.90%) is better than in training (65.88%), while the separation of G4 in training (66.45%) is better than in test (39.39%). In the case of Table 10 for {G3 vs. G4}, the positive predictivity (80.90%) is higher than the negative predictivity (75.16%) of the training set. Then, the positive predictivity (16.78%) is much lower than the negative predictivity (95.14%) of the test set. The prediction of G3 in training (80.90%) is better than that of test (16.78%), and the prediction of G4 in test (95.14%) is better than that of training (75.16%).

In summary, we observed that the predictions of G2 in test for {Gl vs. G2}, {G2 vs. G3}, and {G2 vs. G4} is always better than those in training. The prediction of G3 in training for {G2 vs. G3} and {G3 vs. G4} is better than those of test. Finally, the prediction of G4 for {G2 vs. G4} in training reverses that of {G3 vs. G4} in test. If we emphasize the test results, these results are favorable to G2. This may be due to the size of G2 (non-treatment), which is larger than all other classes. The classification results can change if the sizes of Gl, G3, and G4 increase significantly.

Network Intrusion Detection

Network intrusions are malicious activities that aim to misuse network resources. Although various approaches have been applied to network intrusion detection, such as statistical analysis, sequence analysis, neural networks, machine learning, and artificial immune systems, this field is far from maturity, and new solutions are worthy of investigation. Since intrusion detection can be treated as a classification problem, it is feasible to apply a multiple-criterion classification model to this type of application. The objective of this experiment is to examine the applicability of MCLP and MCQP models in intrusion detection.

KDD99 Dataset

The KDD-99 dataset provided by DARPA was used in our intrusion detection test. The KDD-99 dataset includes a wide variety of intrusions simulated in a military network environment. It was used in the 1999 KDD-CUP intrusion detection contest. After the contest, KDD-99 has become a de facto standard dataset for intrusion detection experiments. Within the KDD-99 dataset, each connection has 38 numerical variables and is labeled as normal or attack. There are four main categories of attacks: denial-of-service (DOS), unauthorized access from a remote machine (R2L), unauthorized access to local root privileges (U2R), surveillance and other probing. The training dataset contains a total of 24 attack types, while the testing dataset contains an additional 14 types (Stolfo, Fan, Lee, Prodromidis, & Chan, 2000). Because the number of attacks for R2L, U2R, and probing is relatively small, this experiment focused on DOS.

Experimental Results of MCLP

Following the heuristic process described in this chapter, training and testing datasets were selected: first, the 'normal' dataset (812,813 records) was divided into 100 intervals (each interval has 8,128 records). Within each interval, 20 records were randomly selected. Second, the 'DOS' dataset (247,267 records) was divided into 100 intervals (each interval has 2,472 records). Within each interval, 20 records were randomly selected. Third, the 2,000 normal and 2,000 DOS records were combined to form a training dataset. Because KDD-99 has over 1 million records, and 4,000 training records represent less than 0.4% of it, the whole KDD-99 dataset is used for testing. Various training and testing datasets can be obtained by repeating this process. Considering the previous high detection rates of KDD-99 by other methods, the across-the-board threshold of 95% was set for both normal and DOS. Since training dataset classification accuracies are all 100%, only testing dataset (10 out of the total 300 results) results are summarized in Table 11 (Kou et al., 2004a). The average predictive accuracies for normal and DOS groups in the testing dataset are 98.94% and 99.56%.

Improvement of MCLP with Ensemble Method

The majority-vote ensemble method demonstrated its superior performance in credit card accounts classification. Can it improve the classification accuracy of network intrusion detection? To answer this question, the majority-vote ensemble was applied to the KDD-99 dataset. Ensemble results are summarized in Table 12 (Kou et al., 2004a). The average predictive accuracies for normal and DOS groups in the testing dataset are 99.61% and 99.78%. Both normal and DOS predictive accuracies have been slightly improved.

Table 11. MCLP KDD-99 classification results

Cross Validation	Testing Set (812813 Normal + 247267 Dos)			
	Normal	Accuracy	DOS	Accuracy
DataSet 1	804513	98.98%	246254	99.59%
DataSet 2	808016	99.41%	246339	99.62%
DataSet 3	802140	98.69%	245511	99.29%
DataSet 4	805151	99.06%	246058	99.51%
DataSet 5	805308	99.08%	246174	99.56%
DataSet 6	799135	98.32%	246769	99.80%
DataSet 7	805639	99.12%	246070	99.52%
DataSet 8	802938	98.79%	246566	99.72%
DataSet 9	805983	99.16%	245498	99.28%
DataSet 10	802765	98.76%	246641	99.75%

Table 12. MCLP KDD-99 classification results with ensemble

Number of Voters	Normal	Accuracy	DOS	Accuracy
3	809567	99.60%	246433	99.66%
5	809197	99.56%	246640	99.75%
7	809284	99.57%	246690	99.77%
9	809287	99.57%	246737	99.79%
11	809412	99.58%	246744	99.79%
13	809863	99.64%	246794	99.81%
15	809994	99.65%	246760	99.79%
17	810089	99.66%	246821	99.82%
19	810263	99.69%	246846	99.83%

Experimental Results of MCQP

A similar MCQP procedure used in credit card accounts classification was used to classify the KDD-99 dataset. A part of the results is summarized in Table 13 (Kou et al., 2004b). These results are slightly better than MCLP.

Improvement of MCQP with Ensemble Method

The majority-vote ensemble was used on MCQP results, and a part of the outputs is summarized in Table 14 (Kou et al., 2004b). The average predictive accuracies for normal and DOS groups in the testing dataset are 99.86% and 99.82%. Although the increase in classification accuracy is small,

Table 13. MCQP KDD-99 classification results

Cross Validation	Testing Set(812813 Normal + 247267 Dos)			
	Normal	Accuracy	DOS	Accuracy
DataSet 1	808142	99.43%	245998	99.49%
DataSet 2	810689	99.74%	246902	**99.85%**
DataSet 3	807597	99.36%	246491	99.69%
DataSet 4	808410	99.46%	246256	99.59%
DataSet 5	810283	99.69%	246090	99.52%
DataSet 6	809272	99.56%	246580	99.72%
DataSet 7	806116	**99.18%**	246229	99.58%
DataSet 8	808143	99.43%	245998	99.49%
DataSet 9	811806	**99.88%**	246433	99.66%
DataSet 10	810307	99.69%	246702	99.77%

Table 14. MCQP KDD-99 classification results with ensemble

NO of Voters	Normal	Accuracy	DOS	Accuracy
3	810126	99.67%	246792	99.81%
5	811419	99.83%	246930	**99.86%**
7	811395	99.83%	246830	99.82%
9	811486	99.84%	246795	99.81%
11	812030	99.90%	246845	99.83%
13	812006	99.90%	246788	99.81%
15	812089	99.91%	246812	99.82%
17	812045	99.91%	246821	99.82%
19	812069	99.91%	246817	99.82%
21	812010	99.90%	246831	99.82%
23	812149	**99.92%**	246821	99.82%
25	812018	99.90%	246822	99.82%

both normal and DOS predictive accuracies have been improved compared with previous 99.54% and 99.64%.

RESEARCH CHALLENGES AND OPPORTUNITIES

Although the above multiple criteria optimization data mining methods have been applied in the real-life applications, there are number of challenging problems in mathematical modeling. While some of the problems are currently under investigation, some others remain to be explored.

Variations and Algorithms of Generalized Models

Given Model 1, if p=2, q=1, it will become a convex quadratic program which can be solved by using some known convex quadratic programming algorithm. However, when p=1, q=2, Model 1 is a concave quadratic program; and when p=2, q=2, we have Model 3 (MCQP), which is an indefinite quadratic problem. Since both concave quadratic programming and MCQP are NP-hard problems, it is very difficult to find a global optimal solution. We are working on both cases for developing direct algorithms that can converge to local optima in classification (Zhang, Shi, & Zhang, 2005).

Kernel Functions for Data Observations

The generalized model in the chapter has a natural connection with known support vector machines (SVM) (Mangasarian, 2000; Vapnik, 2000) since they both belong to the category of optimization-based data mining methods. However, they differ from ways to identify the classifiers. As we mentioned before, while the multiple criteria optimization approaches in this chapter use the overlapping and interior distance as two standards to measure the separation of each observation in the dataset, SVM selects the minority of observations (support vectors) to represent the majority of the rest of the observations. Therefore, in the experimental studies and real applications, SVM may have a high accuracy in the training set, but a lower accuracy in the testing result. Nevertheless, the use of kernel functions in SVM has shown its efficiency in handling nonlinear datasets. How to adopt kernel functions into the multiple criteria optimization approaches can be an interesting research problem. Kou, Peng, Shi, and Chen (2006) explored some possibility of this research direction. The basic idea is outlined.

First, we can rewrite the generalized model (Model 1) similar to the approach of SVM. Suppose the two-classes G_1 and G_2 are under consideration. Then, a $n \times n$ diagonal matrix Y, which only contains +1 or -1, indicates the class membership. A -1 in row i of matrix Y indicates the corresponding record $A_i \in G_1$, and a +1 in row i of matrix Y indicates the corresponding record $A_i \in G_2$. The constraints in Model 1, $A_iX = b + \alpha_i - \beta_i$, $\forall A_i \in G_1$ and $A_iX = b - \alpha_i + \beta_i$, $\forall A_i \in G_2$, are converted as: $Y(\langle A{\cdot}X\rangle - eb) = \alpha - \beta$, where $e = (1,1,\ldots,1)^T$, $\alpha = (\alpha_1,\ldots,\alpha_n)$, and $\beta = (\beta_1,\ldots,\beta_n)^T$. In order to maximize the distance $\frac{2}{\|X\|_2}$ between the two adjusted bounding hyper planes, the function $\frac{1}{2}\|X\|_2$ should also be minimized. Let $s = 2$, $q = 1$, and $p = 1$, then a simple quadratic programming (SQP) variation of Model 1 can be built as:

Model 5: SQP

$$\textit{Minimize } -\frac{1}{2}\|X\|_2 + w_\alpha \sum_{i=1}^{n} \alpha_i - w_\beta \sum_{i=1}^{n} \beta_i$$

Subject to $Y(\langle A{\cdot}X\rangle - eb) = \alpha - \beta$, where $e = (1,1,\ldots,1)^T$, $\alpha = (\alpha_1,\ldots,\alpha_n)^T$ and $\beta = (\beta_1,\ldots,\beta_n)^T \geq 0$.

Using Lagrange function to represent Model 5, one can get an equivalent of the Wolfe dual problem of Model 5 expressed as:

Model 6: Dual of SQP

$$Maximize -\frac{1}{2}\sum_{i=1}^{n}\sum_{j=1}^{n}\xi_i y_i \xi_j y_j (\mathbf{A_i}\cdot\mathbf{A_j})+\delta\sum_{i=1}^{n}\xi_i$$

$$Subject\ to \sum_{i=1}^{n}\xi_i y_i = 0,\ w_\beta \le \xi_i \le w_\alpha,$$

where $w_\beta < w_\alpha$ are given, $1 \le i \le n$.

The global optimal solution of the primal problem if Model 5 can be obtained from the solution of the Wolfe dual problem:

$$\mathbf{X}^* = \sum_{i=1}^{n}\xi_i^* y_i \mathbf{A_i},\ b^* = y_j - \sum_{i=1}^{n}\xi^* y_i(\mathbf{A_i}\cdot\mathbf{A_j}).$$

As a result, the classification decision function becomes:

$$sgn((\mathbf{X}^*\cdot B)-b^*)\begin{cases} > 0, B\in G_1 \\ \le 0,\ B\in G_2 \end{cases},$$

We observe that because the form $(A_i \cdot A_j)$ of Model 6 is inner product in the vector space, it can be substituted by a positive semi-definite kernel $K(A_i, A_j)$ without affecting the mathematical modeling process. In general, a kernel function refers to a real-valued function on $\chi\times\chi$ and for all $A_i, A_j \in \chi$. Thus, Model 6 can be easily transformed to a nonlinear model by replacing $(A_i \cdot A_j)$ with some positive semi-definite kernel function $K(A_i, A_j)$. Use of kernel functions in multiple criteria optimization approaches can extend its applicability to linear inseparable datasets. However, there are some theoretical difficulties to directly introduce kernel function to Model 5. How to overcome them deserves a careful study. Future studies may be done on establishing a theoretical guideline for selection of a kernel that is optimal in achieving a satisfactory credit analysis result. Another open problem is to study the subject of reducing computational cost and improving algorithm efficiency for high dimensional or massive datasets.

Choquet Integrals and Non-Additive Set Function

Considering the r-dimensional attribute vector $a = (a_1,...,a_r)$ in the classification problem, let $P(a)$ denote the power set of a. We use $f(a_1),...,f(a_r)$ to denote the values of each attribute in an observation. The procedure of calculating a Choquet integral can be given as (Wang & Wang, 1997):

$$\int f\,d\mu = \sum_{j=1}^{r}[f(a'_j) - f(a'_{j-1})]\times\mu(\{a'_1, a'_2,..., a'_r\}),$$

where $\{a'_1, a'_2,..., a'_r\}$ is a permutation of $a = (a_1,...,a_r)$. Such that $f(a'_0) = 0$ and $f(a'_1),..., f(a'_r)$ is non-decreasingly ordered such that: $f(a_1) \le ... \le f(a_r)$. The non-additive set function is defined as: μ:P(a)→(-∞,+∞), where $\mu(\varnothing) = 0$. We use μ_i to denote set function μ, where $i = 1,...,2^r$.

Introducing the Choquet measure into the generalized model of an section refers to the utilization of Choquet integral as a representative of the left-hand side of the constraints in Model 1. This variation for non-additive data mining problem is (Yan, Wang, Shi, & Chen, 2005):

Model 7: Choquet Form

Minimize $f(\alpha)$ *and Maximize* $g(\beta)$

Subject to:

$$\int f\,d\mu - \alpha_i + \beta_i - b = 0,\ \forall A_i \in G_1,$$
$$\int f\,d\mu + \alpha_i - \beta_i - b = 0,\ \forall A_i \in G_2,$$

where $\int f\,d\mu$ denotes the Choquet integral with respect to a signed fuzzy measure to aggregate the attributes of a observation f, b is unrestricted, and $\alpha = (\alpha_1,...,\alpha_n)^T$, $\beta = (\beta_1,...,\beta_n)^T$; α_i, $\beta_i \ge 0$, $i = 1,..., n$.

Model 7 results in the replacement of a linear combination of all the attributes $A_i X$ in the left-hand side of constraints with the Choquet integral representation $\int f\,d\mu$. The number of parameters, denoted by μ_i, increases from r to 2^r (r is the number attributes). How to determine the parameters through linear programming framework is not easy. We are still working on this problem and shall report the significant results.

CONCLUSION

As Usama Fayyad pointed out at the KDD-03 Panel, data mining must attract the participation of the relevant communities to avoid re-inventing wheels and bring the field an auspicious future (Fayyad, Piatetsky-Shapiro, & Uthurusamy, 2003). One relevant field to which data mining has not attracted enough participation is optimization. This chapter summarizes a series of research activities that utilize multiple criteria decision-making methods to classification problems in data mining. Specifically, this chapter describes a variation of multiple criteria optimization-based models and applies these models to credit card scoring management, HIV-1 associated dementia (HAD) neuronal damage and dropout, and network intrusion detection as well as the potential in various real-life problems.

ACKNOWLEDGMENT

Since 1998, this research has been partially supported by a number of grants, including First Data Corporation, USA; DUE-9796243, the National Science Foundation of USA; U.S. Air Force Research Laboratory (PR No. E-3-1162); National Excellent Youth Fund #70028101, Key Project #70531040, #70472074, National Natural Science Foundation of China; 973 Project #2004CB720103, Ministry of Science and Technology, China; K.C. Wong Education Foundation (2001, 2003), Chinese Academy of Sciences; and BHP Billiton Co., Australia.

REFERENCES

Bradley, P.S., Fayyad, U.M., & Mangasarian, O.L. (1999). Mathematical programming for data mining: Formulations and challenges. *INFORMS Journal on Computing, 11*, 217-238.

Bauer, E., & Kohavi, R. (1999). an empirical comparison of voting classification algorithms: Bagging, boosting, and variants. *Machine Learning, 36*, 105-139.

C 5.0. (2004). Retrieved from http://www.rulequest.com/see5-info.html

Charnes, A., & Cooper, W.W. (1961). *Management models and industrial applications of linear programming* (vols. 1 & 2). New York: John Wiley & Sons.

Dietterich, T. (2000). *Ensemble methods in machine learning.* In Kittler & Roli (Eds.), Multiple classifier systems (pp. 1-15). Berlin: Springer-Verlag (Lecture Notes in Pattern Recognition 1857).

Fayyad, U.M., Piatetsky-Shapiro, G., & Uthurusamy, R. (2003). Summary from the KDD-03 Panel: Data mining: The next 10 years. *ACM SIGKDD Explorations Newsletter, 5*(2), 191-196.

Fisher, R.A. (1936). The use of multiple measurements in taxonomic problems. *Annals of Eugenics, 7*, 179-188.

Freed, N., & Glover, F. (1981). Simple but powerful goal programming models for discriminant problems. *European Journal of Operational Research, 7*, 44-60.

Freed, N., & Glover, F. (1986). Evaluating alternative linear programming models to solve the two-group discriminant problem. *Decision Science, 17*, 151-162.

Han, J.W., & Kamber, M. (2000). *Data mining: Concepts and techniques.* San Diego: Academic Press.

He, J., Liu, X., Shi, Y., Xu, W., & Yan, N. (2004). Classifications of credit cardholder behavior by using fuzzy linear programming. *International Journal of Information Technology and Decision Making, 3*, 633-650.

Hesselgesser, J., Taub, D., Baskar, P., Greenberg, M., Hoxie, J., Kolson, D.L., & Horuk, R. (1998). Neuronal apoptosis induced by HIV-1 gp120 and the Chemokine SDF-1alpha mediated by the Chemokine receptor CXCR4. *Curr Biol, 8*, 595-598.

Kaul, M., Garden, G.A., & Lipton, S.A. (2001). Pathways to neuronal injury and apoptosis in HIV-associated dementia. *Nature, 410*, 988-994.

Kou, G., & Shi, Y. (2002). *Linux-based Multiple Linear Programming Classification Program: (Version 1.0.)* College of Information Science and Technology, University of Nebraska-Omaha, USA.

Kou, G., Liu, X., Peng, Y., Shi, Y., Wise, M., & Xu, W. (2003). Multiple criteria linear programming approach to data mining: Models, algorithm designs and software development. *Optimization Methods and Software, 18*, 453-473.

Kou, G., Peng, Y., Yan, N., Shi, Y., Chen, Z., Zhu, Q., Huff, J., & McCartney, S. (2004a, July 19-21). Network intrusion detection by using multiple-criteria linear programming. In *Proceedings of the International Conference on Service Systems and Service Management*, Beijing, China.

Kou, G., Peng, Y., Chen, Z., Shi, Y., & Chen, X. (2004b, July 12-14). A multiple-criteria quadratic programming approach to network intrusion detection. In *Proceedings of the Chinese Academy of Sciences Symposium on Data Mining and Knowledge Management*, Beijing, China.

Kou, G., Peng, Y., Shi, Y., & Chen, Z. (2006). *A new multi-criteria convex quadratic programming model for credit data analysis.* Working Paper, University of Nebraska at Omaha, USA.

Kuncheva, L.I. (2000). Clustering-and-selection model for classifier combination. In *Proceedings of the 4th International Conference on Knowledge-Based Intelligent Engineering Systems and Allied Technologies* (KES'2000).

Kwak, W., Shi, Y., Eldridge, S., & Kou, G. (2006). Bankruptcy prediction for Japanese firms: Using multiple criteria linear programming data mining approach. In *Proceedings of the International Journal of Data Mining and Business Intelligence.*

Jiang, Z., Piggee, C., Heyes, M.P., Murphy, C., Quearry, B., Bauer, M., Zheng, J., Gendelman, H.E., & Markey, S.P. (2001). Glutamate is a mediator of neurotoxicity in secretions of activated HIV-1-infected macrophages. *Journal of Neuroimmunology, 117*, 97-107.

Lam, L. (2000). *Classifier combinations: Implementations and theoretical issues.* In Kittler & Roli (Eds.), Multiple classifier systems (pp. 78-86). Berlin: Springer-Verlag (Lecture Notes in Pattern Recognition 1857).

Lee, S.M. (1972). *Goal programming for decision analysis.* Auerbach.

Lindsay, P.H., & Norman, D.A. (1972). *Human information processing: An introduction to psychology.* New York: Academic Press.

LINDO Systems Inc. (2003). *An overview of LINGO 8.0.* Retrieved from http://www.lindo.com/cgi/frameset.cgi?leftlingo.html;lingof.html

Lopez, A., Bauer, M.A., Erichsen, D.A., Peng, H., Gendelman, L., Shibata, A., Gendelman, H.E., & Zheng, J. (2001). The regulation of neurotrophic factor activities following HIV-1 infection and immune activation of mononuclear phagocytes. In *Proceedings of Soc. Neurosci. Abs.*, San Diego, CA.

Mangasarian, O.L. (2000). Generalized support vector machines. In A. Smola, P. Bartlett, B. Scholkopf, & D. Schuurmans (Eds.), Advances in large margin classifiers (pp. 135-146). Cambridge, MA: MIT Press.

Olson, D., & Shi, Y. (2005). *Introduction to business data mining*. New York: McGraw-Hill/ Irwin.

Opitz, D., & Maclin, R. (1999). Popular ensemble methods: An empirical study. *Journal of Artificial Intelligence Research, 11*, 169-198.

Parhami, B. (1994). *Voting algorithms. IEEE Transactions on Reliability, 43*, 617-629.

Peng, Y., Kou, G., Chen, Z., & Shi, Y. (2004). Cross-validation and ensemble analyses on multiple-criteria linear programming classification for credit cardholder behavior. In *Proceedings of ICCS 2004* (pp. 931-939). Berlin: Springer-Verlage (LNCS 2416).

Shi, Y., & Yu, P.L. (1989). *Goal setting and compromise solutions*. In B. Karpak & S. Zionts (Eds.), Multiple criteria decision making and risk analysis using microcomputers (pp. 165-204). Berlin: Springer-Verlag.

Shi, Y. (2001). *Multiple criteria and multiple constraint levels linear programming: Concepts, techniques and applications*. NJ: World Scientific.

Shi, Y., Wise, W., Luo, M., & Lin, Y. (2001). *Multiple criteria decision making in credit card portfolio management*. In M. Koksalan & S. Zionts (Eds.), Multiple criteria decision making in new millennium (pp. 427-436). Berlin: Springer-Verlag.

Shi, Y, Peng, Y., Xu, W., & Tang, X. (2002). Data mining via multiple criteria linear programming: Applications in credit card portfolio management. *International Journal of Information Technology and Decision Making, 1*, 131-151.

Shi, Y, Peng, Y., Kou, G., & Chen, Z. (2005). Classifying credit card accounts for business intelligence and decision making: A multiple-criteria quadratic programming approach. *International Journal of Information Technology and Decision Making, 4*, 581-600.

Shibata, A., Zelivyanskaya, M., Limoges, J., Carlson, K.A., Gorantla, S., Branecki, C., Bishu, S., Xiong, H., & Gendelman, H.E. (2003). Peripheral nerve induces macrophage neurotrophic activities: Regulation of neuronal process outgrowth, intracellular signaling and synaptic function. *Journal of Neuroimmunology, 142*, 112-129.

Stolfo, S.J., Fan, W., Lee, W., Prodromidis, A., & Chan, P.K. (2000). Cost-based modeling and evaluation for data mining with application to fraud and intrusion detection: Results from the JAM project. In *Proceedings of the DARPA Information Survivability Conference*.

Vapnik, V.N. (2000). T*he nature of statistical learning theory* (2nd ed.). New York: Springer.

Wang, J., & Wang, Z. (1997). Using neural network to determine Sugeno measures by statistics. *Neural Networks, 10*, 183-195.

Weingessel, A., Dimitriadou, E., & Hornik, K. (2003, March 20-22). An ensemble method for clustering. In *Proceedings of the 3rd International Workshop on Distributed Statistical Computing*, Vienna, Austria.

Yan, N., Wang, Z., Shi, Y., & Chen, Z. (2005). *Classification by linear programming with signed fuzzy measures*. Working Paper, University of Nebraska at Omaha, USA.

Yu, P.L. (1985). *Multiple criteria decision making: Concepts, techniques and extensions*. New York: Plenum Press.

Zenobi, G., & Cunningham, P. (2002). An approach to aggregating ensembles of lazy learners that supports explanation. *Lecture Notes in Computer Science, 2416*, 436-447.

Zhang, J., Shi, Y., & Zhang, P. (2005). S*everal multi-criteria programming methods for clas-*

sification. Working Paper, Chinese Academy of Sciences Research Center on Data Technology & Knowledge Economy and Graduate University of Chinese Academy of Sciences, China.

Zheng, J., Thylin, M., Ghorpade, A., Xiong, H., Persidsky, Y., Cotter, R., Niemann, D., Che, M., Zeng, Y., Gelbard, H. et al. (1999). Intracellular CXCR4 signaling, neuronal apoptosis and neuropathogenic mechanisms of HIV-1-associated dementia. *Journal of Neuroimmunology, 98*, 185-200.

Zheng, J., Zhuang, W., Yan, N., Kou, G., Erichsen, D., McNally, C., Peng, H., Cheloha, A., Shi, C., & Shi, Y. (2004). Classification of HIV-1-mediated neuronal dendritic and synaptic damage using multiple criteria linear programming. *Neuroinformatics, 2*, 303-326.

Zimmermann, H.-J. (1978). Fuzzy programming and linear programming with several objective functions. *Fuzzy Sets and Systems, 1*, 45-55.

This work was previously published in Research and Trends in Data Mining Technologies and Applications, edited by D. Taniar, pp. 242-275, copyright 2007 by IGI Publishing, formerly known as Idea Group Publishing (an imprint of IGI Global).

Chapter II
Making Decisions with Data: Using Computational Intelligence Within a Business Environment

Kevin Swingler
University of Stirling, Scotland

David Cairns
University of Stirling, Scotland

ABSTRACT

This chapter identifies important barriers to the successful application of computational intelligence (CI) techniques in a commercial environment and suggests a number of ways in which they may be overcome. It identifies key conceptual, cultural and technical barriers and describes the different ways in which they affect both the business user and the CI practitioner. The chapter does not provide technical detail on how to implement any given technique, rather it discusses the practical consequences for the business user of issues such as non-linearity and extrapolation. For the CI practitioner, we discuss several cultural issues that need to be addressed when seeking to find a commercial application for CI techniques. The authors aim to highlight to technical and business readers how their different expectations can affect the successful outcome of a CI project. The authors hope that by enabling both parties to understand each other's perspective, the true potential of CI can be realized.

INTRODUCTION

Computational intelligence (CI) appears to offer new opportunities to a business that wishes to improve the efficiency of their operations. It appears to provide a view into the future, answering questions such as, "What will my customers buy?", "Who is most likely to file a claim on an insurance policy?", and "What increase in demand will follow an advertising campaign?" It can filter good prospects from bad, the fraudulent from the genuine and the profitable from the loss-making.

These abilities should bring many benefits to a business, yet the adoption of these techniques has been slow. Despite the early promise of expert systems and neural networks, the application of computational intelligence has not become mainstream. This might seem all the more odd when one considers the explosion in data warehousing, loyalty card data collection and online data driven commerce that has accompanied the development of CI techniques (Hoss, 2000).

In this chapter, we discuss some of the reasons why CI has not had the impact on commerce that one might expect, and we offer some recommendations for the reader who is planning to embark on a project that utilizes CI. For the CI practitioner, this chapter should highlight cultural and conceptual business obstacles that they may not have considered. For the business user, this chapter should provide an overview of what a CI system can and cannot do, and in particular the dependence of CI systems on the availability of relevant data.

Given the right environment the technology has been shown to work effectively in a number of fields. These include financial prediction (Kim & Lee, 2004; Trippi & DeSieno, 1992; Tsaih, Hsu, & Lai, 1998), process control (Bhat & McAvoy, 1990; Jazayeri-Rad, 2004; Yu & Gomm, 2002) and bio-informatics (Blazewicz & Kasprzak, 2003). This path to successful application has a number of pitfalls and it is our aim to highlight some of the more common difficulties that occur during the process of applying CI and suggest methods for avoiding them.

BACKGROUND

Computational intelligence is primarily concerned with using an analytical approach to making decisions based on prior data. It normally involves applying one or more computationally intensive techniques to a data set in such a way that meta-information can be extracted from these data. This meta-information is then used to predict or classify the outcome of new situations that were not present in the original data. Effectively, the power of the CI system derives from its ability to generalize from what it has seen in the past to make sensible judgements about new situations.

A typical example of this scenario would be the use of a computational intelligence technique such as a neural network (Bishop, 1995; Hecht-Neilsen, 1990; Hertz, Krogh, & Palmer, 1991) to predict who might buy a product based on prior sales of the product. A neural network application would process the historical data set containing past purchasing behaviour and build up a set of weighted values which correlate observed input patterns with consequent output patterns. If there was a predictable consistency between a buyer's profile (e.g., age, gender, income) and the products they bought, the neural network would extract the salient aspects of this consistency and store it in the meta-information represented by its internal weights. A prospective customer could then be presented to the neural network which would use these weights to calculate an expected outcome as to whether the prospect is likely to become a customer or not (Law, 1999).

Although neural networks are mentioned above, this process is similar when used with a number of different computational intelligence approaches. Even within the neural network field, there are a large number of different approaches that could be used (Haykin, 1994). The common element in this process is the extraction and use of information from a prior data set. This information extraction process is completely dependent upon the quality and quantity of the available data. Indeed it is not always clear that the available data are actually relevant to the task at hand — a difficult issue within a business environment when a contract has already been signed that promises to deliver a specific result.

BEING COMMERCIAL

This chapter makes two assumptions. The first is that the reader is interested in applying CI techniques to commercial problems. The second is that the reader has not yet succeeded in doing so to any great extent. The reader may therefore be a CI practitioner who thoroughly understands the computational aspects and is having difficulties with the business aspects of selling CI, or a business manager who would like to use CI but would like to be more informed about the requirements for applying it. In this chapter we offer some observations we have made when commercializing CI techniques, in the hope that the reader will find a smoother route to market than they might otherwise have taken.

If you are hoping to find commercial application for your expertise in CI, then it is probably for one or more of the following reasons:

- You want to see your work commercially applied.
- Commercialization is stipulated in a grant you have won.
- You want to earn more money.

Many technologists with an entrepreneurial eye will have heard the phrase, "When you have invented a hammer, everything looks like a nail." Perhaps the most common mistake made by any technologist looking to commercialize their ideas for the first time is to concentrate too much on the technology and insufficiently on the needs of their customers (Moore, 1999). The more tied you are to a specific technique, the easier this mistake is to make. It is easy to concentrate on the technological aspects of an applied project, particularly if that is where your expertise lies.

CONCEPTUAL, CULTURAL, AND TECHNICAL BARRIERS

We believe that computational intelligence has a number of barriers that impede its general use in business. We have broken these down into three key areas: conceptual, cultural and technical barriers. On the surface, it may appear that technical barriers would present the greatest difficulties, however, it is frequently the conceptual and cultural barriers that stop a project dead in its tracks. The following sections discuss each of these concepts in turn. We first discuss some of the main foundations of CI under the heading of "Conceptual Barriers," this is followed by a discussion of the business issues relating to CI under the topic of "Cultural Barriers" and we finish off by covering the "nuts and bolts" of a CI project in a section on "Technical Barriers."

Conceptual Barriers

CI offers a set of methods for making decisions based on calculations made from data. These calculations are normally probabilities of possible outcomes. This is not a concept that many people are familiar with. People are used to the idea of a computer giving definitive answers—the value of sales for last year, for example. They are less comfortable with the idea that a computer can make a judgement that may turn out to be wrong.

The end user of a CI system must understand what it means to make a prediction based on data, the effect of errors and non-linearity and the requirements for the right kind of data if a project is to be successful. Analysts will understand these points intuitively, but if managers and end users do not understand them, problems will often arise.

Core Concepts

In this section, we will define and explain some of the mathematical concepts that everybody involved in a CI project will need to understand. If you are reading this as a CI practitioner, it may seem trivial and somewhat obvious. This unfortunately is one of the first traps of applying CI—there will be people who do not understand these concepts or perhaps have an incomplete understanding, which may lead them to expect different outcomes. These differences in understanding must be resolved in order for a project to succeed. We highlight these mathematical concepts because they are what makes CI different from the type of computing many people find familiar. They are conceptual barriers because their consequences have a material impact on the operation of a CI-based system.

Systems, Models, and The Real World

First, let us define some terms in order to simplify the text and enhance clarity. A system is any part of the real world that we can measure or observe. Generally, we will want to predict its future behaviour or categorize its current state. The system will have inputs: values we can observe and often control, that lead to outputs that we cannot directly control. Normally the only method available to us if we want to change the values of the outputs is to modify the inputs. Our goal is usually to do this in a controlled and predictable manner.

In the purchasing example used above, our inputs would be the profile of the buyer (their age, gender, income, etc.) and the outputs would be products that people with a given profile have bought before. We could then run a set of possible customers through the model of the system and record those that are predicted to have the greatest likelihood of buying the product we are trying to sell.

Given that a CI system is generally derived from data collected from a real-world system, it is important to determine what factors or variables affect the system and what can safely be ignored. It is often quite difficult to estimate in advance all the factors or variables that may affect a system and even if it were, it is not always possible to gather data about those factors.

The usual approach, forced on CI modelers through pragmatism, is to use all the variables that are available and then exclude variables that are subsequently found to be irrelevant. Time constraints frequently do not allow for data on further variables to be collected. It is important to acknowledge that this compromise is present since a model with reduced functionality will almost certainly be produced. From a business point of view, it is essential that a client is made aware that the limitations of the model are attributable to the limitations of their data rather than the CI technique that has been used. This can often be a point of conflict and therefore needs to be clarified at the very outset of any work.

Related to this issue of collecting data for all the variables that could affect a system is the collection of sufficient data that span the range of all the values a variable might take with respect to all the other variables in the system. The goal here is to develop a model that accurately links the patterns in the input data to corresponding output patterns and ideally this model would be an exact match to the real-world system. Unfortunately, this is rarely the case since it is usually not possible to gather sufficient data to cover all the possible intricacies of the real-world system.

The client will frequently have collected the data before engaging the CI expert. They will have done this without a proper knowledge of what is likely to be required. A significant part of the CI practitioner's expertise is concerned with the correct collection of the right data. This is a complex issue and is discussed in detail in Baum and Haussler (1989).

A simple example of this might be the collection of temperature readings for a chemical process. Within the normal operation of this process, the temperature may remain inside a very stable range, barely moving by a few degrees. If regular recordings of the system state are being made every 5 seconds then the majority of the data that are collected will record this temperature measurement as being within its stable range. An analyst may however be interested in what happens to the system when it is perturbed outside its normal behaviour or perhaps what can be done to make the system optimal. This may involve temperature variations that are relatively high or low compared to the norm. Unless the client is willing to perturb their system such that a large number of measurements of high and low temperatures can be obtained then it will not be possible to make queries about how the system will react to novel situations.

This lack of relevant data over all the "space" that a system might cover will lead to a model that is only an approximation to the real world. The model has regions where it maps very well to the real world and produces accurate predictions, but it will also have regions where data were sparse or noisy and its approximations are consequently very poor.

Inputs and Outputs

Input and output values are characterized by variables — a variable describes a single input or output, for example "temperature" or "gender." Variables take values — temperature might take values from 0 to 100 and gender would take the values "male" or "female." Values for a given variable can be numeric like those for a temperature range or symbolic like those of "gender." It is rare that a variable will have values that are in part numeric and in part symbolic. The general approach in this case is to force the variable to be regarded as symbolic if any of its values are symbolic. Fuzzy systems can impose an order on symbolic data, for example we can say that "cold" is less than "warm" which is less than "hot." This enables us to combine the two concepts.

Numbers have an order and allow distances to be calculated between them, symbolic variables do not, although they may have an implied scale such as "small," "medium" or "large." Ignoring the idea of creating an artificial distance metric for symbolic variables, a computational intelligence system cannot know, for example, that blue and purple are closer than blue and yellow. This information may be present in the knowledge of a user, but it is not obvious from just looking at the symbolic values "blue" and "yellow."

Coincidence and Causation

If two things reliably coincide, it does not necessarily follow that one caused the other. Causation cannot be established from data alone. We can observe that A always occurs when B occurs, but we cannot say for sure that A causes B (or indeed, that B causes A). If we observe that B always follows A, then we can rule out B causing A, but we still can't conclude that A causes B from the data alone. If A is "rain" and B is "wet streets" then we can infer that there is a causal effect, but if A is "people sending Christmas cards" and B is "snow falling" then we know that A does not cause B nor B cause A, yet the two factors are associated. Generally, however, if A always occurs when B occurs, then we can use that fact to predict that B will occur if we have seen A. Spotting such co-occurrences and making proper use of them is at the heart of many CI techniques.

Non-Linearity

Consider any system in which altering an input leads to a change in an output. Take the relationship between the price of a product and the demand for that product. If an increase in price of $1 always leads to a decrease in demand of 50 units regardless of the current price then the relationship is said to be linear. If, however, the change in demand following a $1 increase varies depending on the current price, then the system is non-linear. This is the standard demand curve and is an example of non-linearity for a single input variable.

Adding further input variables can introduce non-linearity, even when each individual variable produces a linear effect if it alone is changed.

This occurs when two or more input variables interact within the system such that the effect of one is dependent upon the value of the other (and vice versa). An example of such a situation would be the connection between advertising spend, price of the product and the effect these two input variables might have on the demand for the product. For example, adding $1 to the price of the product during an expensive advertising campaign may cause less of a drop in demand compared with the same increase when little has been spent on advertising.

Non-linearity has a number of major consequences for trying to predict a future outcome from data. Indeed, it is these non-linear effects that drove much of the research into the development of the more sophisticated neural networks. It is also this aspect of computational intelligence that can cause significant problems in understanding how the system works. A client will frequently request a simplified explanation of how a CI system is deriving its answer. If the CI model requires a large number of parameters (e.g., the weights of a neural network) to capture the non-linear effects, then it is usually not possible to provide a simplified explanation of that model. The very act of simplifying it removes the crucial elements that encode the non-linear effects.

This directly relates to one of the more frequently requested requirements of a CI system — the decision-making process should be traceable such that a client can look at a suggested course of action and then examine the rationale behind it. This can frequently lead to simple, linear CI techniques being selected over more complex and effective non-linear approaches because linear processes can be queried and understood more easily.

A further consequence of non-linearity is that it makes it impossible to answer a question such as "How does *x* affect *y*?" with a general all encompassing answer. The answer would have to become either, "It depends on the current value of *x*" in the case of *x* having a simple non-linear relationship with *y*, and "It depends on *z*" in cases where the presence of one or more other variables introduce non-linearity.

Here is an example based on a CI system that calculates the risk of a person making a claim on a motor insurance policy. Let us say we notice that as people grow older, their risk increases, but that it grows more steeply once people are over 60 years of age. That is a non-linearity as growing older by one year will have a varying effect on risk depending on the current age.

Now let us assume that the effect of age is linear, but that for males risk gets lower as they grow older and for females the risk gets higher with age. Now, we cannot know the effect of age without knowing the gender of the person in question. There is a non-linear effect produced by the interaction of the variables "age" and "gender." It is possible for several inputs to combine to affect an output in a linear fashion. Therefore, the presence of several inputs is not a sufficient condition for non-linearity.

Classification

A classification system takes the description of an object and assigns it to one class among several alternatives. For example, a classifier of fruit would see the description "yellow, long, hard peel" and classify the fruit as a banana. The output variable is "class of fruit," the value is "banana." It is tempting to see classification as a type of prediction. Based on a description of an object, you predict that the object will be a banana. Under normal circumstances, that makes sense but there are situations where that does not make sense, and they are common in business applications of CI.

A CI classification system is built by presenting many examples of the descriptions of the objects to be classified to the classifier-building algorithm. Some algorithms require the user to specify the classes and their members in this data. Other algorithms (referred to as clustering algorithms) work out suitable classes based on groups of objects that are similar enough to each other but different enough from other things to qualify for a class of their own.

A common application of CI techniques in marketing is the use of an existing customer database to build a CI system capable of classifying new prospects as belonging to either the class "customer" or "non-customer." Classifying a prospect as somebody who resembles a customer is not the same as predicting that the person will become a customer. Such systems are built by presenting examples of customers and non-customers. When they are being used, they will be presented with prospective customers (i.e., those who do not fall into the class of customer at the moment since they have not bought anything). Those prospects that are classified by the CI system as "customer" are treated as good prospects as they share sufficient characteristics with the existing customers.

It must be remembered, however, that they currently fall into the non-customer category, so the use of the classification to predict that they would become customers if approached is erroneous. What the system will have highlighted is that they have a greater similarity to existing customers than those classified as "non-customer." It does not indicate that they definitely will become a "customer."

For example, if such a system were used to generate a mailing list for a direct-mail campaign, you would choose all the current non-customers who were classified as potential customers by the CI system and target them with a mail shot. If a random mailing produced a 1% response rate and you doubled that to 2% with your CI approach, the client should be more than satisfied. However, if you treated your classification of customers as a prediction that those people would respond to the mailing, you would still have been wrong on 98% of your predictions.

Prospect list management is increasingly seen as an important part of Customer Relationship Management (CRM) and it is in that aspect that CI can offer real advantages. Producing a list of 5,000 prospects and predicting that they will all become customers is a sure way of producing scepticism in the client at best, and at worst of failing to deliver.

Dealing with Errors and Uncertainty

Individual predictions from a CI system have a level of error associated with them. The level of error may depend on the values of the inputs for the current situation, with some situations being more predictable than others. This lack of certainty can be caused by noise in the data, inconsistencies in the behaviour of the system under consideration or by the effects of other variables that are not available to the analysis. Dealing with this uncertainty is an important part of any CI project. It is important both in technical terms—measuring and acting on different levels of certainty—and conceptual levels—ensuring that the client understands that the uncertainty is present. (See Jepson, Collins, & Evans, 1993; Srivastava & Weigend, 1994 for different methods for measuring errors.)

We have stated that a classification can be seen as a label of a class that a new object most closely resembles, as opposed to being a prediction of a class of behaviour. A consequence of this is that a CI system can make a prediction or a classification that turns out to be wrong. In the broadest sense, this would be defined as an error but could also be seen as a consequence of the probabilistic nature of CI systems. For example, if a CI system predicts that an event will occur with a probability of 0.8 and that event does not occur for a given prediction, then the prediction and its associated probability could still be seen as being correct. It is just that in this instance the most probable outcome did not occur. In order to validate the system, you must look at all the results for the all the predictions. If a CI model assigns a probability of 0.8 to an event, it should occur 8 times out of 10 for the system to be valid but you should still expect it to misclassify 2 out of 10 events.

For example, if a given insurance claim is assigned a probability of being fraudulent of 0.8 then one would expect 8 out of 10 identical claims to be fraudulent. If this turned out not to be the case, for example only 6 out 10 turned out to be fraudulent, then the CI system would be considered to be wrong.

Returning to the customer-prospecting example, it is clear that the individual cost of a wrong classification in large campaigns is small. If we have made it clear that the prospects were chosen for looking most like previous customers and that no predictions are made about a prospect actually converting, the job of the CI system becomes to increase the response rate to a campaign.

There are many cases where it is necessary to introduce the concept of the CI system being able to produce an "I don't know" answer. Such cases are defined as any prediction or classification with a confidence score below a certain threshold. By refusing to make a judgement on such cases, it is possible to reduce the number of errors made in all other cases.

The authors have found that neural network based systems are very useful for the detection of fraudulent insurance claims. A system was developed that could detect fraudulent claims with reasonable accuracy. However, the client did not want to investigate customers whose claims looked fraudulent but were not. By introducing the ability of the system to indicate when it was uncertain about a given case, we were able to significantly reduce the number of valid claims that were investigated.

The two aspects that had to be considered when looking at the pattern of errors within the above example were the cost of a false positive and the cost of a false negative. An example of a false positive would be a situation where an insurance fraud detection system classified a claim as "positive" for fraud (i.e., fraudulent) but subsequent investigation indicated the claim to be valid. In the case of a false negative, the insurance fraud system might indicate that a claim is "negative" for fraud when in fact it was actually fraudulent. In the latter case you would not know that you had paid out on a fraudulent claim unless you explicitly investigated every claim while validating the fraud detection system.

False positives and false negatives have a cost associated with them in any specific application. The key to dealing with these errors lies in the cost-benefit ratio for each type of error. A false positive in the above case may cost two days work for an investigator. A false negative (i.e., paying out on a missed fraudulent claim) may cost many thousands of dollars.

Interpolation vs. Extrapolation

Many users want a model that they can use to make predictions about uncharted territory. This involves either interpolation within the current model or extrapolation into regions outside the data set from which the model was built. This might happen in a case where the user asks the system to make a prediction for the outcome of a chemical process when one of the input variables, such as temperature, is higher than any example provided in the recorded data set.

Without a measure of the non-linearity in the system, it can be difficult to estimate how accurate such predictions are likely to be. For example, interpolation within a data-rich area of the variable space is likely to produce accurate results unless the system is highly non-linear. Conversely, interpolation within a data-poor area is likely to produce almost random answers unless the system is very linear in the region of the interpolation. The problem with many computational systems is that it is often not obvious when the model has strayed outside its "domain knowledge."

A good example comes from a current application being developed by the authors. We are using a neural network to predict sales levels of newly released products to allow distributors and retailers to choose the right stocking levels. The effect on sales of the factors that we can measure is non-linear, which means that we do not know how those factors would lead to sales levels that were any higher than those we have seen already. The system is constrained to predicting sales levels up to the maximum that it has already seen. If a new product is released in the future, and sells more than the best selling product that we have currently seen, we will fall short in our prediction.

In the case of interpolation, the simplest method for ensuring that non-linearity is accurately modelled is to gather as much data as possible. This is because the more data we have, the more likely it is that areas of non-linearity within the system will have sample data points indicating the shape of the parameter space. If there were insufficient data in a non-linear part of the system, then a CI method would tend to model the area as though it were linear.

In the extreme, you only need two data points to model a linear relationship. As soon as a line becomes a curve then we need a multitude of data points along the curve to map out its correct shape. Figure 1 (a) shows a simple case of identifying a linear relationship in a system with two variables. With only two points available, the most obvious conclusion to draw would be that the system is linear. Figure 1 (b) highlights what would happen if we were to obtain more data points. Our initial assumptions would be shown to be potentially invalid. We would now have a case for suspecting that the system is non-linear or perhaps very noisy. A CI model would adapt to take account of the new data points and produce an estimate of the likely shape of the curve that would account for the shape of the points (Figure 1 (c)).

It can be seen from Figure 1 (c) that if we had interpolated between the original two points shown in Figure 1 (a) then we would have made an incorrect prediction. By ensuring that we had adequate data, the non-linearity of the system would be revealed and the CI technique would adapt its model accordingly.

Figure 1 (a) A simple linear system derived from 2 points. (b) The addition of further data reveals non-linearity. (c) A CI system fits a curve to the available data. (d) The shape of the estimated curve showing how further data produces a new shape — extrapolation would fail in these regions.

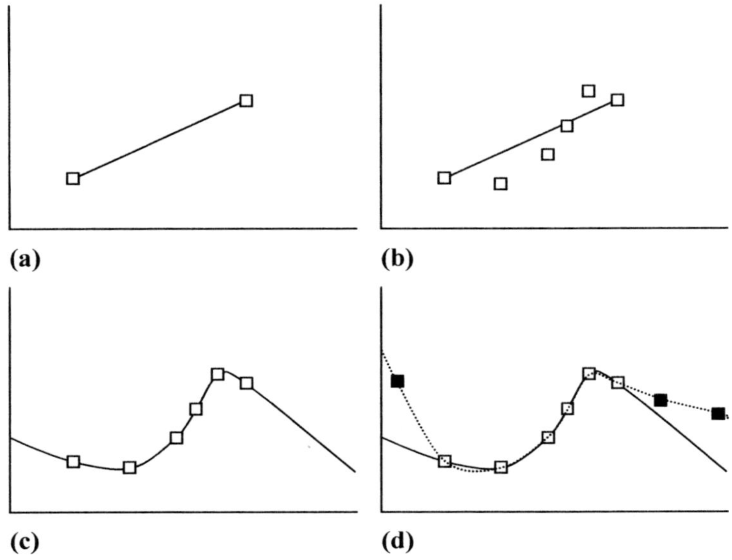

Related to this concept is the possibility of extrapolating from our current known position in order to make predictions about areas outside the original data set used to build the model. Extrapolation of the linear system in Figure 1 (a) would be perfectly acceptable if we knew the system was actually linear. However, if we know the system is non-linear, this approach becomes very error prone. An example of the possible shape of the curve is shown in Figure 1 (c), however, we have no guarantee that this is actually where the curve goes. Further data collection in the extremes of the system (shown by black squares) might reveal that the boundaries of the curve are actually quite a different shape to the one we have extrapolated (Figure 1 (d)). While we remained in the data-rich central area of the curve, our prediction would have remained accurate. However, as soon as we went to the extremes, errors would have quickly crept in.

Given that we have the original data set at our disposal, it is possible to determine how well sampled a particular region is that we wish to make a prediction in. This should enable us to provide a measure of uncertainty about the prediction itself. With regards to extrapolation, we usually know what the upper and lower bounds are for the data used to build the model. We will therefore know that we have set a given input variable to a value outside the range on which the data used to build the model was limited to. For anything but the simplest of systems, this should start ringing alarm bells. It is important that a client using a CI system understands the implications of what they are asking for under each of these situations and where possible, steer away from trying to use such information.

Generalization

This leads us to the concept of generalization—an important issue in the development of an actual CI system. Generalization is concerned with avoiding the construction of a CI system that is very accurate when tested with data that has been used to build it, but performs very poorly when presented with novel data. With regard to the previous section, generalization deals with the ability of a non-linear system to accurately interpolate between points from the data used to construct it.

An idealized goal for a CI system is that it should aim to produce accurate predictions for data that it has not seen before. With a poorly constructed CI system that may have been built with unrepresentative data, the system is likely to perform well when making predictions in the region of this unrepresentative data and very poorly when tested with novel data that is more representative of the typical operating environment. In simple terms, the system attempts to build a predictive system that very closely follows all the observed historical data to the detriment of new data.

If all the data used to build a system completely captured the behaviour of the system then generalization would not be an issue. This is almost never the case, as it is very difficult to capture all the data describing the state of a system and furthermore data usually have some degree of noise associated with them. The CI practitioner will understand these limitations and will attempt to minimize their effects on the performance of the CI system. For the business manager interested in applying CI with the assistance of the practitioner, this will generally present itself as a need for a significant amount of data in an attempt to overcome the noise within it and ensure that a representative sample of the real-world system has been captured.

Cultural Barriers

CI's apparent power lies in its ability to address issues at the heart of a business: choosing prospects, pricing insurance, or warning that a machine needs servicing. These are high-level decisions that a business trusts experts to perform. Can you go into a business and challenge the expertise of their marketing team, their underwriters, or their engineers in the same way that production line robots have replaced car assembly workers? We look at these cultural barriers and the ways in which they have been successfully overcome. Whether you are an external consultant selling to a client or an internal manager selling an idea to the board, you will need to understand how to win acceptance for this new and challenging approach if your project is to succeed.

People who are experts at their job do not like to think that a computer can do it better. In general, computers are regarded as dumb tools—there to help the human experts with the tedious aspects of their work. Robots and simple machines have successfully replaced a lot of manual labor. There have been barriers to this replacement—protests from unions and doubts about quality for example—but automation of manual labor is now an integral aspect of the industrialized world.

Computational intelligence might be vaunted as offering a modern computational revolution where machines are able to replace human decision-making processes. This replacement process should free up people to focus on special cases that require thought and knowledge of context that the computer may be lacking. Given these positive aspects, there are still many reasons why this shift might not come about. In the first instance, there is the position of power held by the people to be replaced. The people who make the decisions are less than happy with the idea that they might be replaceable and that they might be called upon to help build the systems that might replace them. Manual workers have little say in the running of an organization. However, marketing executives and underwriters are higher up a company's decision-making chain—replacing them with a computer is consequently a more difficult prospect.

Next there is the issue of trust. I might not believe that a machine can build a car, but show me one that does and I have to believe you. If I do not believe that a computer can understand my customers better than me, you can show me an improved response to a mail campaign for a competing company, but I will still believe that my business is different and it will require a lot of evidence before I will change my mind.

Related to the issue of trust is that of understanding. This is a problem on two levels—first people do not always understand how they themselves do something. For example, we interviewed experts in spotting insurance fraud, who said things like, "You can just tell when a claim is dodgy—it doesn't look right." You can call it intuition or experience, but it is hard to persuade somebody that it is the result of a set of non-linear equations served up by their subconscious. The brain is a mysterious thing and people find the idea that in some areas it can be improved upon by a computer very hard to swallow.

The second problem is that people have difficulty believing that a computer can learn. If a person does not understand the concepts of computer learning and how it is possible to use data to make a computer learn, then it is hard for them to make the conceptual leap required to believe that a computer could be good at something that they see as a very human ability.

Here is an example to illustrate the point. A printing company might upgrade from an old optical system to a complete state-of-the-art digital system. In the process they would replace the very core of their business with a new technology, perhaps with the result that their old skills become obsolete. A graphic designer, however, would not want to buy a system that could automatically produce logos from a written brief, no matter how clever the technology.

Our experience has shown that many of these problems can be overcome if the right kind of simplification is applied to the sales pitch. That is not to say that technical details should be avoided or that buyers should be considered stupid. It means choosing the right level of technical description and, more importantly, setting the strength of claims being made about the technology on offer.

We shall use the task of building a CI system for use in motor insurance as an example. We developed a system that could calculate the risk associated with a new policy better than most underwriters. It could spot fraud more effectively than most claims handlers and it could choose prospective customers for direct marketing better than the marketing department. Insurance could be revolutionized by the use of CI (Viaene, Derrig, Baesens, & Dedene, 2003), but the industry has so far resisted.

An insurance company would never replace its underwriters, so if we are to help them with a CI system, it must be clearly positioned as a tool—something that helps them do their job better without doing it for them. Even though you could train a neural network to predict the probability associated with a new policy leading to a claim better than the underwriter can, it would do too much of his or her job to be acceptable.

Our experience has shown us that approximately 90% of motor insurance policies carry a similar, low probability of leading to a claim. There are 5% that have a high risk associated with them and 5% that have a very low risk. A system for spotting people who fall into the interesting 10% in order to avoid the high risks and increase the low risk policies would leave the underwriters still doing their job on the majority of policies and give them an extra tool to help avoid very high risks. The CI system becomes the basis of a portfolio management system and the sale is then about better portfolio management and not about intelligent computing—a much easier prospect to sell.

Within the context of the insurance fraud example, investigators spent a considerable amount of their time looking at routine cases. Each case took a brief amount of time to review but, due to the large number of them, this took up the majority of their time. If you put forward the argument that the investigators would be better spending their time on the more complex cases where their skills could truly be used, then you can make a case for installing a CI system that does a lot of the routine work for them and only presents the cases that it regards as suspicious.

Another barrier to the successful commercialization of CI techniques is, to put it bluntly, a lack of demand. It is easy to put this lack of demand down to a lack of awareness, but it should be stated with more strength than that: CI is not in the commercial consciousness. Perhaps if prospective customers understood the power of CI techniques, then they would be easily sold on the idea. To an extent, of course, that is true. But to find the true reason behind a lack of demand, we must look at things from a customer's point of view. Will CI be on the customer's shopping list? Will there be a budget allocated? Are there pressing reasons for a CI system to be implemented? If the answer to these questions is no, then there is no demand. There is only, at best, the chance to persuade a forward-thinking visionary in the company who has the time, resources and security to risk a CI approach.

To use our e-commerce example again, a company building an online shop will need to worry about secure servers, an e-commerce system, order processing, delivery and promotion of the site. Those things will naturally be on their shopping list. An intelligent shop assistant to help the customer choose what to buy might be the only thing that would make a new e-commerce site stand out. It might be a perfect technical application of CI, and it might double sales, but it will not be planned, nor budgeted for. That makes the difference between you having to sell and the customer wanting to buy.

Technical Barriers

It has been our experience that the most common and fundamental technical barrier to most CI applications concerns access to source data of the correct type and quality. Obviously, if there is no data available relating to a given application, then no data-driven CI technique will be of use. A number of more common problems arise, however, when a client initially claims to have adequate data.

Is the client able to extract the data in electronic form? Some database systems actually do not have a facility for dumping entire table contents, compelling the user to make selections one-at-a-time. Some companies still maintain paper-only storage systems and some companies have a policy against data leaving their premises. It is also well worth remembering that the appropriate data will not only need to be available at the time of CI system development, but at run time, too. A typical use of CI in marketing is to make predictions about the buying behaviour of customers. It is easy to append lifestyle data to a customer database off-line ready for analysis, but will that same data be available online when a prediction is required for any given member of the population as a whole?

Does the data reflect the task you intend to perform and does it contain the information required to do so? Ultimately, finding the answer to this question is the job of the CI expert, but this is only true when the data appears to reflect the application well. It can be worth establishing early on that the data at least appears useful.

There are also technical aspects of a CI project that will have an impact on the contractual arrangements between you and the client. These are consequences of the fact that it is not always possible to guarantee the success of a CI project since the outcome depends on the quality of the data.

If the client does not have suitable data but is willing to collect some, it is important to be clear about what is to be collected and when that data will be delivered. If your contract with the client sets out a time table, be sure that delays in the data collection (which are not uncommon) allow your own milestones to be moved. Be clear that your work cannot start until the data are delivered. You may also want to be clear that the data must meet a certain set of criteria.

You also need to make it very clear what the client is buying. Most clients will be used to the idea that if they have a contract with (say) a software company to develop a bespoke solution, then that solution will be delivered, working as agreed upon in the specification. If it is not, then the contract will usually allow for payment to be reduced or withheld. It should be made clear to the client that their data, and whether or not it contains the information required to allow the CI approach to work, will be the major contributing factor to the success or otherwise of the project. The client must understand that success cannot be guaranteed. It has been our experience that the client often does not see it this way — the failure of the CI model to accurately predict who their customers are is seen as a failure of CI, not their carefully collected data.

Another consequence of the lack of available data at the start of a project is the difficulty it presents if you plan to demonstrate your approach to a prospective client. You can generate mock data that carries the information you hope to find in your client's database, but this proves little to the client as it is clearly invented by you. You can talk about (and possibly even demonstrate) what you have done for similar, anonymous clients, but each company's situation is usually different and CI models are very specific to each customer. The difficulty of needing data therefore remains.

There are many specific technical problems including choosing the right CI technique and using it to produce the best results. Each CI technique has its own particular requirements and issues. It is beyond the scope of this chapter to cover such topics—we have focussed on the elements that occur generally across the diverse set of CI approaches. Further chapters in this book address technique specific issues.

FUTURE TRENDS

We believe that CI technology is currently at a stage of development where weaknesses in the techniques are not the major barriers to immediate commercial exploitation. We have identified what we consider to be the main cultural, conceptual and technical barriers to commercialization of CI and the reader may have noticed that the technical barriers did not include any shortcomings of the CI methods themselves.

There is a large gap between the power of the techniques available and the problems that are currently being solved by those techniques. Unusually, however, it is the technology that is ahead. One can easily imagine impressive applications of CI techniques that are yet to be perfected—Web agents that can write you an original essay on any topic you choose, robot cars capable of negotiating the worst rush hour traffic at high speeds, and intelligent CCTV cameras that can recognize that a crime is taking place and alert the police. None of these applications are possible today and they are likely to remain difficult for a long time to come. The small improvements to the techniques that are possible in commercially viable time scales will not bring about a step change in the types of applications to which the techniques may be applied.

Our view of the near future of the commercial exploitation of CI, therefore, is concentrated on the methods of delivery of existing techniques and not the development or improvement of those techniques. Of course, the development of CI techniques is important, but it is the commercialization that must catch up with the technology, and not the converse. The consequence of this observation is, we believe, that the near future of the commercial exploitation of CI techniques requires little further technical research. The current techniques can do far more than they are being asked to do.

We expect to see a shift away from selling the idea of the techniques themselves and towards selling a product or service improved by the techniques without reference to those techniques. The search engine Google is a good example. People do not care about the clever methods behind it. They just know it works as a very good search engine. Another good example of underplayed technology comes from the world of industrial control. Most industrial control is done using a technique known as PID. Many university engineering departments have produced improvements to the PID controller and very few of them have found their way into an industrial process. One reason for this is that everybody in the industry understands and trusts PID controllers. Nobody wants to open the Pandora's box of new and challenging techniques that might fragment the industry and its expertise.

One company developed an improvement to the PID controller and did not even admit to its existence. They simply embedded it in a new product and sold it as a standard PID controller. It worked just that bit better than all the others. Nobody really knew why it was better, but it was. The controller sold very well, nobody was threatened by the new technique, and there was no technical concept to sell. It just worked better.

An alternative and related example is the use of CI systems to spot fraudulent credit card behaviour. It is simply not practical for investigators to analyze every single credit card transaction. A CI system can be used to monitor activity for each user and determine when it has become unusual. At this point an investigator is alerted who can contact the owner of the card to verify their spending behaviour. People are generally not aware that CI systems are behind such applications, and for all practical purposes, this does not matter. The important issue is the benefits they bring rather than their technical sophistication.

The authors have put this approach into practice. Having spent several years selling CI technology to direct marketing agencies with little success, they have recently launched a Web-based direct marketing system that is driven by CI techniques. The service allows clients to upload their current customer database to a Web server. It then appends lifestyle data to the names in the database, which is then used to generate a new list of prospects for the client to download.

The primary selling point of the service is that it is easy to use and inexpensive (the techniques are automated). These are far easier concepts to sell because they are clearly demonstrable—the client can see our prices and visit our Web site to see how easy the process is to use. Having got a foot in the door of the mailing list market, our system quietly uses some very straightforward CI techniques to produce prospect lists that yield response rates up to four times better than the industry standard.

Our approach is proving successful and it is based on the following points:

- We selected a market where there was clearly money to be made from delivering an improvement to the existing, inefficient norm.
- The main selling point of the product is not technical, thus all problems associated with explaining and selling the CI concept are avoided.
- We deliver a service that the customer needs, already has budgeted for, and understands perfectly.
- The data we receive are always in the same format (names and addresses) and we provide all the additional data required. Consequently, we never have a problem with data quality.

This approach has a number of advantages. It removes the need for the client to worry that they are taking a risk by using a new technology. It removes the need for us to try and sell the idea of the technology, and it allows us to sell to a mature market.

SUMMARY

We have seen that there are a number of barriers to the successful commercialization of CI techniques. There is a lack of awareness and understanding from potential customers. Their mathematical nature and the fact that the success of a project depends on the quality of the data it uses can make the concept hard to sell. The lack of awareness also means that companies are not actively looking for CI solutions and are consequently unlikely to have budgets in place with which to buy them.

CI techniques face cultural barriers to their adoption as they could potentially replace existing human expertise. The existing human experts are often in a powerful position to prevent even the risk of this replacement and their unwillingness to change should not be underestimated. We have also touched upon technical barriers, such as accessing the correct data both at design and run-time, and the problems of specifying, demonstrating and prototyping a system based on data.

We have suggested a number of approaches designed to overcome the barriers discussed in this chapter. These approaches can be summarized by the notion of putting yourself in your prospective customer's position. Ask yourself what the customer needs, not what you can offer. Think about how much change a customer is likely to accept and whether or not they could cope with that change. Ask yourself whether you are making more work for the customer or making their life easier. Think about whether the customer is likely to have a budget for what you offer. If not, can you present it as something they do have budget for? Find out what level of technical risk the customer is likely to be comfortable with. Are they early adopters or conservative followers?

We believe that the future success of CI will rely on keeping your customer on board and giving them what they want, not impressing them with all the clever tricks that you can perform.

The key element is for both you and the client to maintain the same point of view of the problem you are both trying to solve. This will primarily mean that if you are the provider of the CI solution, you will need to adapt your perspective to fit that of the client. It is, however, important that the client understands the conceptual limits of CI as discussed in the early parts of this chapter. In order to maintain a positive working relationship with a client, it is important that they understand both the benefits and limitations of computational intelligence and therefore know, at least in principle, what can and cannot be done.

REFERENCES

Baum, E.B., & Haussler, D. (1989). What net size gives valid generalisation? *Neural Computation, 1*(1), 151-160.

Bhat, N., & McAvoy, T. J. (1990). Use of neural nets for dynamic modelling and control of chemical process systems. *Computer Chemical Engineering, 14*(4/5), 573-583.

Bishop, C. M. (1995). *Neural networks for pattern recognition.* Oxford, UK: Oxford University.

Blazewicz, J., & Kasprzak, M. (2003). Determining genome sequences from experimental data using evolutionary computation. In G. G. Fogel & D. W. Corne (Eds.), *Evolutionary computation in bioinformatics* (pp. 41-58). San Francisco: Morgan Kaufmann.

Haykin, S. (1994). *Neural networks, a comprehensive foundation.* New York: Macmillan.

Hecht-Nielsen, R. (1990). *Neurocomputing.* Reading, MA: Addison Wesley.

Hertz, J., Krogh, A., & Palmer, R. G. (1991). *Introduction to the theory of neural computation.* Redwood City, CA: Addison Wesley

Hoss, D. (2000). The e-business explosion: Strategic data solutions for e-business success. *DM Review, 10*(8), 24-28.

Jazayeri-Rad, H. (2004). The nonlinear model-predictive control of a chemical plant using multiple neural networks. *Neural Computing and Applications, 13*(1), 2-15.

Jepson, B., Collins, A., & Evans, A. (1993). Post-neural network procedure to determine expected prediction values and their confidence limits. *Neural Computing and Applications, 1*(3), 224-228.

Kim, K., & Lee, W. B. (2004). Stock market prediction using artificial neural networks with optimal feature transformation. *Neural Computing and Applications, 13*(3), 255-260.

Law, R. (1999). Demand for hotel spending by visitors to Hong Kong: A study of various forecasting techniques. *Journal of Hospitality and Leisure Marketing, 6*(4), 17-29.

Moore, G. (1999). *Crossing the chasm: Marketing and selling high-tech products to mainstream customers.* Oxford, UK: Capstone.

Srivastava, A. N., & Weigend, A. S. (1994). Computing the probability density in connectionist regression. In M. Marinara & G. Morasso (Eds.), *Proceedings ICANN, 1* (pp. 685-688). Berlin: Springer-Verlag.

Trippi, R. R., & DeSieno, D. (1992). Trading equity index futures with a neural-network. *Journal of Portfolio Management, 19,* 27-33.

Tsaih, R., Hsu, Y., & Lai, C. C. (1998). Forecasting S & P 500 stock index futures with a hybrid AI system. *Decision Support Systems, 23*(2), 161-174.

Viaene, S., Derrig, R. A., Baesens, B., & Dedene, G. (2003). A comparison of state-of-the-art classification techniques for expert automobile insurance claim fraud detection. *Journal of Risk and Insurance*, *69*(3), 373-421.

Yu, D. L., & Gomm, J. B. (2002). Enhanced neural network modelling for a real multi-variable chemical process. *Neural Computing and Applications*, 10(4), 289-299.

This work was previously published in Business Applications and Computational Intelligence, edited by K. E. Voges, & N. K. L. Pope, pp. 19-37, copyright 2006 by IGI Publishing, formerly known as Idea Group Publishing (an imprint of IGI Global).

Chapter III
Data Mining Association Rules for Making Knowledgeable Decisions

A.V. Senthil Kumar
CMS College of Science and Commerce, India

R. S. D.Wahidabanu
Govt. College of Engineering, India

ABSTRACT

This chapter describes two techniques used to explore frequent large itemsets in the database. In the first technique called "closed directed graph approach," the algorithm scans the database once making a count on possible 2-itemsets from which only the 2-itemsets with a minimum support are used to form the closed directed graph which explores possible frequent large itemsets in the database. In the second technique, dynamic hashing algorithm, large 3-itemsets are generated at an earlier stage which reduces the size of the transaction database after trimming and the cost of later iterations will be less. Furthermore the authors hope that these techniques help researchers not only to understand about generating frequent large itemsets, but also assist with the understanding of finding association rules among transactions within relational databases.

INTRODUCTION

Recently, with the advent of the vast growth in applications of computers, large amounts of transaction data are stored in databases. This has occurred in all areas of human endeavors, from the mundane (such as supermarket transaction data, credit card usage records, telephone call details, and government statistics) to the more exotic (such as images of astronomical bodies, molecular databases, and medical records). Little wonder, then that interest has grown in the possibility of tapping these data, of extracting from them information that might be of value to the owner of the database. The discipline concerned with this task has become known as data mining.

Data mining, *the extraction of hidden predictive information from large databases,* is a powerful new technology with great potential to help companies focus on the most important information in their data warehouses. Data mining tools predict future trends and behaviors, allowing businesses to make proactive, knowledge-driven decisions. The automated, prospective analyses offered by data mining move beyond the analyses of past events provided by retrospective tools typical of decision support systems. Data mining tools can answer business questions that traditionally were too time consuming to resolve. They scour databases for hidden patterns, finding predictive information that experts may miss because it lies outside their expectations.

Most companies already collect and refine massive quantities of data. Data mining techniques can be implemented rapidly on existing software and hardware platforms to enhance the value of existing information resources, and can be integrated with new products and systems as they are brought online. When implemented on high performance client/server or parallel processing computers, data mining tools can analyze massive databases to deliver answers to questions such as, "Which clients are most likely to respond to my next promotional mailing, and why?"

Data mining techniques are the result of a long process of research and product development. This evolution began when business data was first stored on computers, continued with improvements in data access, and more recently, generated technologies that allow users to navigate through their data in real time. Data mining takes this evolutionary process beyond retrospective data access and navigation to prospective and proactive information delivery. Data mining is ready for application in the business community because it is supported by three technologies that are now sufficiently mature:

- Massive data collection
- Powerful multiprocessor computers
- Data mining algorithms

Commercial databases are growing at unprecedented rates. A recent research survey conducted by GoldenGate Software in San Francisco shows that almost half of the data warehouses are growing between 10 and 50 percent annually. In some industries, such as retail, these numbers can be much larger. The accompanying need for improved computational engines can now be met in a cost-effective manner with parallel multiprocessor computer technology. Data mining algorithms embody techniques that have existed for at least 10 years, but have only recently been implemented as mature, reliable, understandable tools that consistently outperform older statistical methods.

Data mining is the analysis of (often large) observational data sets to find unsuspected relationships and to summarize the data in novel ways that are both understandable and useful to the data owner. The relationships and summaries derived through a data mining exercise are often referred to as models or patterns. Examples include linear equations, rules, clusters, graphs, tree structures, and recurrent patterns in time series. Association rules are among the most popular representations for local patterns in data mining.

BACKGROUND

There are quite a few rules that are available for analyzing data transformation for making intelligent decision. The association rule is by far the most useful method in this respect, which is described next.

Association Rule

An association rule is a simple probabilistic statement about the co-occurrence of certain events in a database, and is particularly applicable to sparse transaction data sets. For the sake of simplicity all variables are assumed to be binary. An association rule takes the following form:

If A=1 AND B=1 THEN C=1 with probability ρ, where A, B, and C are binary variables and ρ = ρ(C=1|A=1,B=1) that is, the conditional probability that C=1 given that A=1 and B=1. The conditional probability ρ is sometimes referred to as the "accuracy" or "confidence" of the rule, and ρ(A=1, B=1 C=1) is referred to as the "support." This pattern structure or rule structure is quite simple and interpretable, which helps explain the general appeal of this approach. Typically the goal is to find all rules that satisfy the constraint that the accuracy ρ is greater than some threshold ρ , and the support is greater than some threshold ρ_s (for example, to find all rules with support greater than 0.05 and accuracy greater than 0.8). Such rules comprise a relatively weak form of knowledge; they are really just summarizes of co-occurrence patterns in the observed data, rather than strong statements that characterize the population as a whole. Indeed, in the sense that the term "rule" usually implies a causal interpretation (from the left to the right hand side), the term "association rule" is strictly speaking a misnomer since these patterns are inherently correlational, but need not be causal.

The general idea of finding association rules originated in applications involving "market-basket data". These data are usually recorded in a database in which each observation consists of an actual basket of items (such as grocery items), and the variables indicate whether or not a particular item was purchased. One can think of this type of data in terms of a data matrix of *n* rows (corresponding to baskets) and *p* columns (corresponding to grocery items). Such a matrix can be very large, with *n* in the number of millions and *p* in the tens of thousands, and is generally very sparse, since a typical basket contains only a few items. Association rules were invented as a way to find simple patterns in such data in a relatively efficient computational manner.

Details about who calls whom, how long they are on the phone, and whether a line is used for fax as well as voice can be invaluable in targeting sales of services and equipment to specific customers. But these tidbits are buried in masses of numbers in the database. By delving into its extensive customer-call database to manage its communications network, a regional telephone company identified new types of unmet customer needs. Using its data mining system, it discovered how to pinpoint prospects for additional services by measuring daily household usage for selected periods. For example, households that make many lengthy calls between 3 p.m. and 6 p.m. are likely to include teenagers who are prime candidates for their own phones and lines. When the company used target marketing that emphasized convenience and value for adults—"Is the phone always tied up?"—hidden demand surfaced. Extensive telephone use between 9 a.m. and 5 p.m. characterized by patterns related to voice, fax, and modem usage suggests a customer has business activity. Target marketing offering those customers "business communications capabilities for small budgets" resulted in sales of additional lines, functions, and equipment.

The ability to accurately gauge customer response to changes in business rules is a powerful competitive advantage. A bank searching for new ways to increase revenues from its credit card operations tested a nonintuitive possibility: Would credit card usage and interest earned increase significantly if the bank halved its minimum required payment? With hundreds of gigabytes of data representing two years of average credit card balances, payment amounts, payment timeliness, credit limit usage, and other key parameters, the bank used a powerful data mining system to model the impact of the proposed policy change on specific customer categories, such as customers consistently near or at their credit limits who make timely minimum or small payments. The bank discovered that cutting minimum payment requirements for small, targeted customer categories could increase average balances and extend indebtedness periods, generating more than $25 million in additional interest earned.

Merck-Medco Managed Care is a mail-order business which sells drugs to the country's largest health care providers: Blue Cross and Blue Shield state organizations, large HMOs, U.S. corporations, state governments, and so forth. Merck-Medco is mining its one terabyte data warehouse to uncover hidden links between illnesses and known drug treatments, and spot trends that help pinpoint which drugs are the most effective for what types of patients. The results are more effective treatments that are also less costly. Merck-Medco's data mining project has helped customers save an average of 10-15% on prescription costs.

Association Rule Mining to Find Frequent Itemsets

The task in association rule discovery is to find all rules fulfilling given prespecified frequency and accuracy criteria. This task might seem a little daunting, as there is an exponential number of potential frequent sets in the number of variables of the data, and that number tends to be quite large in market basket applications. Fortunately, in real data sets it is the typical case that there will be relatively few frequent sets (for example, most customers will buy only a small subset of the overall universe of products).

If the data set is large enough, it will not fit into main memory. Thus researchers aimed at methods that read the data as few times as possible. Algorithms to find association rules from data typically divide the problem into two parts: (1) find the frequent itemsets and then (2) form the rules from the frequent sets. If the frequent sets are known, then finding association rules is simple. If a rule $X \Rightarrow B$ has frequency at least s, then the set X must by definition have frequency at least s. Thus, if all frequent sets are known, researchers can generate all rules of the form $X \Rightarrow B$, where X is frequent, and evaluate the accuracy of each of the rules in a single pass through the data.

Studies on mining association rules have evolved from techniques for discovery of functional dependencies (Mannila & Raiha, 1987), strong rules (Agrawal & Srikant, 1994), classification rules (Han, Cai, & Cercone, 1993; Quinlan, 1992), casual rules (Michalski & Tecuci, 1994), clustering (Fisher, 1987), and so forth to disk-based, efficient methods for mining association rules in large set of transaction data (Agrawal, Imelinski, & Swami, 1993; Agrawal & Srikant,1994; Agrawal & Srikant, 1995; Mannila, Toivonen, & Verkamo, 1994; Park, Chen, & Yu, 1995). Discovery of association rules is an important class of data mining and aims at deciphering interesting relationships among attributes in the data (Houtsma & Swami, 1995; Michalski & Tecuci, 1994; Quinlan, 1992). To achieve this, efficient algorithms are to be implemented to conduct mining on these data. As a base, for given database of sales transactions, one could like to decipher all transactions among items such that the presence of some items in a transaction will imply the presence of some items in the same transaction. The problem of mining association rules on the basis of database was first explored in Agrawal et al. (1993). Various algorithms have been proposed to discover the large itemsets (Agrawal & Srikant, 1994; Houtsma & Swami, 1995; Jong Soo Park, Ming-Syan Chen, & Philip S. Yu., 1997).

The main limitation of almost all proposed algorithms (Agrawal et al., 1993; Agrawal & Srikant, 1994; Agrawal & Srikant, 1995; Mannila, Toivonen, & Verkamo, 1994; Park, Chen & Yu, 1995) is that they make repeated passes over the disk-resident database partition, incurring high I/O overheads. Moreover, these algorithms use complicated hash structures which entails additional overhead in maintaining and searching them, and they typically suffer from poor cache locality (Zaki, Parthasarathy, & Li, 1997). The problem with *Partition*, even though it makes only two scans, is that, as the number of partitions increase, the number of locally possible frequent itemsets increases. While this can be reduced by randomizing the partition selection, but the results from sampling experiment (Toivonen et al., 1996; Zaki, Parthasarathy, Li, & Ogihara, 1997) indicate that the randomized partitions will have a large number of possible frequent itemsets in common. Construction of directed graphs with a single pass over the database reduces high I/O overheads (Senthil Kumar & Wahidabanu, 2006). *Partition* can thus spend a lot of time in performing redundant computation (Zaki, Parthasarathy, Li, & Ogihara, 1997). Approaches using only general-purpose DBMS systems and relational algebra operations have been studied (Holsheimer, Kersten, Mannila, & Toivonen, 1995; Houtsma & Swami, 1995), but these do not compare favorably with the specialized approaches. A number of parallel algorithms have also been proposed (Han, Karypis, & Kumar, 1997; Houtsma & Swami, 1995). In this aspect, to find efficient algorithms the author proposes two forms of approaches in the next section that would overcome the drawbacks of currently available approaches.

TECHNIQUES USED FOR FINDING FREQUENT ITEMSETS

The goal of the techniques described in this section is to detect relationships or associations between specific values of categorical variables in large data sets. Closed directed graph approach technique scans a database only once making a count on possible 2-itemsets from which only the 2-itemsets with a minimum support are used to form the closed directed graph which explores possible frequent large itemsets in the database and the dynamic hashing algorithm technique generates large 3-itemsets at an earlier stage which reduces the size of the transaction database after trimming and consequently iterations will be less.

Closed Directed Graph Approach

Various algorithms used for discovering large itemsets make multiple passes over the data. During the first pass, a count is made to find the support of individual items. As a result, the support of individual items is used to determine which of them are *large*, that is, have minimum support. In each subsequent iterations, the set of itemsets found to be large in the previous pass is used as the base for generating new potentially large itemsets, called *candidate* itemsets. A count is made to find the actual support for these candidate itemsets during the pass over the data. The candidate itemsets which are actually large, is identified at the end of the pass, which forms a base for the next pass. The same process is repeated until no other new large itemsets are found (Agrawal et al., 1994; Jong Soo Park et al., 1997).

The construction of closed directed graph requires a single scan of the transaction database. Our method begins by generating all the 2-subsets in the database and performing a single pass of the database to find the counts of support of 2-itemsets. The frequent 2-itemsets with the minimum support are used to generate a directed graph from which only the closed portion of the directed graph is used to identify the frequent large itemsets. To illustrate this, a transaction database has been considered as shown in Figure 1.

Figure 1. Database

Transaction ID	Items
100	b e
200	a b c e
300	b c e
400	a c d

At first each possible 2-itemsets from the database are being taken as shown in Figure 2. Initially, a counter is set for each identified 2-itemsets. During the scanning of the entire database, the support of each candidate 2-itemsets is counted and stored in their respective counters.

After the scanning of the database is over, the value in each counter specifies the total number of counts for that particular 2-itemset. The total number of *support* for each possible 2-itemsets for the given database is shown in Figure 3.

To discover the possible frequent large itemsets, directed graphs are constructed using the items in the set of possible 2-itemsets with minimum support. A directed graph is one of the prevailing techniques to depict associations. A directed graph $G = \{V,E\}$ consists of a finite set V, together with a subset $E \subseteq V x V$. The elements of V are the *vertices* of the graph, and the elements of E are the *edges* of the graph. An edge of a directed graph is an ordered pair $[u,v]$ where u and v are the vertices of the graph. We say that the vertex v is *adjacent* to the vertex u, and the vertex u is *adjacent* to the vertex v. Moreover, it is said that the edge is *incident* from the vertex u and *incident* to the vertex v. An association graph can quickly turn into a tangled display with as few as a dozen rules. Suppose that the minimum support is 2, the directed graphs for the possible 2-itemsets with the minimum support will be as shown in Figure 4. The value of the edges represent the number of counts of each pair from the possible 2-itemsets.

After directed graphs for possible frequent 2-itemsets with minimum support have been constructed, all the directed graphs in Figure 4 are used to construct a single directed graph as follows (see Figure 5(i) and subsequently Figures 5(ii) to Figure 5(iv) as emerged):

In Figure 5(*iv*) the nodes of items b, c and e are closed. Only the node of item {a} is not closed. The edges of the closed nodes are used for the identification of frequent large itemsets. In the Figure 5(*iv*) itemset {b c e} represents the possible

Figure 2. Generation of possible 2-itemsets

{b e}		{b e}
{a b c e}	→	{{a b},{a c},{a e},{b c},{b e},{c e}}
{b c e}		{{b c},{b e},{c e}}
{a c d}		{{a c},{a d},{c d}}

Figure 3. Support for possible 2-itemsets

			counts
		{a b}	1
		{a d}	1
{b e}		**{a c}**	**2**
{{a b},{a c},{a e},{b c},{b e},{c e}}	→	{a e}	1
{{b c},{b e},{c e}}		**{b c}**	**2**
{{a c},{a d},{c d}}		**{b e}**	**3**
		{c d}	1
		{c e}	**2**

frequent k-itemset and itemsets {b c}, {b e}, and {c e} represents the possible frequent 2-itemsets for the given database.

Dynamic Hashing Algorithm

In Jong Soo Park et al. (1997) DHP algorithm, it has been shown that, the amount of data that has to be scanned during the large itemset discovery is a performance-related issue. Reducing the number of transactions to be scanned and trimming the number of items in each transaction improves the data mining efficiency in later stages.

The proposed algorithm DHA uses a technique of dynamic hashing to filter out unnecessary itemsets for next candidate itemsets generation. When a transaction database is considered, frequent 2-itemsets will not be much useful for improving the sales. Considering this, DHA accumulates the information only about 3-itemsets in advance in such a way that all possible 3- itemsets of each transaction after some pruning are hashed to a dynamic hash table.

In Jong Soo Park et al. (1997) DHP generates a much smaller C_2, so that the step for determining L_2 will be less expensive than which is spent in

Figure 4. Directed graphs for the possible frequent 2-itemsets with minimum support

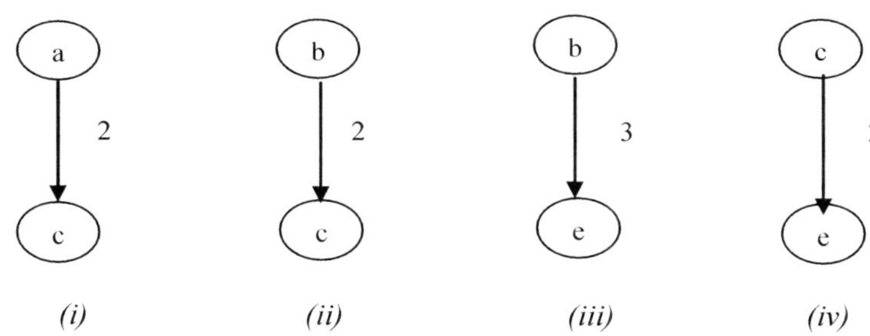

Figure 5. Formation of closed directed graph. (i): Directed graph 4(i) is used as the base. (ii): Since there is no node for item b in Figure 5(i), a new node for item b is added newly and directed graph 4(ii) is added to Figure 5(i) as shown. (iii). Since there is no node for item e in Figure 5(ii), a new node for item e is added newly and directed graph 4(iii) is added to Figure 5(ii) as shown. (iv). Since already nodes for items c and e exists in Figure 5(iii), directed graph 4(iv) is added to Figure 5(iii) as shown.

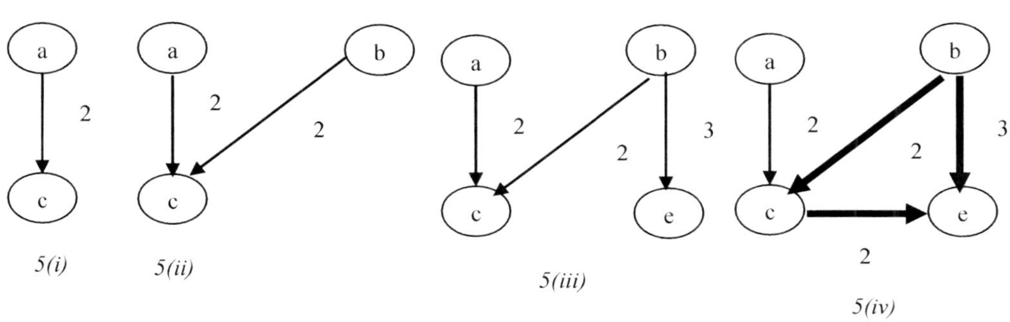

Apriori algorithm. Since, frequent 2-itemsets will not be much useful for the improvement of sales, DHA algorithm generates frequent 3-itemsets in advance, which in turn reduces the cost spent for generating k-itemsets in subsequent passes. Also the time and cost spent for generating frequent 2-itemsets will also be reduced.

The proposed algorithm uses dynamic hash technique to filter out unnecessary itemsets for next candidate itemset generation and also reduces the size of the database as a minimum for the next large itemsets generation. Initially, during the scanning of the entire database, the support of candidate k-itemsets is counted and DHA accumulates information only about *k+1*-itemsets (*where hash value is equal to 0*) in advance in such a way that all efficient *k+1*-itemsets of each transaction after some pruning are assigned to 0 and for each k+1-itemsets this number will be incremented by 1, and are hashed to the dynamic hash table. The buckets in the dynamic hash table consists of only *k+1* itemsets whose hash value is equal to 0. Instead of hashing all the k+1 itemset into a hash entry whose value is larger than or equal to *support s*, DHA hashes only the efficient k+1-itemsets whose hash value is equal to 0 in the dynamic hash table. Since, the number of keys, *m*, is not fixed but varies with time, DHA algorithm uses a dynamic hash table.

DHA algorithm is divided into 3 parts. In Part 1, the algorithm simply counts item occurrences to determine the large 1-itemsets and makes a dynamic hash table (i.e., DH_3) in which only the efficient 3-itemsets (*whose hash value is equal to 0*) are filtered and assigned a value starting from 0 for the first efficient 3-itemset and the value increases for the next efficient 3-itemsets by 1 and hashed into a dynamic hash table. In Part 2, based on the dynamic hash table (i.e., DH_k) generated in the previous pass, a set of candidate itemsets C_k is generated, which determines the set of large *k* itemsets L_k, reducing the size of database as a minimum for the next large itemsets and makes a dynamic hash table for candidate large *k+1* itemsets. Part 3 is basically same as Part 2 except that it does not employ a dynamic hash table. Since DHA is particularly powerful to determine large itemsets in early stages, thus improving the performance bottleneck. The size of C_k decreases significantly in later stages, thus rendering little justification its further filtering. This is the very reason that Part 2 will be used for early iterations, and Part 3 will be used for later iterations. It is noted that in Part 3 procedure apriori_gen to generate C_{k+1} from C_k is essentially the same as the method used by algorithm Apriori in [4] in determining candidate itemsets, and authors have omitted the details on it. Part 3 is used in DHA only for the completeness of this method.

However, this algorithm can be explained with an example as shown in Figure 6. The transaction database D as shown in Figure 6 is used for the discussion of the large itemsets. In the first pass of the algorithm, all the transactions of the database are scanned to count the support of the 1-itemsets to form the candidate set of large 1-itemsets, that is, $C_1 = \{\{A\}, \{B\}, \{C\}, \{D\}, \{E\}, \{F\}, \{G\}\}$. For this purpose, a hash tree for C_1 is built on the fly for the purpose of efficient counting. For each item in the database, it is checked with the items in the hash table. If the item is already present in the hash table, then the corresponding count of the item is incremented by one. If the item is not present in the hash table, then the new item is inserted into the hash table and the count is initialized as one. For each transaction, after occurrences of all the 1-subsets are counted, all the 3-subsets of this transaction are generated and only the efficient 3-subsets are hashed into the dynamic hash table DH_3 in such a way that each efficient 3-subsets generated using the hash function are assigned a count starting from 0 for the first 3-subset generated by using the hash function and for each 3-subset generated thereafter, the count will be incremented by one. Given the dynamic hash table, to filter out 3-itemsets from $L_1 * L_1$, the number of occurrences of 3-itemsets in dynamic hash table are compared with a mini-

Figure 6. Example of a dynamic hash table and generation of C_3 and C_4

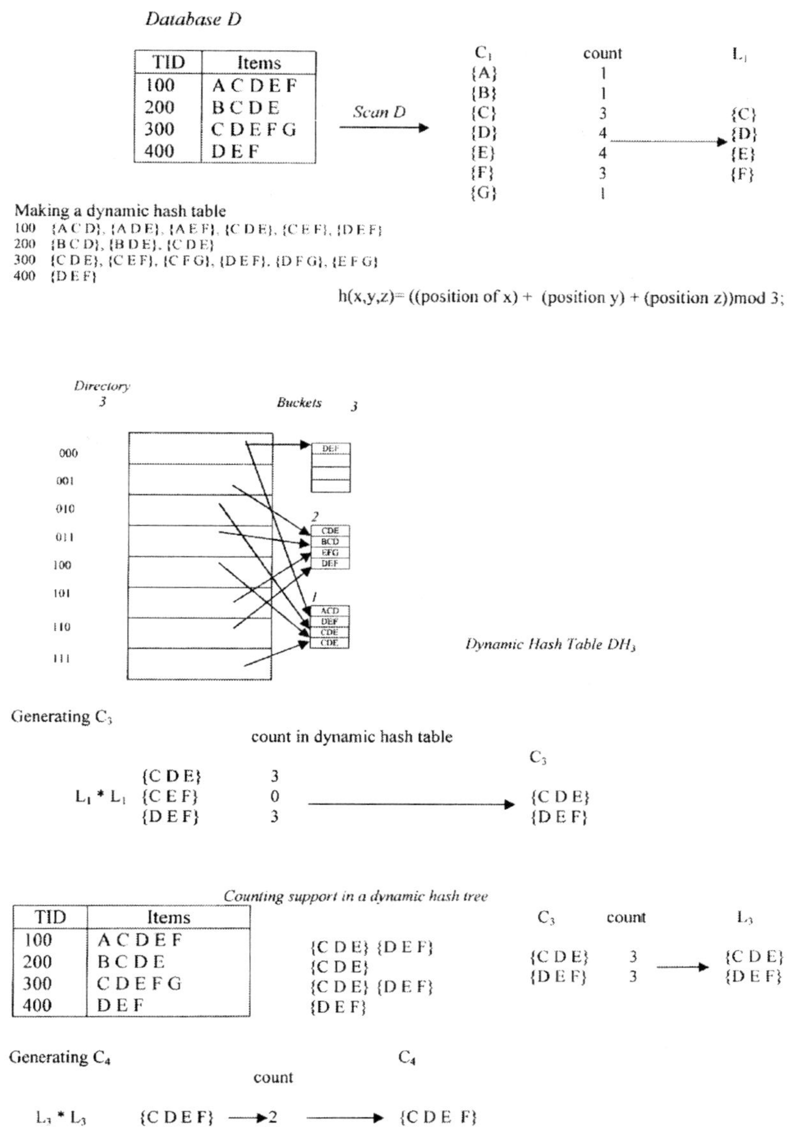

mum support equal to 2, thus forming C_3= {{C D E},{D E F}}.

FUTURE TRENDS

It can be stated that, in the short-term, the results of data mining will be in profitable taste, even if in mundane, business related areas. Moreover, micromarketing campaigns will explore new niches and foremost, advertising will target potential customers with new precision tools.

In the medium term, data mining may be as common and easy to use as e-mail. Users may use these tools to find the best airfare to New York, root out a phone number of a long-lost classmate, or find the best prices on lawn mowers.

However, the long-term prospects are truly exciting. Imagine intelligent agents turned loose on medical research data or on subatomic particle data. Computers may reveal new treatments for diseases or new insights into the nature of the universe. There are potential dangers, though, as discussed next.

What if every telephone call you make, every credit card purchase you make, every flight you take, every visit to the doctor, every warranty card you send in, every employment application you fill out, every school record you have, your credit record details, every Web page you visit— was all collected together? Much would be known about you! This is an all-too-real possibility. This kind of information is already stored in a database. Remember that phone interview you gave to a marketing company last week? All of your replies went into a database. Remember that loan application you filled out? It has been dumped in a database. Is there too much information about too many people for anybody to make sense of? Not with data mining tools running on massively parallel processing computers! Would any one feel comfortable about someone (or lots of someones) having access to all this data about you? And remember, all this data do not have to reside in one physical location; as the net grows, information of this type becomes more available to more people.

All these demand justified nature of data mining, their implications (social, ethical, cultural, economical) in the society, and technically speaking, a more robust set of data mining algorithm defining all these parameters and filtering all those preconditions.

CONCLUSION

Closed directed graph approach first constructs directed graphs for the possible 2-itemsets with minimum support and these directed graphs are used to construct a single directed graph from which only closed portion of the directed graph is used in identifying the possible frequent large itemsets. This method scans the database only once so that the I/O cost for retrieving the possible frequent large itemsets will be less. Since the number of keys, *m*, is not fixed and it varies with time a dynamic hash table is used in DHA algorithm to store the efficient large candidate 3-itemsets. The proposed algorithm also prunes the transactions, which do not contain any frequent itemsets, and trims the nonfrequent items from the transactions at the initial stage itself. The expected storage utilization will be greater than that of the DHP algorithm (Jong Soo Park et al., 1997). Considering frequent 2-itemsets will not be much useful for the sales improvement, frequent 3-itemsets are generated in the initial stage which in turn also reduces the cost and time used for generating frequent 2-itemsets. However, both the techniques explained above helps is discovering interesting association relationships among huge amounts of data which in turn helps in marketing, decision making and business management.

REFERENCES

Agrawal, R., Imelinski, T. & Swami, A (1993). Mining association rules between sets of items in large data bases. In *Proceedings of the ACM SIGMOD,* (pp. 207-216).

Agrawal, R. & Srikant, R. (1994). Fast algorithms for mining association rules in large databases. In *Proceedings of the 20th International Conference Very Large Data Bases,* (pp.478-499).

Agrawal, R. & Srikant, R. (1995). Mining sequential patterns. In *Proceedings of the 1995 International Conference Data Engineering,* Taipei, Taiwan.

Fisher, D. (1987). Improving inference through conceptual clustering. In *Proceedings of the 1987 AAAI Conference* (pp. 461-465). Seattle, Washington.

Han, J., Cai, Y., & Cercone, N., (1993). Data-driven discovery of quantitative rules in relational databases. *IEEE Trans. Knowledge and Data Engineering, 5*, 29-40.

Han, E. H., Karypis, G., & Kumar, V. (1997). Scallable parallel data mining for association rules. In *Proceedings of the ACM SIGMOD Conference Management of Data.*

Holsheimer, M., Kersten, M., Mannila, H., & Toivonen, H. (1995). A perspective on databases and data mining. In *Proceedings of the 1st International Conference Knowledge Discovery and Data Mining.*

Houtsma, M. & Swami, A. (1995). Set-oriented mining for association rules in relational databases. In *Proceedings of the 11th International Conference Data Engineering*, (pp.25-33).

Jong Soo Park, Ming-Syan Chen, & Philip S. Yu. (1997). Using a hash-based method with transaction trimming for mining association rules. *IEEE Transactions on Knowledge and Data Engineering, 9*(5), 813-825.

Mannila, H. & Raiha, K. J. (1987). Dependency inference. In *Proceedings of the 1987 International Conference Very Large Data Bases,* (pp. 155-158). Brighton, England.

Mannila, H., Toivonen, H., & Verkamo, I. (1994). Efficient algorithms for discovering association rules. In *Proceedings of the AAAI Workshop, Knowledge Discovery in Databases.*

Michalski, R. S. & Tecuci, G., (1994). *Machine learning, A multistrategy approach* (Vol. 4). Morgan Kaufmann.

Park, J. S., Chen, M. S. & Yu, P. S. (1995). An effective hash based algorithm for mining association rules. In *Proceedings of the 1995 ACM-SIGMOD International Conference Management of Data,* San Jose, California.

Parthasarathy, S., Zaki, M. J., & Li, W. (1997). *Application driven memory placement for dynamic data structures* (Tech. Rep. URCS TR 653). University of Rochester.

Quinlan, J. R. (1992). *Programs for machine learning.* Morgan Kaufmann.

Senthil Kumar, A. V. & Wahidabanu, R. S. D. (2006). Directed graph approach for association rule mining. In *Proceedings of the 2nd International Conference ICTS,* Indonesia.

Toivonen, H. (1996). Sampling large databases for association rules. In *Proceedings of the 22nd VLDB Conference.*

Zaki, M. J., Parthasarathy, S., Ogihara, M. & Li, W. (1997). New algorithms for fast discovery of association rules. In *Proceedings of the 3rd International Conference on Knowledge Discovery and Data Mining.*

Zaki, M. J., Parthasarathy, S., Li, W., & Ogihara, W. (1997). Evaluation of sampling for data mining of association rules. In *Proceedings of the 7th International. Workshop Research Issues in Data Engineering.*

Zaki, M. J.,Parthasarathy, S., & Li, W. (1997). A localized algorithm parallel association mining. In *Proceedings of the 9th ACM Symposium Parallel Algorithms and Architectures.*

Section II
Tools, Techniques, Methods

Chapter IV
Image Mining:
Detecting Deforestation Patterns Through Satellites

Marcelino Pereira dos Santos Silva
Rio Grande do Norte State University, Brazil

Gilberto Câmara
National Institute for Space Research, Brazil

Maria Isabel Sobral Escada
National Institute for Space Research, Brazil

ABSTRACT

Daily, different satellites capture data of distinct contexts, which images are processed and stored in many institutions. This chapter presents relevant definitions on remote sensing and image mining domain, beyond referring to related work on this field and to the importance of appropriate tools and techniques to analyze satellite images and extract knowledge from this kind of data. The Amazonia deforestation problem is discussed, as well INPE's effort to develop and spread technology to deal with challenges involving Earth observation resources. An image mining approach is presented and applied on a case study, detecting patterns of change on deforested areas of Amazonia. The purpose of the authors is to present relevant technologies, new approaches and research directions on remote sensing image mining, demonstrating how to increase the analysis potential of such huge strategic data.

INTRODUCTION

Motivation

Data acquisition and storage technology progress has led to a huge amount of data stored in repositories, which grow fast. Among increasing and relevant data acquired and processed, there is a strategic segment: satellite images, also known as remote sensing images.

The search for less expensive and more efficient ways to observe Earth motivated man in developing remote sensing satellites. They are currently the most significant source of new data about the planet, and remote sensing image databases are the fastest growing archives of spatial information. The variety of spatial and spectral resolutions for remote sensing images ranges from IKONOS 1-meter panchromatic images to the next generation of polarimetric radar imagery satellites. Given the widespread availability of remotely sensed data, many government and private institutions have built large remote sensing image archives.

The US National Satellite Land Remote Sensing Data Archieve, managed by USGS EROS Data Center, hosts 1.400 terabytes of satellite data gathered during 40 years. Satellites, like Terra and Aqua (NASA), generate 3 terabytes of images every day. The Brazil's National Institute for Space Research (INPE) holds more than 130 terabytes of image data, covering 30 years of remote sensing activities which are available on a database with free online access.

Actual society problems demand smart exploration of the vast and growing remote sensing data. There is a need for understanding relevant data and use it effectively and efficiently. Although valuable information is contained in image repositories, the volume and complexity of this data makes difficult (generally impossible) for human beings extract strategic information (knowledge) without appropriate tools (Piatetsky-Shapiro, Djeraba, Getoor, Grossman, Feldman & Zaki, 2006).

Data mining research has enabled powerful tools, new technologies and challenging techniques for relevant data domains. However, large image datasets need specific analysis resources and smart techniques and methodologies. The availability of huge remote sensing image repositories demands appropriate resources to explore this data.

A vast remote sensing database is a collection of landscape snapshots, which supplies a single opportunity to understand how, when and where changes occurred in the world. When such rich data is not analyzed, or it is done inefficiently, relevant information to understand complex processes and help solving challenging problems is wasted.

General Perspective and Objectives of the Chapter

In this chapter, which extends previous work (Silva, Câmara, Souza, Valeriano, & Escada, 2005), the authors intend to present relevant definitions on remote sensing and image mining domain, beyond presenting related work on this field and the importance of appropriate tools and techniques to explore satellite images and extract strategic knowledge from this kind of data.

They also discuss the Amazonia deforestation problem to demonstrate, through an image mining process, the strength of this approach to identify patterns and fight against the increase of affected areas in this forest. Developed technologies to support the process will be presented, providing an overview of methodologies, tools and techniques involved in research efforts.

Future trends and conclusion will bring reflection elements to consider classical and new mining resources to deal with challenging demands, citing limitations and also revealing directions to new research initiatives and relevant problems.

REMOTE SENSING AND IMAGE MINING

Broad Definitions

The first operational remote sensing satellite (LANDSAT-1) was launched in 1972, since then there has been a large worldwide experience in data gathering, processing and analysis of remotely sensed data. According to Canada Centre for Remote Sensing (2003), *remote sensing* is the science (and to some extent, art) of acquiring information about the Earth surface without actually being in contact with it. In other words, remote sensing is a field of applied sciences for information acquisition of the Earth surface through devices that perform the sensing and recording of the reflected or emitted energy, followed by processing, analysis, and application of this information. Such devices are called remote sensors, which are boarded on remote sensing aircrafts or satellites—also called Earth observation satellites. Images obtained through remote sensing acquisition and processing are used in many fields, once information from these remote sensing images is strongly demanded in many areas, including government, economy, infrastructure, and hydrology (e.g., security and social purposes, crop forecasting, urban planning, water resources monitoring).

In the image acquisition process, four concepts are fundamental: *spatial, spectral, radiometric and temporal resolution*. The *spatial resolution* defines the detail level of an image, that is, if a sensor has a spatial resolution of 20m then each pixel represents an area of 20m x 20m. The *spectral resolution* determines the sensor capability to define short intervals of wavelength; the finer the spectral resolution, the narrower the wavelength range for a particular channel or band. The *radiometric resolution* of an imaging system describes its ability to discriminate very slight differences in energy; the finer the radiometric resolution of a sensor, the more sensitive it is to detect small differences in reflected or emitted energy. The *temporal resolution* determines the necessary time for the sensor revisit a specific target and image the exact same area, that is, the time required to complete one entire orbit cycle; if a sensor is able to obtain an image of an area each 16 days, then its temporal resolution is this period (Canada Centre for Remote Sensing, 2003).

Before getting into remote sensing image mining, it is necessary to state *spatial data mining*, which refers to the extraction of knowledge, spatial relationships, or other interesting but not explicit patterns stored in spatial databases. Such mining approaches integrate spatial database and data mining issues, bringing valuable resources to understand facts and processes represented in spatial data, discovering spatial relationships, building up spatial knowledge bases, and revealing spatial patterns and processes contained in spatial repositories. Applications of the technology include, beyond remote sensing, geographic information systems, medical imaging, geomarketing, navigation, traffic control, enviromental studies, and many other areas where spatial data are used (Han & Kamber, 2001).

Remote sensing image mining deals specifically with the challenge of capturing patterns, processes, and agents present in the geographic space, in order to extract specific knowledge to understand or to make decisions related to a set of relevant topics, including land change, climate variations and biodiversity studies. Events like deforestation patterns, weather change correlations and species dynamics are examples of precious knowledge contained in remote sensing image repositories.

The *Amazonia forest*, located in South America, has 6,500,000 km^2, involving seven frontier countries. Brazil holds 63.4% of South America Amazonia, which extends to the following Brazilian states: Mato Grosso, Tocantins, Maranhão, Amazonas, Pará, Acre, Amapá, Rondônia and Roraima. Since it is the world's largest tropical forest, deforestation in the Amazonia rainforest is an important contributor to global land change. According to INPE's estimates, close to 200,000 km^2 of forest were cut in Brazilian Amazonia in the period from 1995 to 2005 (INPE, 2005). INPE uses LANDSAT and CBERS images to provide yearly assessments of the deforestation in Amazonia. Given the extent of the deforestation on tropical forests, figuring out the processes and its agents are important issues for setting up public policies that can help preserve the environment (Figure 1).

Figure 1. Amazonia deforestation (source: Isabel Escada - INPE)

Related Work

Nagao and Matsuyama (1980) developed, at Kyoto University (Japan), the first high level vision system for aerial image interpretation. The system processing modules operated on a common dataset. The analysis process was divided in the following steps: smoothing, when the images were processed to remove noise and spots on boundaries; segmentation, when elementary regions were extracted through a basic region growing algorithm; global exam of the scene, to estimate object domains using image metadata; detailed area analysis, when object detection subsystems analyzed a knowledge base to find specific objects; communication among object detection subsystems, in order to control the analysis flow managing the information on databases, resolve conflicts among detection subsystems and correct segmentation problems.

GeoMiner (Han, Koperski & Stefanovic, 1997), developed at Simon Fraser University (Canada), is a prototype of spatial data mining system, with resources to characterize spatial data through rules, compare, associate, classify and group datasets, analyze patterns and perform mining tasks in differente levels. The prototype has a language for mining tasks of spatial data (GMQL), beyond visualization tools for data and spatial mining results. GeoMiner is integrated to data warehousing technology, and it is able to access different spatial database servers.

SPIN! project (May & Savinov, 2002), developed by the Fraunhofer Institute for Autonomous Intelligent Systems (Germany), is focused on producing a spatial data mining system that integrates Geographic Information Systems and data mining in a open, extensible and tightly coupled framework. The project prioritizes issues like scalability, security, multiuser access, robustness, and platform independence. Its functionality levels include data access and management, interactive thematic mapping for statistic data visualization, detection and explanation of spatial clusters and spatial events.

ADaM, a NASA's project with the University of Alabama at Huntsville (USA), is a set of scientific data and image mining tools (Rushing, Ramachandran, Nair, Graves, Welch & Lin, 2005). Its resources include pattern recognition, image processing, optimization, association rule mining, among others. The system is a set of components that may be put together to perform complex tasks. A focus of the project is the efficient implementation of critical performance components, keeping each component of the system as independent as possible, in order to enable the use of appropriate module subsets to specific applications, including linking to third party software.

Position about the Technology

Such initiatives, among other important projects, led by institutions and researchers of different countries since 1980, demonstrates the relevance, strength and demand for efficient and robust approaches, once the mining process on image repositories demand a strong commitment with efficiency and robustness. The huge volume of the datasets need an efficient hardware and software infrastructure. The relativity of values, the spatial complexity, and the multitude of interpretations require robust implementations, competent domain specialists and experient data analysts for the mining task performances.

However, still a limited capacity is available for extracting information from large remote sensing image databases. Currently, most image processing techniques are designed to operate on a single image, and there are few algorithms and techniques for handling multitemporal images. This situation has lead to a "knowledge gap" in the process of deriving information from images and digital maps (MacDonald, 2002). This "knowledge gap" has arisen because there are currently very few techniques for image data mining and information extraction in large image data sets, and thus researchers are failing to exploit the huge remote sensing data archives.

Although there has been a large research effort in content-based image retrieval (CBIR) techniques (Rui, Huang & Chang, 1999; Smeulders, Worring, Santini, Gupta & Jain, 2000; Wang, Khan & Breen, 2002), the specific problem of mining remote sensing image databases has received much less attention. Proposals such as VISIMINE (Aksoy, Koperski, Tusk, & Marchisio, 2004) and KIM (Schröder, Rehrauer, Seidel& Datcu, 2000) are focused on clustering methods that operate on the feature space, the multidimensional space which is created by the different spectral bands of a remote sensing image. These techniques are useful for distinguishing spectral signatures of different land cover types, such as finding areas which are classified as "lakes," "cities" or "forests."

Nevertheless, in remote sensing image mining, one of the most important challenges is tracking patterns of land use change. A large remote sensing image database is a collection of snapshots of landscapes, which provide a unique opportunity for understanding how, when, and where changes take place in the world. Extensive fieldwork also indicates that the different actors involved in land cover change (e.g., small-scale farmers, large plantations, cattle ranchers) can be distinguished by their different spatial patterns of land use (Lambin, Geist & Lepers, 2003). Furthermore, these patterns evolve in time; new small farms will be created and large farms will increase their agricultural area at the expense of the forest. In these and related situations, patterns of land use change will have similar spectral signatures and image mining techniques based on clustering in the feature space will not be able to distinguish between them. Therefore, tracking the temporal evolution of patterns in remote sensing imagery requires methods that are different from standard content-based image retrieval (CBIR) systems. A typical CBIR system uses a query image as the source and images in the database as targets, and query results are a set of images sorted by feature similarities with respect to the source (Chen, Wang & Krovetz, 2003). When searching for patterns in remote sensing image databases, a different approach is necessary. Instead of similarity searches between image pairs, a system for mining remote sensing image databases must be able to do similarity searches between patterns found in different images. Therefore, mining remote sensing image databases is searching for patterns of change, not searching for internal content.

CHALLENGES AND TECHNOLOGICAL STRATEGIES ON DEFORESTATION ISSUE

Brazil's Challenge: Monitor and Decrease Amazonia Deforestation

The *land cover* describes the physical state of the land surface, which may be forest, water, buildings, and so on. Changes on this cover may be caused by climate variations, changes on river courses, and so on. However, most changes on land cover are attributed to human activities. Such modifications implies on changes on the extension (area increase or decrease) of a specific type of coverage. The *land use*, influenced by human activities and enviromental processes and features, is related to the purpose to which it is used, like agriculture, habitation, mining, leisure, among others. Land use changes occur in several spatial levels and in different periods, characterizing the environment and human dynamics on territorial segments (Briassoulis, 2004).

Desertification, climate change, biodiversity loss—among others—can imply in severe consequences to the environment and consequently to humans. The modification of forest and crop areas for urban use is an important land change, due to serious implications. The causes and consequences of land use and cover change, its social, economics and enviromental impacts have motivated different research projects. One of them is (Lambin, 1999), which emphasizes that land cover change is an important global change factor, interacting with climate, ecosystem processes, biochemical cycles, biodiversity and even with human activities. The key issues of the project deal with land cover patterns, change processes, human response to changes, integrated global and local models, development of databases about Earth surface, biophysics processes and fundamental factors. This approach aims to increase the understanding, and get new knowledge about interactive land change.

The Amazonia case is characterized by the complexity, dimension, and interests involved in the issues concerning land change (Becker, 1997). Alves (2002) presents an investigation on spatio-temporal deforestation dynamics of the Amazonia, using remote sensing images to analyze deforestation spatial patterns on 1970's, and between 1991 and 1997. This work brings valuable information: the deforested area increased from 10,000,000 ha (1970's) to 59,000,000 ha in 2000; an intensification on the deforestation rate on 1970's and 1980's was caused by the federal government politics, which included huge highway infrastructures, and a roadside colonization of 100 km along the extended highways; analyzing the images and the patterns, it is clear that beyond of the roadside deforestation along main roads and development areas, there is still the merging of little deforested areas, what originates large ones.

Once the fast deforestation process causes land degradation, social tension and irregular urbanization, faster the precise identification of areas with these tendencies, higher the chances of preventing, managing and reducing the consequences of the processes. Daily, different satellites capture data belonging to this context, which images are available to many institutions. Image mining tools can, in fact, increase the analysis potential of such huge strategic data.

Developed Technologies at INPE Concerning Image Analysis and Mining

Researchers of the Brazil's National Institute for Space Research (INPE) has been studying the structural patterns on Amazonia, holding a wide know-how on the forest issues. Moreover, the historical development process is also a research topic at INPE, which maintains a rich dataset of remote sensing images that provide an extensive spatiotemporal perspective of the Amazonia territory. In addition, the Institute experience on image processing and analysis, as well the development of methodologies and software tools, supplies important elements to keep building up image analysis and mining technologies. In this context, relevant ones developed at INPE are: SPRING, TerraLib, CBERS, PRODES and DETER, which are freely available on Internet.

SPRING (www.dpi.inpe.br/spring) is a state-of-the-art geographic information system (GIS) and remote sensing image processing system with an object-oriented data model which provides the integration of raster and vector data representations in a single environment (Figure 2). SPRING main features include: an integrated GIS for environmental, socioeconomic and urban planning applications; a multiplatform system, including support for Windows and Linux; a widely accessible freeware for the GIS community with a quick learning curve. The software is a mechanism of diffusion of the knowledge developed by INPE and its partners with the introduction of new algorithms and methodologies (Câmara, Souza, Freitas & Garrido 1996).

TerraLib (www.dpi.inpe.br/terralib) is a GIS classes and functions library, available from the Internet as open source, allowing a collaborative environment and its use for the development of multiple GIS tools. Its main objective is to enable the development of a new generation of GIS applications, based on the technological advances on spatial databases. TerraLib is free software developed by INPE and its partners. The main motivation for this project is the current lack of either public or commercial GIS libraries that provide components for the diversity of GIS data and algorithms, especially when viewed upon the latest advances in geographical information science. On a practical side, TerraLib enables quick development of custom-built geographical applications using spatial databases. As a research tool, TerraLib is aimed at providing a rich and powerful environment for the development of GIScience research, enabling the implementation of GIS prototypes that include new concepts such as spatio-temporal data models, geographical ontologies and advanced spatial analysis techniques (Câmara, Vinhas, Souza, Paiva, Monteiro, Carvalho, 2001).

The *CBERS* program (http://www.cbers.inpe.br/en/index_en.htm), a joint effort of Brazil and China, embodied the development and construction of two remote sensing satellites that carry on-board imaging cameras and additionally a repeater

Figure 2. SPRING: image processing and geographic information system

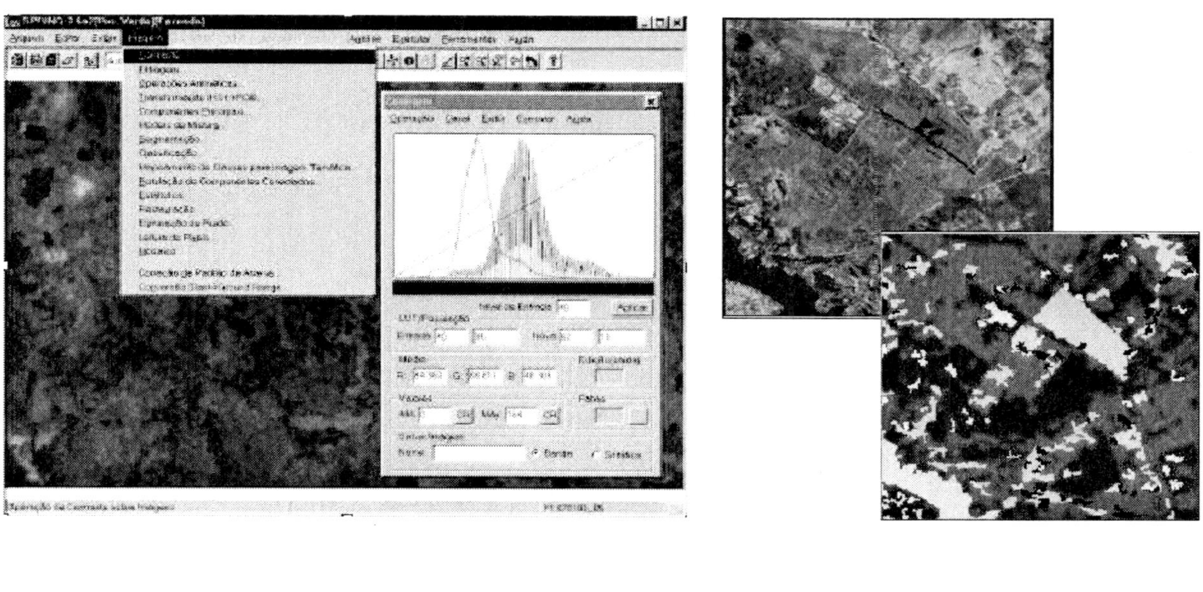

for the Brazilian System of Environmental Data Collection. CBERS-1 and CBERS-2 are identical in their technical structure, space mission and payload (on-board equipment like cameras, sensors, computers, among other equipment designed for scientific experiments). CBERS-1 was launched by the Chinese Long March 4B launcher from the Taiyuan Launch Base on October 14, 1999. CBERS-2 was launched on October 21, 2003 from the Taiyuan Satellite Launch Center in China (Figure 3). CBERS-2 was integrated and tested in the integration and test laboratory of INPE. The CBERS satellite has a set of sensors—WFI (wide field imager), CCD (charge coupled device high resolution imaging camera), IRMSS (infrared multispectral scanner)—with a high potential to meet multiple application requirements including: forestry alteration, signs of recent fires, monitoring of agricultural development, support for crop forecasting, identification of anthropic anomalies, analysis of natural recurrent events, mapping of land use, urban sprawling, identification of water-continent borders, coast studies and management, reservoir monitoring, acquisition of stereoscopic images for proper cartographic analysis, support for soil survey and geology, generation of support material for educational activities. In 2002, both governments decided to expand the initial agreement by including CBERS-3 and 4, which must be launched, respectively, in 2008 and 2012. The program objectives are: build a family of remote sensing satellites to support the needs of users in Earth resource applications, and improve the industrial capabilities of space technology in Brazil.

Since 1988 INPE has been monitoring Brazilian Amazonia using satellite images, producing estimations on annual deforestation rates of the forest through the *PRODES* project (Amazônia deforestation calculation program). From 2002 on, these estimations are being generated by image digital classification with PRODES methodology (www.obt.inpe.br/prodes/). The main advantage of this approach is the precision of georeferenced deforestation polygons, enabling a multitemporal geografic database. Using deforestation increments identified on each image, the annual rates are estimated for August 1st of the reference year. For the 2003/2004 period, the deforestation rates were obtained from 207 LANDSAT images; INPE estimates that the deforestation from August 2003 to August 2004 was 27.429 km^2. For the 2004/2005 period, the deforestation rates were obtained from 211 LANDSAT classified images; INPE estimates that the deforestation from August 2004 to August 2005 was 18.793 km^2. The Institute estimates a deforested area of 13,100 km^2 for the 2005/2006 period. PRODES digital results of 2000 until 2004

Figure 3. CBERS-2: Launch and Web interface of image catalog (source: www.cbers.inpe.br)

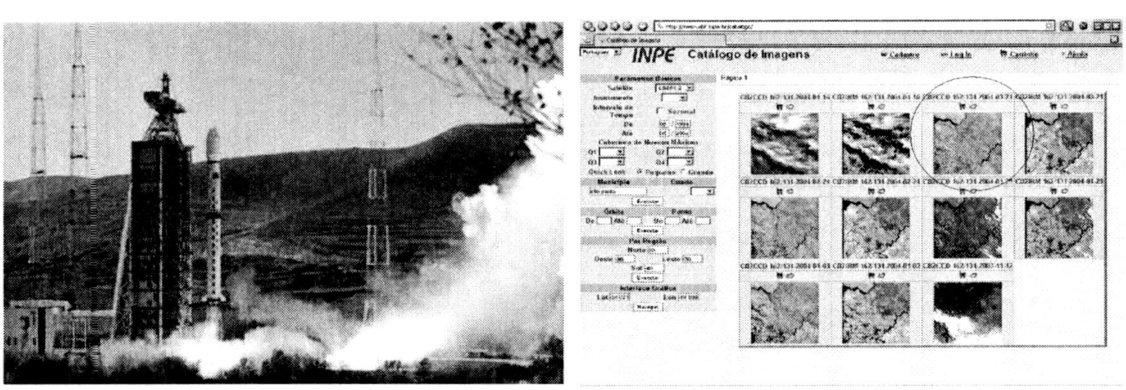

are available on SPRING databases containing LANDSAT satellite images, thematic map of the deforestation of the year, thematic map of the accumulated deforestation, and the shapefiles of the year with polygons of deforestation increment of the year, forest, total accumulated deforestation until the previous year, clouds and non-forest. From 2005 on, it is also available on the shapefile of the deforestation thematic map of the year for each LANDSAT image, and the shapefile of the mosaic of all images.

The *DETER* system (deforestation detection on real time) uses sensors with high observation frequency to reduce cloud cover limitations during the process of detecting deforestation increments (www.obt.inpe.br/deter). The instruments used are the MODIS sensor, aboard TERRA and AQUA satellites (NASA), with a spatial resolution of 250 m and temporal resolution (Brazil) of three to five days, and the WFI sensor, aboard CBERS-2, with a spatial resolution of 260 m and temporal resolution of five days. These resolutions enable the detection of recent deforested areas superior to 0,25 km^2. The results of the methodology—which produces information in almost real time about regions where new deforestation areas occur—allow DETER supplies environment surveillance institutions with periodic information about deforestation events (Figure 4). The goal of the system is not the estimation of total deforestated area in Amazonia, once estimations obtained through DETER are error-prone due to the spatial resolution of MODIS and WFI. The system is concerned on supplying recent and updated information to support government actions against the forest destruction using a higher temporal resolution of the sensors. The system is an INPE project, part of a federal plan of reducing Amazonia deforestation.

INPE's Effort to Spread Earth Observation Technologies

There is a need for global land observation advance, once the world is changing rapidly. Global land observation is a crucial need for the world, and Earth observation (EO) systems are a public good. INPE's effort on advanced policies of development of state-of-the-art software, hardware, methodologies and products relies on the need of building capacity in EO to supply the wide demand of the area.

Build capacity in Earth observation implies on removing the barriers to make all sectors of society use publically funded EO data. Three relevant obstacles are: lack of data (much EO data is expensive or unavailable), lack of tools (once

Figure 4. Amazonia deforestation process detected by DETER

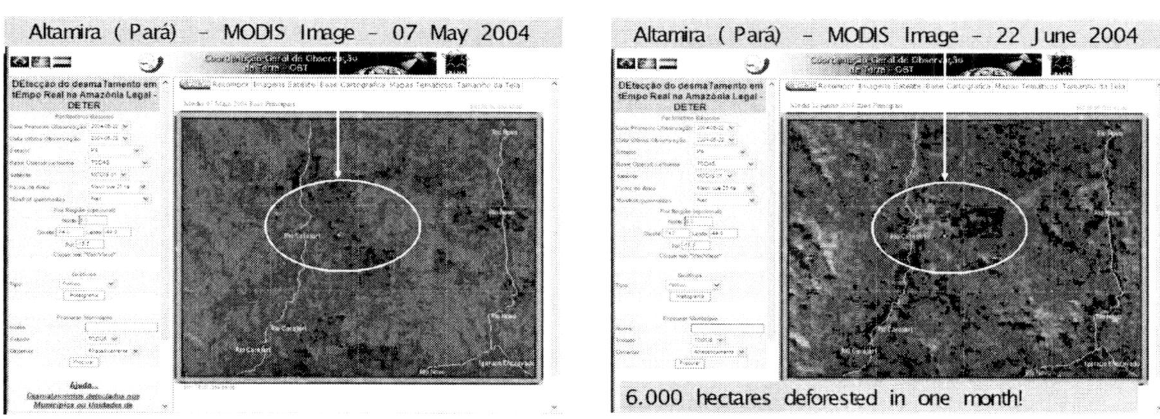

good software is required to explore EO data), and lack of expertise (it is necessary to build capacity at a massive scale). INPE's approach to overcome such barriers are: make EO data free, produce good open source software for EO data handling, and provide open access to on-line training and to scientific literature.

The Internet has reduced the cost of data distribution to very close to zero, and society responds very quickly to open availability of free data and good on the Web. CBERS images received in Brazil are freely available on the Internet for Brazilian and Latin American users, and CBERS images received in China are freely available on the Internet for Chinese users (www.cbers.inpe.br). Free EO data and free EO technology create new users and new applications, increasing the need for other types of EO data. Private companies, for example, state the free CBERS data benefit: enables new business development, facilitates trial uses for new clients, creates jobs by reducing cost of data buys, increases work quality by adding data previously unavailable, and eases the planning of new applications.

Commercial EO market in many countries does not have enough income to research and development investment, once it is still a small size market. To let it grow, it is necessary to supply improvements on information extraction through high-quality software. Concerning the tool challenge, INPE developed GIS and image-processing softwares (TerraLib and SPRING) available free on the Internet, providing good software for EO data handling (www.dpi.inpe.br/spring; www.dpi.inpe.br/terralib).

The research system on EO in the developed world discourages the production of training material, once academic institutions in US and Europe graduate qualified personnel and there are good books on GIS and remote sensing (unfortunately, these books are in English and are expensive). Developing countries need innovative responses, especially good training material and on-line books. Brazilian experience is overcoming the expertise challenge releasing free books online, a three-volume set: *Introduction to GIS*, *Spatial Analysis*, and *Spatial Databases* (http://www.dpi.inpe.br/gilberto/livros.html).

INPE is focusing on the "white-box" model: *results = people + data + software*. This means support for people learning by doing and using, timely and free geospatial datasets, and adequate data analysis and integration softwares. The results: an enormous demand for remote sensing data in developing countries, a relevant increase on the number of users of Earth observation data due to free online data access, and the success of CBERS data policy that has been extremely well-received by government and society in Brazil.

DETECTING DEFORESTATION PATTERNS THROUGH SATELLITES

Patterns of Change in Remote Sensing Image Databases

Given a large remote sensing image database, researchers would like to explore the database with questions such as: What are the different land use patterns present in the database? When did a certain land use pattern emerge? What are the dominant land use patterns for each region? How do patterns emerge and change over time? The answer to these and similar questions requires the availability of data mining techniques which are able to perform searches for patterns found in different images. Silva (2006) approached this problem by using spatial patterns as a mean of describing relevant semantic features of an image.

The primary consideration is that the instruments onboard remote sensing satellites capture energy at different parts of the electromagnetic spectrum, which is then converted into digital imagery. These instruments are not designed for a specific application, but are a compromise between sensor technology and requirements from different user communities. As a result, remote sensing images have a structural description which is independent of the application domain that a scientist employs to extract information. The *image domain* and the *application* domain are distinguished, as shown in Figure 5:

- **Spatial patterns:** The geometric structures that can be extracted from the images using techniques for feature extraction, segmentation, and image classification. They must be identified and labeled according to a typology which expresses their semantics. Examples of such patterns include corridor-like regions and regular-shaped polygons representing patterns of the mined data.
- **Application concepts:** The different classes of spatial objects, which are associated to a specific domain. For example, in deforestation assessments, concepts include large-scale agriculture, small-scale agriculture, cattle ranching, and wood logging.

To associate structures found in the image to concepts in the application, there is a *structural classifier*, which is able to relate the same structures to different application domains. This strategy differs from most remote sensing image database mining systems, such as KIM (Schröder et al., 2000) and VISIMINE (Aksoy et al., 2004), which implicitly assume that there is one "best fit" for associating semantic concepts in the user domains to image-derived structures. In this approach, different structural classifiers will produce different associations between spatial patterns and the user domain concepts, and each association is valid within a given application context. In other words, there are many ways to bridge the "sensory gap" and a "best fit" should not be searched. For each type of application, there will be an appropriate structural classifier.

In what follows, the methodology for image mining is described and applied to the problem of mining patterns in INPE's remote sensing image database. In this context, the application domain is concerned with describing land use change in tropical forests using remote sensing satellites.

Methodology for Mining Land Use Patterns on Remote Sensing Images

The methodology for image mining in large remote sensing databases uses the application-dependent structural classifier, as outlined previously. The methodology consists of three steps:

- Definition of a spatial pattern typology according to the user's application domain (Figure 6).
- Building a reference set of spatial patterns. This reference set is built using a prototypical set of images. Landscape objects are identified and labeled: the identification employs image segmentation and the labeling is performed according to the spatial pattern typology (Figure 7).
- Mining the database using a structural classifier (guided by the application concepts of the domain), matching the reference set of spatial patterns to the landscape objects

Figure 5. Overview of pattern mining process

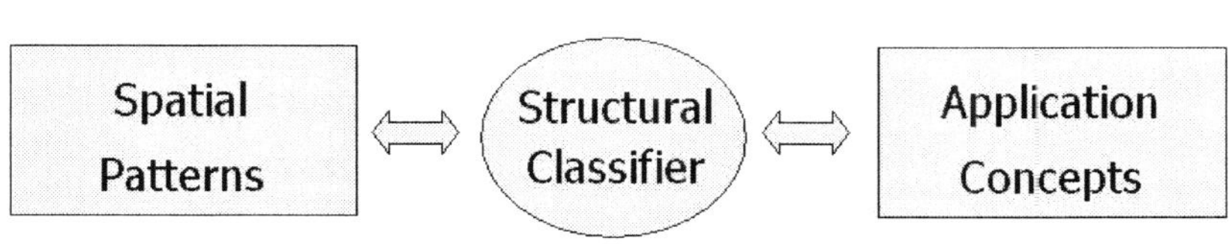

identified in images, thus revealing the spatial configurations present in each image (Figure 9).

Defining a Spatial Pattern Typology

The first phase of the methodology calls for the definition of a spatial pattern typology which is associated to a given application domain (Escada, Monteiro, Aguiar, Carneiro & Câmara, 2005). In order to illustrate the proposal, a typology defined for mapping different types of land use change in tropical forests will be used.

When using remote sensing images for understanding the forces driving changes in tropical forests, the assumption is that the expression of change is captured by changes in land use. Extensive fieldwork also indicates that the different actors involved in land use change (small-scale farmers, large plantations, cattle ranchers) can be distinguished by their different patterns of land use (Lambin, Geist & Lepers, 2003). They propose a typology of the land use patterns in terms of deforestation processes (see Figure 6): corridor (commonly associated with riverside and roadside colonization), diffuse (generally related to smallholder subsistence agriculture), fishbone (typical of planned settlement schemes), and geometric (frequently linked to large-scale clearings for modern sector activities).

Three spatial patterns typology of Lambin will be used (corridor, diffuse, geometric), relating them to the structures of landscape objects in order to obtain the spatial patterns, through a cognitive assessment process, in which a human specialist associates landscape objects to spatial patterns typology elements.

Building a Reference Dataset of Spatial Patterns

To represent the structures detected in remote sensing images, the concept of a landscape object will be introduced. A landscape object is a structure detected in a remote sensing image by means of an image segmentation algorithm. Landscape objects can be associated to different types of spatial patterns.

To build a reference set of *spatial patterns* (Figure 7), a set of prototypical landscape objects is obtained, which are extracted from a set of sample images. Segmentation algorithms are used to partition the image into regions which are spatially continuous, disjoint and homogenous. Recent surveys (Meinel & Neubert, 2004) indicate that region-growing approaches are well suited for producing closed and homogenous regions. In this proposal, it is adopted the region-growing segmentation algorithm developed by INPE (Bins, 1996), and implemented in the SPRING

Figure 6. Spatial patterns of tropical deforestation (from left to right): corridor, diffuse, fishbone, and geometric (Source: Lambin, Geist & Lepers, 2003)

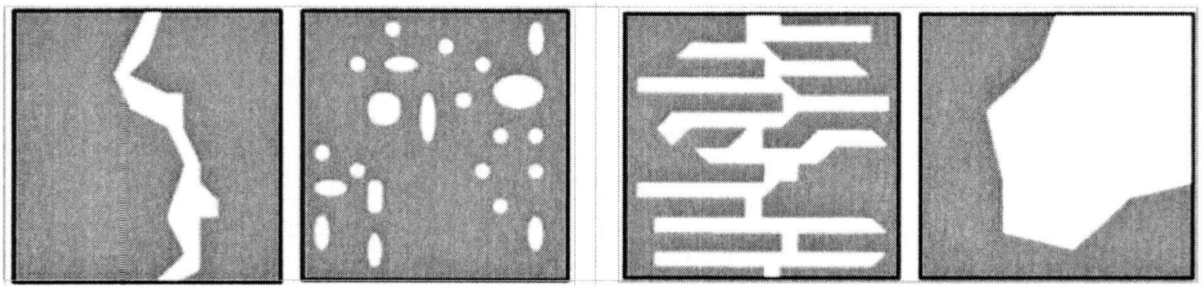

software system (Câmara, 1996). This algorithm has been extensively validated for extracting land use patterns in tropical forests (Shimabukuro et al., 1998) and has been very favorably reviewed in a survey (Meinel & Neubert, 2004).

SPRING's region growing algorithm works as follows (Figure 8) (Bins, 1996): (a) the image is first segmented into atomic cells of one or few pixels; (b) each segment is compared with its neighbors to determine if they are similar or not. If similar, they are merged and the mean gray level of the new segment is updated; (c) the segment continues growing by comparing it with all the neighbors until there is no remaining joinable region, at which point the segment is labeled as a completed region; (d) the process moves to the next uncompleted cell, repeating the entire sequence until all cells are labeled. The algorithm requires two parameters: a similarity threshold value, and an area threshold value.

Mining the Database Using a Structural Classifier

Once the reference set of *spatial patterns* is built, the next phase will use it to mine *spatial configura-*

Figure 7. Building a reference set of spatial patterns

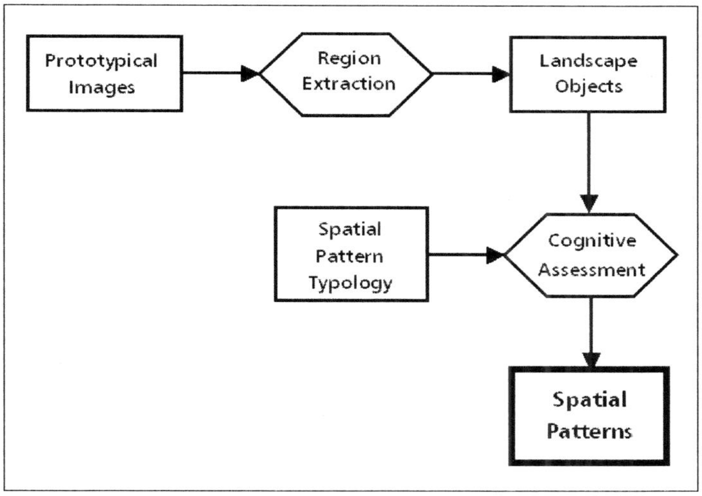

Figure 8. Example of a segmentation process

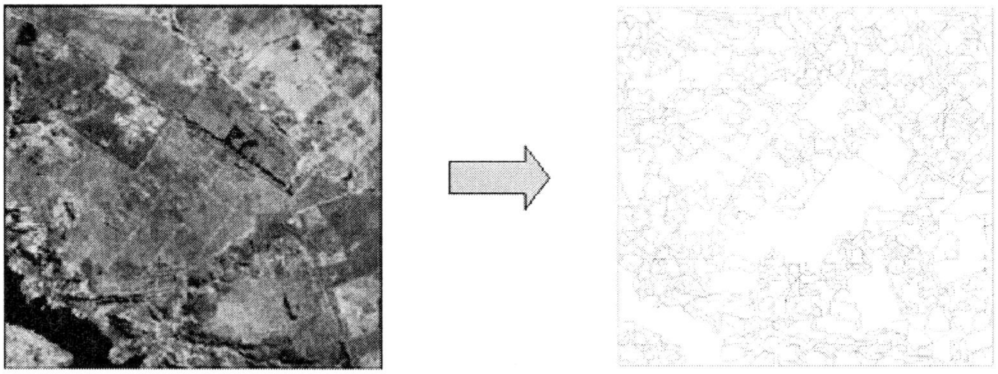

Figure 9. Obtaining spatial configurations

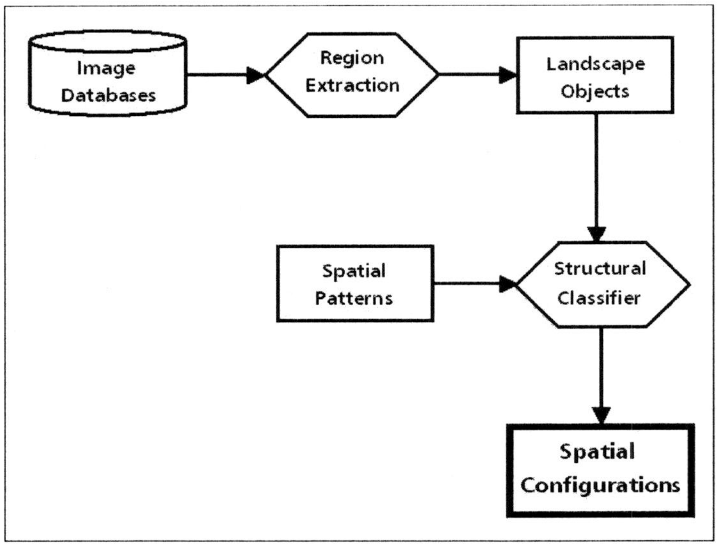

tions from image databases. The *structural classifier* enables the association between landscape objects extracted from images and the reference set of *spatial patterns* (Figure 9).

The structural classifier must be able to distinguish between different spatial patterns. It uses the C4.5 classifier (Quinlan, 1993), a classification method based on a decision tree. It predicts the value of a categorical attribute (Witten & Frank, 1999) based on noncategorical attributes. The categorical attribute is the pattern type and the noncategorical attributes are a set of numerical attributes that characterize each pattern.

To select the attributes that distinguish the different types of land use patterns, the concepts from landscape ecology (Turner, 1989) are used. Landscape ecology is based on the notion that environmental patterns strongly influence ecological processes. One of the key components of landscape ecology theory is the definition of metrics that characterize geometric and spatial properties of categorical map patterns (McGarigal, 2002). The pattern metrics used in landscape ecology include metrics of spatial configuration that operate at the patch level. Patches form the building blocks for categorical maps and within-patch heterogeneity is ignored. Patch metrics refer to the spatial character and arrangement, position, or orientation of patches within the landscape. The pattern metrics proposed by the FRAGSTATS (Spatial Pattern Analysis Program for Categorical Maps) software (McGarigal & Marks, 1995) are used, which include:

- *Perimeter* (m) and *area* (ha).
- *Para* (perimeter-area ratio): A measure of shape complexity.
- *Shape* (shape index): Patch perimeter divided by the minimum perimeter possible for a maximally compact patch of the corresponding patch area.
- *Frac* (fractal dimension index): Two times the logarithm of patch perimeter (m) divided by the logarithm of patch area (m^2).
- *Circle* (related circumscribing circle): 1 minus patch area (m^2) divided by the area (m^2) of the smallest circumscribing circle.

- *Contig* (contiguity index): Equals the average contiguity value for the cells in a patch.

The landscape ecology metrics are fed into the C4.5 classification algorithm to distinguish the different types of spatial patterns. After this classifier is properly trained, it can be used to label the landscape objects found in other images. Therefore, for each image in the database, this procedure identifies the number and location of the different types of spatial patterns. A specific set of spatial patterns found in an image is referred as a *spatial configuration.*

By identifying the spatial configurations of different images, the user will be able to evaluate the emergence and evolution of different types of change. Each spatial pattern is associated to a different type of land use change. Therefore, the comparison between spatial configurations of images in different locations and between spatial configurations of images at the same location in different times will allow new insights into the processes and actors that bring about change.

Table 1. Land use change in tropical forests

Landscape object	Land use change
Corridor pattern	Roadside colonization Riverside deforestation
Diffuse pattern	Smallholder agriculture Small deforestation increments
Geometric pattern	Large farms

Case Study: Image Mining for Deforestation Patterns

Controlling deforestation on Amazon rain forest is a difficult challenge for Brazil, once the causes of deforestation include economic, social and political factors, and the current pace of land use change is substantial, with a deforested area of about 200,000 km^2 during the decade 1995-2005. The situation demands fast and effective actions for reducing this pace of devastation. In order to monitor the extremely fast process of land use change in Amazonia, it is very important that INPE be able to use its huge data archive to the maximum extent possible. In this context, the image mining methodology was used to achieve a better understanding of the processes of land use change in Amazonia.

A case study was developed using Landsat TM images (225/64, 226/64, 226/65, 225/65) of 1997, 2000, 2001, 2002 and 2003, which cover the region of São Félix do Xingu in the state of Pará. This is a region with many violent land conflicts and one of the largest annual rates of deforestation in Amazônia (INPE, 2005). The main land use activity developed in São Felix do Xingu is cattle ranching, which holds around 10% of the cattle of Pará state (Américo, Vieira, Veiga & Araujo, in press). Deforestation in the region has two main agents: migrants, that have settled in small areas, and large cattle ranchers, many of whom have occupied land illegally (Escada, Vieira, Amaral, Araújo, Veiga, Aguiar, & Veiga, 2005). The images

Figure 10. Spatial patterns representing corridor, diffuse, and geometric patterns

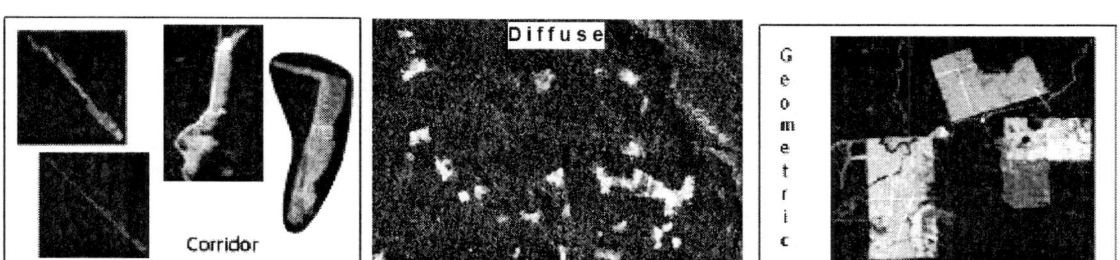

and deforestation data were provided by PRODES Project (INPE, 2005). The application concepts for this task are guided by the land use change domain in tropical forests (Table 1).

Building Spatial Patterns

According to the image mining methodology, landscape objects were extracted from prototypical images. Then, a human specialist, through cognitive assessment, obtained *spatial patterns* based on the spatial patterns typology of tropical deforestation (Figure 6). Spatial patterns are presented in Figure 10.

Obtaining Spatial Configurations

The *structural classifier*, using the *spatial patterns*, extracted *spatial configurations* from the set of images just mentioned. Results are presented below.

In a first case, it is necessary to answer the following question: "What's the behavior of large farmers in São Félix do Xingu during 1997-2003 period? Is the area of new large farms increasing?"

Figure 11. Large farms dynamic in São Félix do Xingu

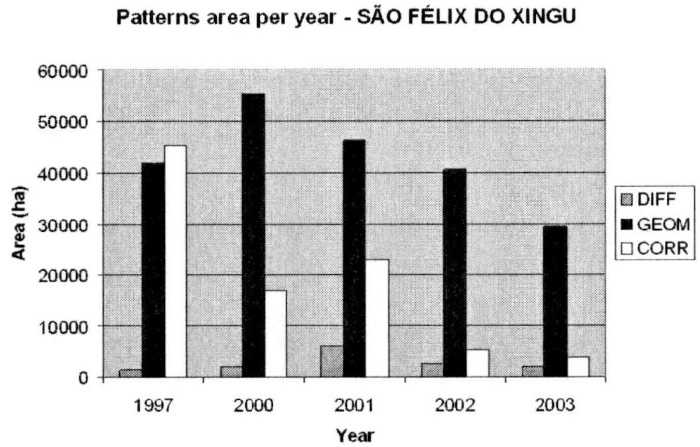

Figure 12. Diffuse pattern in São Félix do Xingu 1997-2003

Observing the evolution of the corresponding *spatial configuration* (geometric patterns - GEOM) in Figure 11, it was possible to conclude that, "in 2000, this kind of deforestation reached a peak of 55,000 ha, but decreased in the following years. In 2003, the deforestation area associated to large farms decreased to 29,000 ha. This indicates that large farms are reducing their contribution to deforestation."

There is a second question: "What's the distribution of smallholder agriculture and small deforestation increments in São Félix do Xingu area during the years 1997-2003?" Observing Figure 12, it is possible to conclude: "the distribution of this land use pattern (diffuse) in this period was mainly concentrated in the northeast and southeast of this area."

The next question is: "In São Félix do Xingu region, is there any dominant land use change pattern?" Observing Figure 13, the conclusion is: "Diffuse pattern represented 61% of total occurrences of land use changes in 2001, indicating an increase in smallholder agriculture / small increments in deforested areas in that year."

FUTURE TRENDS

A consortium of Earth observation satellites for global land monitoring, a network of cooperating ground stations, EO data free on the Internet with global weekly coverage, satellite sensor resolution improvements and the availability of web services to perform image mining tasks will provide necessary resources for new applications and a wide range of demands, specially in developing countries. Moreover, hardware and software performance increase will support mining processes on huge and improved image datasets, allowing a more intensive and extensive use of satellite image mining in strategic fields like forestry and reservoir monitoring, agricultural expansion, soil survey, analysis of natural phenomena, and urban studies.

Future research directions in remote sensing image mining include tracking individual trajectories of change. Patterns found in one map are linked to those in earlier and later maps, thus enabling a description of the trajectory of change of each landscape object. The current method aggregates landscape objects of the same type. A

Figure 13. Diffuse patterns in São Félix do Xingu

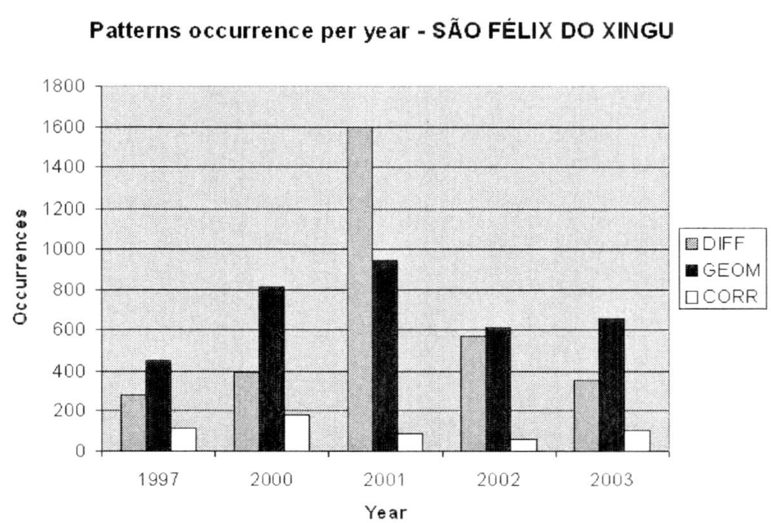

more sophisticated approach would be to describe the evolution of each landscape object, including operations such as merging of adjacent regions. This description would allow the image-mining tool to describe when two irregular areas of land use (associated to small settlers) were merged. It would also show when the merged region was extended with a regular pattern (suggesting that a large cattle ranch had been established). This description could increase even more the ability to understand the land use changes that are detectable in remote sensing image databases.

CONCLUSION

This chapter presents relevant issues on satellite image mining, describing a method for mining patterns of change that enables extracting spatial arrangements from remote sensing image databases. It addresses the problem of describing land use change. It combines techniques from data mining, digital image processing and landscape ecology to identify patterns in images of distinct dates. The method points out that patch metrics can be used to identify agents of land use change. Images of distinct dates enabled the detection of pattern changes, which are extremely valuable when assessing, managing or preventing deforestation processes.

This methodology enables associating land change objects to causative agents, and it can assist the environmental community to respond to the challenge of understanding and modeling relevant issues in a rapidly changing world. The results from the case study show that image-mining techniques are a step forward in understanding and modeling land use and cover change. The proposed method also enables a more effective use of the large land remote sensing image databases available in agencies such as USGS, ESA and INPE.

The remote sensing image-mining process is an interactive one; once it demanded the sample selection, model building and rating, context evaluation, return to specific points of the process, among others. During experiments, the result evaluation in different phases demonstrated the need of new prototype objects, better model calibration, or even adjustments on the spatial pattern typology. Once provided such topics, relevant results were obtained and validated through extensive fieldwork.

Taking into account the heterogeneity of the Amazonia context, a relevant (and expected) question is the fact that the model training and application must be performed in spatially similar regions, that is, train the structural classifier in a specific region and apply it in another region, with different spatial features, causes the generation of inconsistent results. Another methodology limitation concerns the quantity and quality of prototype objects used to generate the model for structural classification. If the number of elements or their description ability to distinguish patterns is not appropriate, the generated model (decision tree) will classify inconsistently many objects. The methodology also demands a proper spatial pattern typology, which must characterize the spatial patterns and the semantic aspects that must be detected during the process.

The mining process requires a domain specialist, due to the intense Amazonia dynamics, especially on the prototype object selection and during the spatial configuration interpretation. Further experiments are necessary to improve the method, to test alternatives for image segmentation algorithms and for pattern classifiers. The limitations of the current method are also associated to the two-dimensional nature of land use maps. An extension of the method would combine spatial information (patch metrics) with spectral information (pixel and region trajectories in multitemporal images).

Uncle Scrooge principle states that, "a penny saved is a penny earned." However, the anti-Uncle Scrooge principle reveals that, "a pixel saved is a penny wasted." Why is that so? Because "value comes from use." Coherent EO programs can supply strategic components for the enormous demand of remote sensing data, expertise, and analysis tools in developing countries. This work resources may help to leverage the power of detecting, evaluating and reducing the pace of Amazonia deforestation, once INPE holds know-how and a wide spatiotemporal coverage of the forest. Moreover, the present technology can be ported to provide solutions to a broad range of image mining applications.

REFERENCES

Américo, M. C. S., Vieira, I. C. G., Veiga, J. B. & Araujo, R. (in press) Pecuária e Amazônia: Estratégias sociais e reestruturação do território nas frentes pioneiras: Rodovia PA-279 e região da Terra do Meio no Pará [Cattle ranching and Amazonia: Social strategies and territory reorganization in new frontiers – PA-279 and Terra do Meio region in Pará state]. In R. Araujo & P. Lena (Eds.), *Alternativas de desenvolvimento sustentável na Amazônia: Experiências recentes* [Alternatives of sustainable development in Amazônia: Recent experiences].

Aksoy, S., Koperski, K., Tusk, C. & Marchisio, G. (2004). Interactive training of advanced classifiers for mining remote sensing image archives. In *Proceedings of the ACM International Conference on Knowledge Discovery and Data Mining* (pp. 773-782). Seattle, Washington.

Alves, D. S. (2002). Space-time dynamics of deforestation in Brazilian Amazonia. *International Journal of Remote Sensing, 23*(14).

Bins, L., Fonseca, L. & Erthal, G. (1996). Satellite imagery segmentation: A region growing approach. In *Proceedings of the 8th Brazilian Symposium on Remote Sensing* (pp.1-4).

Becker, B. (1997). *Amazonia.* São Paulo: Atica.

Briassoulis, H. (2004). *Analysis of land use change: Theoretical and modeling approaches.* Retrieved April 8, 2008, from http://www.rri.wvu.edu/WebBook/Briassoulis

Câmara, G., Souza, R., Freitas, U. & Garrido, J. (1996). SPRING: Integrating Remote Sensing and GIS with object-oriented data modelling. *Computers and Graphics, 15*(6), 13-22.

Câmara, G., Vinhas, L., Souza, L., Paiva, L., Monteiro, A., Carvalho, M. & Raoult, B. (2001). Design patterns in GIS development: The Terralib experience. In *Proceedings of the III Brazilian Symposium in Geoinformatics, GeoInfo 2001*, Rio de Janeiro.

Canada Centre for Remote Sensing (2003). *Fundamentals of remote sensing.* Remote Sensing Tutorial (pp. 5-44). Retrieved April 8, 2008, from www.ccrs.nrcan.gc.ca/ccrs/learn/tutorials/fundam/fundam_e.html

Chen, Y., Wang, J. Z. & Krovetz, R. (2003). CLUE: Cluster-based retrieval of images by unsupervised learning. In K. A. Meraim, I. Bloch (Eds.), In *Proceedings of the IEEE Seventh International Symposium on Signal Processing and its Applications* (pp. 202-231).

Escada, M. I. S., Monteiro, A. M., Aguiar, A. P., Carneiro, T. & Câmara, G. (2005). Análise de padrões e processos de ocupação para a construção de modelos na Amazônia [Analysis of land use patterns and processes for the construction of models in Amazonia]. In *Proceedings of the XII Brazilian Symposium on Remote Sensing* (pp. 2973-2983), Goiania, Brazil.

Escada, M. I. S., Vieira, I. C. G., Amaral, S., Araújo, R., Veiga, J. B. D., Aguiar, A. P. D., Veiga, I., Oliveira, M., Pereira, J. L. G., Filho, A. C., Fearnside, P. M., Venturieri, A., Carriello, F., Thales, M., Carneiro, T. S., Monteiro, A. M. V., & Câmara, G. (2005). Padrões e processos de ocupação nas novas fronteiras da Amazônia: O interflúvio do Xingu/Iriri [Land use patterns and processes in Amazonian new frontiers: The Xingu/Iriri region]. *Estudos Avançados* [Advanced Studies], *19*, 9-23.

Han, J., Koperski, K. & Stefanovic, N. (1997). GeoMiner: A system prototype for spatial data mining. In *Proceedings of the ACM SIGMOD International Conference on Management of Data* (pp. 553-556).

Han, J. & Kamber, M. (2001). Data mining - Concepts and tecniques. In D. Cerra & H. Severson (Eds.) (pp.405-412). San Diego, CA: Morgan Kaufmann Publishers.

INPE, National Institute for Space Research (2005). *PRODES project - Monitoring the Brazilian Amazon forest using satellites.* National Institute for Space Research. Retrieved April 8, 2008, from http://www.obt.inpe.br/prodes

Lambin, E. (1999). *Land-use and land-cover change implementation strategy.* Retrieved April 8, 2008, from http://www.geo.ucl.ac.be/LUCC/lucc.html

Lambin, E. F., Geist, H. J. & Lepers, E. (2003). Dynamics of land use and land cover change in Tropical Regions. *Annual Review of Environment and Resources, 28*(1) 205-241.

MacDonald, J. (2002). The Earth observation business and the forces that impact it. In D. Couts (Ed.), *Earth observation business network 2002.* Vancouver, CA: MacDonald Dettwiler.

May, M. & Savinov, A. (2002). An integrated platform for spatial data mining and interactive visual analysis. In *Proceedings of the International Conference on Data Mining Methods and Databases for Engineering* (pp. 90-101).

McGarigal, K. & Marks, B. (1995). *FRAGSTATS: Spatial pattern analysis program for quantifying landscape structure.* USDA Forestry Service Technical Report PNW-351, Washington, DC.

McGarigal, K. (2002). Landscape pattern metrics. In A.H. El-Shaarawi & W.W. Piegorsch (Eds.), *Encyclopedia of environmentrics* (pp. 1135-1142). Sussex, England: John Wiley & Sons.

Meinel, G. & Neubert, M. (2004). A comparison of segmentation programs for high resolution remote sensing data. *International Archives of Photogrammetry and Remote Sensing, 35*(1), 1097-1105.

Nagao, M. & Matsuyama, T. (1980). *A structural analysis of complex aerial photographs.* New York: Plenum Press.

Piatetsky-Shapiro, G., Djeraba, C., Getoor, L., Grossman, R., Feldman, R. & Zaki, M. (2006). What are the grand challenges for data mining? - KDD-2006 Panel Report. *SIGKDD Explorations, 8*(2), 70-77.

Quinlan, R. (1993). *Programs for machine learning.* San Francisco: Morgan Kaufmann.

Rui, Y., Huang, T. S. & Chang, S. F. (1999). Image retrieval: Current techniques, promising directions and open issues. *Journal of Visual Communication and Image Representation, 10*(1), 39-62.

Rushing, J., Ramachandran, R., Nair, U. J., Graves, S. J., Welch, R. & Lin, A. (2005). ADaM: A data mining toolkit for scientists and engineers. *Computers and Geosciences, 31*(5), 607-618.

Schröder, M., Rehrauer, H., Seidel, K. & Datcu, M. (2000). Interactive learning and probabilistic retrieval in remote sensing image archives. *IEEE Transactions on Geoscience and Remote Sensing, 23*(1), 2288-2298.

Shimabukuro, Y. et al. (1998). Using shade fraction image segmentation to evaluate deforestation in Landsat thematic mapper images of the Amazon region. *International Journal of Remote Sensing 19*(3), 535-541.

Silva, M. P. S., Câmara, G., Souza, R. C. M., Valeriano, D. M. & Escada, M. I. S. (2005). Mining patterns of change in remote sensing image databases. J. Han & B. Wah (Eds.), In *Proceedings of the Fifth IEEE International Conference on Data Mining* (pp. 362-369).

Silva, M. P. S. (2006). Mineração de Padrões de Mudança em Imagens de Sensoriamento Remoto [Mining patterns of change in remote sensing images] (Unpublished doctoral thesis). São José dos Campos: National Institute for Space Research (INPE).

Smeulders, A.W.M., Worring, M., Santini, S., Gupta, A. & Jain, R. (2000). Content-based image retrieval at the end of the early years. *IEEE Transactions on Pattern Analysis and Machine Intelligence, 22*(1), 1349-1380.

Turner, M. G. (1989). Landscape ecology: The effect of pattern on process. *Annual Review of Ecology and Systematics, 20*, 171-197.

Wang, L., Khan, L. & Breen, C. (2002). Object boundary detection for ontology-based image classification. In *Proceedings of the Third ACM International Workshop on Multimedia Data Mining* (pp. 51-61).

Witten, I. H. & Frank, H. (1999). *Data mining: Practical machine learning tools and techniques with Java implementations*. San Francisco: Morgan Kaufmann.

Chapter V
Machine Learning and Web Mining: Methods and Applications in Societal Benefit Areas

Georgios Lappas
Technological Educational Institution of Western Macedonia, Kastoria Campus, Greece

ABSTRACT

This chapter reviews research on machine learning and Web mining methods that are related to areas of social benefit. It shows that machine learning and Web mining methods may provide intelligent Web services of social interest. The chapter reveals a growing interest for using advanced computational methods, such as machine learning and Web mining, for better services to the public, as most research identified in the literature has been conducted during the last years. The chapter objective is to help researchers and academics from different disciplines to understand how Web mining and machine learning methods are applied to Web data. Furthermore it aims to provide the latest developments on research that is related to societal benefit areas.

INTRODUCTION

The Web is constantly becoming a central part of social, cultural, political, educational, academic, and commercial life and contains a wide range of information and applications in areas that are of societal interest. Web mining is the field of data mining that is related to the discovery of knowledge from the Web. The Web can be considered as a tremendously large and rich in content knowledge base of heterogeneous entries without any well specified structure, which proportionally makes the Web at least as complex as any known complex database and perhaps the largest knowledge repository. The vast information that surrounds the Web does not come only from the content of Websites, but is also related to usage of Web pages, navigation paths and networking between the links of Web-pages. All these properties establish the Web as a very challenging area for the machine learning community to apply their methods usually for extracting new knowledge, discovering interesting patterns and enhancing the efficiency of Websites by providing user-demand content and design.

Web mining is a relatively new area, broadly interdisciplinary, attracting researchers from: computer science fields like artificial intelligence, machine learning, databases, and information retrieval specialists; from business studies fields like marketing, administrative and e-commerce specialists; and from social and communication studies fields such as social network analyzers, pedagogical scientists, and political science specialists. Herrera-Viedma and Pasi (2006) denote that due to the complexity of Web research there is a requirement for the use of interdisciplinary approaches like statistics, databases, information retrieval, decision theory, artificial intelligence, cognitive social theory and behavioral science. As a relatively new area there is a lot of confusion when comparing research efforts from different point of views (Kosala & Blockeel, 2000) and therefore there is a need for surveys that record and aggregate efforts done by independent researchers, provide definitions and explain structures and taxonomies of the field from various points of view.

The overall objective of this chapter is to provide a review of different machine learning approaches to Web mining and draw conclusions on their applicability in societal benefit areas. The novelty of this review is that it focuses on Web mining in societal benefit areas. There exist similar work related to Web mining in (Baldi, Frasconi, & Smyth, 2003; Chakrabarti, 2003; Chen & Chau, 2004; Pal, Talwar & Mitra, 2002). Baldi et al. (2003) cover research and theory on aspects of Internet and Web modeling at the information level based on mathematical, probabilistic, and graphical treatment. Chakrabarti focuses on studies that connect users to the information they seek from the Web providing lots of programs with pseudocode. Chen and Chau provide an extended review of how machine-learning techniques for traditional information retrieval systems have been improved and adapted for Web mining applications. Pal et al. (2002) present an overview of machine learning techniques with focusing on a specific Web mining category, the Web content mining that will be described in next section. This work is differentiated from the aforementioned related work as the chapter particularly focuses on Web mining and machine learning that may help and benefit societal areas in ways of extracting new knowledge, providing support for decision making and empowering valuable management of societal issues. This survey aims to help researchers and academics from different disciplines to understand Web mining and machine learning methods. Thus, it is aimed at a relatively broad audience and tries to provide them with a different and more open view on Web research. Therefore this work addresses researchers from both computer science and other than computer science disciplines with the intention: (a) for computer science researchers, to provide them with the latest developments on the theory and applications of Web mining, focusing also to the need for Web mining applications in societal beneficial areas, and (b) for researchers from other than computer science disciplines, to draw their attention to existing machine learning methods that may help them to seek for more effective results in their Web research.

Later in the chapter, some background to the different perspectives of Web mining has been provided with a short review on machine learning methods. Afterwards, a study on related machine learning methods applied to Web mining have been put forward, which is followed by applications related to societal benefit areas. Finally it discusses current trends and future challenges on machine learning and Web mining.

WEB MINING OVERVIEW

The word *mining* means extracting something useful or valuable, such as mining gold from the earth. The expectation of useful or valuable information discovery from the Web is enclosed in the term "Web mining." Definitional, Web mining refers to the application of data mining techniques to the World Wide Web (Cooley, Mobasher & Srivastava, 1997), or else is the area of data mining that refers to the use of algorithms for extracting patterns from resources distributed in the Web. Over the years, Web mining has been extended to denote the use of data mining and other similar techniques to discover resources, patterns and knowledge from the Web and Web-related data (Chen & Chau, 2004).

According to the different sources of data analysis, Web mining is divided into three mining categories. Figure 1 shows the Web mining taxonomy and the sources of data that are used in mining. Moreover, it displays Web mining categories and the related areas of research interest:

a. *Web content mining* focus on the discovery of knowledge from the content of Web

Figure 1. Taxonomy of Web mining according to source of target data

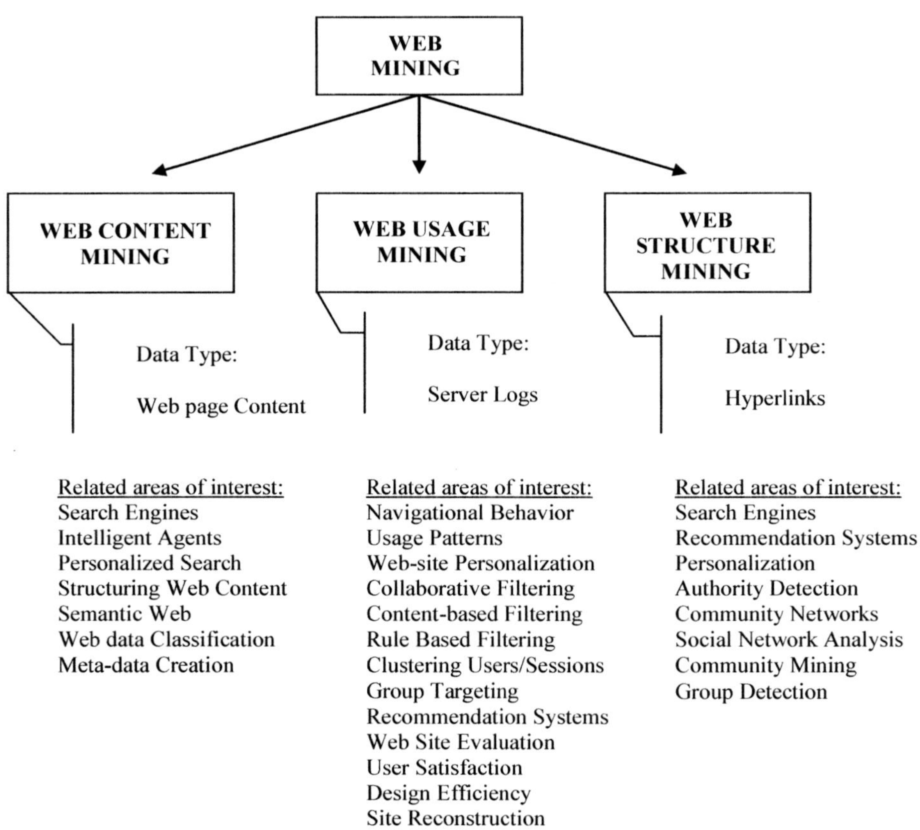

pages and therefore the target data consist of multivariate type of data contained in a Web page as text, images, multimedia, and so forth.

b. *Web usage mining* focus on the discovery of knowledge from user navigation data when visiting a Website. The target data are requests from users recorded in special files stored in the Website's servers called log files.

c. *Web structure mining* deals with the connectivity of Websites and the extraction of knowledge from hyperlinks of the Web.

The above taxonomy is now broadly used in Web mining (Scime, 2005) and has the origins from Coley et al. (1997) who introduced Web content mining and Web usage mining and Kosala and Blockeel (2000), who added Web structure mining.

A well-known problem related to Web content mining, is experienced by any Web user trying to find relevant Web pages that interests the user from the huge amount of available pages. Current search tools suffer from low precision due to irrelevant results (Chakrabarti, 2000). Lawrence and Giles (1999) raise issues related to search engine problems. Search engines are not able to index all pages resulting in imprecise and incomplete searches due to information overload. The overload problem is very difficult to cope as information on the Web is immensely and grows dynamically raising scalability issues.

Moreover, myriad of text and multimedia data are available on the Web prompting the need for intelligence techniques for developing automatic mining using artificial intelligence tools. Such automatic mining is performed by intelligent systems called "intelligent agents" or "agents" that search the Web for relevant information using domain characteristics and user profiles to organize and interpret the discovery information. Agents may be used for intelligent search, for classification of Web pages, and for personalized search by learning user preferences and discovering Web sources meeting these preferences.

Web content mining is more than selecting relevant documents on the Web. Web content mining is related to information extraction and knowledge discovery from analyzing a collection of Web documents. Related to Web content mining is the effort for organizing the semistructured Web data into structured collection of resources leading to more efficient querying mechanisms and more efficient information collection or extraction. This effort is the main characteristic of the "Semantic Web" (Berners-Lee, Hendler & Lassila, 2001), which is the next Web generation. Semantic Web is based on "ontologies," which are metadata related to the Web page content that make the site meaningful to search engines. Sebastiani's study (2002) may be used as a source for Web content mining.

Web usage mining tries to find patterns of navigational behavior from users visiting a Website. These patterns of navigational behavior can be valuable when searching answers to questions like: How efficient is our Website in delivering information? How the users perceive the structure of the Website? Can we predict users next visit? Can we make our site meeting user needs? Can we increase user satisfaction? Can we targeting specific groups of users and make Web content personalized to them?

Answer to these questions may come from the analysis of the data from log files stored in Web servers. A log file is usually a large file that contains all requests of all users to the Website as they arrive in time. Log files may have various formats according to the information stored. The most common format uses information about user IP, date and time of request, type of request (for example get a Web page), a code denoting whether the request has be successfully served or failed, and number of bytes transferred to user. However, Web usage mining should not be confused with tools that analyze log files in order to provide statistics about the site such us:

page hits, times of visits, hits per hour or per day or per month, and so forth. While this information might be interesting or valuable for Website owners, they have low data analysis. Web usage mining is more sophisticated as it refers to find users access behavior (Levene & Loizou, 1999) and usage patterns (Buchner, Mulvenna, Anand & Hughes, 1999). It has become a necessity task to provide Web administrators with meaningful information about users and usage patterns for improving quality of Web information and service performance (Eirinaki & Vazirgiannis, 2003; Spiliopoulou & Pohle, 2001; Wang, Abraham & Smith, 2005). Successful Websites may be those that are customized to meet user preferences both in the presentation of information and in relevance of the content that best fits the user.

In this context, Website personalization is the process of customizing the content and structure of a Website to the specific needs of each user taking advantage of user's navigational behavior (Eirinaki & Vazirgiannis, 2003). Recommendation systems support Website personalization by tracking user's behavior and recommending similar items to those liked in the past (content-based learning), or by inviting users to rate objects and state their preferences and interests so that recommendations can be offered to them based on other users rates with similar preferences (collaborative filtering), or by asking questions to the user and providing services tailored to user needs according to the user's answers (rule-based filtering).

On the other hand, content-based filtering is the most common method for Web personalization from server log files and has attracted considerable attention from researchers (Mobasher, Jain, Han, & Srivastava, 1996; Mobasher, Cooley, & Srivastava, 1999; Ngu & Wu, 1997; Spiliopoulou, Pohle, & Faulstich, 1999; Srivastava, Cooley, Deshpande, & Tan, 2000; Wolfang & Lars, 2000) for constructing user models that represent the behavior of users. Such systems usually apply classification methods or clustering algorithms on Web usage data.

Along this perspective, a common methodology for discovering usage and user behavior patterns consists of the following steps: reconstructing user sessions, that is, the navigational sequence of Web-pages of a user in the site; comparing them with other user's sessions; and clustering or classifying the sessions to extract knowledge of navigational behavior. Extracted usage and user behavior patterns may be used in targeting specific groups of users; in various recommendation systems; and in evaluation and reconstruction of the Website to meet design efficiency issues and user satisfaction requirements. Detailed surveys on Web usage mining are presented by Faca and Lanzi (2005), and by Srivastava et al. (2000).

Subsequently, Web structure mining is closely related to analyzing hyperlinks and link structure on the Web for information retrieval and knowledge discovery. Web structure mining can be used by search engines to rank the relevancy between Websites and to classify them according to their similarity and relationship (Kosala & Blockeel, 2000). Google search engine, for instance, is based on PageRank algorithm (Brin & Page, 1998), which states that the relevance of a page increases with the number of hyperlinks to it from other pages, and in particular of other relevant pages. Personalization and recommendation systems based on hyperlinks are also studied in Web structure mining.

Web structure mining is used for identifying "authorities," which are Web pages that are pointed to by a large set of other Web pages (Desikan, Srivastava, Kumar & Tan, 2002) that make them candidates of good sources of information. Web structure mining is also used for discovering "social networks on the Web" by extracting knowledge from similarity links. The term is closely related to "link analysis" research, which has been developed in various fields over the last decade such as computer science and mathematics for graph-theory, and social and communication sciences for social network analysis (Foot, Schneider, Dougherty, Xenos & Larsen, 2003; Park, 2003; Wasserman & Faust, 1994).

This method is based on building a graph out of a set of related data (Badia & Kantardzic, 2005) and to apply social network theory (Wasserman & Faust, 1994) to discover similarities. Thus, a social network is modeled by a graph, where the nodes represent individuals whereas an edge between two nodes represents a direct relationship between the individuals. Recently Getoor and Diehl (2005) introduce the term "link mining" to put special emphasis on the links as the main data for analysis and provide an extended survey on the work that is related to link mining.

A new term, namely, community mining is a major research area on social networks that emphasizes on discovering groups of individuals, who by sharing the same properties form a specific community on the Web. Domain applications related to Web structure mining of social interest are: criminal investigations and security on the Web, digital libraries where authoring, citations and cross-references form the community of academics and their publications etc. Detailed survey on Web structure mining can be found in Desikan et al. (2002) and Getoor & Diehl (2005).

The taxonomy previously described is based on the characteristics of the source data. Usually when working with one of the three data sources (Web content, log files, hyperlinks), researchers might think the corresponding category. However, this is not strict and might combine source data and target application as for example they can use hyperlinks to predict Web content (Mladenic & Grobelnik, 1999). Another example is "Web community" (Zhang,Yu & Hou, 2005), a term closer to Web structure mining, however, is used for the analysis and construction of "Web communities" not only from hyperlinks, but also from Web document content and user access logs. Mobasher, Dai, Luo, Sung, and Zhu (2000) combine Web usage mining and Web content mining for creating user content profiles. Web usage data combined with ontologies and semantics for improving Web personalization are currently proposed in various systems (Berendt, 2002; Dai & Mobasher, 2003; Oberle, Berendt, Hotho & Gonzalez, 2003; Spiliopoulou & Pohle, 2001).

MACHINE LEARNING OVERVIEW

Machine learning is the basic method for most data mining approaches and therefore will be also an important method in Web mining research. It is a broad field of artificial intelligence investigating the use of algorithms acting as intelligent learning methods in computer systems to gain experience, so that this experience can be used when making decisions based on previous learned tasks. The machine learning methods cover a wide range of learning methods, where some of them have been inspired from nature. Neural networks are inspired from human brain and its neurons for the learning, information storing and information retrieval capability. Genetic algorithms and evolutionary algorithms are inspired from Darwin's theory for the surviving characteristics of the fittest in a population that evolves in time. Other machine learning methods are designed to reach to a decision by asking simple yes/no questions following a path from a tree based graph (decision trees) or to derive rules that find interesting associations and/or correlation relationships among large set of data items (association rules).

Representatives of machine learning methods are: artificial neural networks (ANN), self-organizing maps, Hopfield network, genetic algorithms, evolutionary algorithms, fuzzy systems, rough sets, rule-based systems, support vector machines, decision trees, Bayesian and probabilistic models. Describing in details each of these methods will overpass the chapter. The reader can find many textbooks that describe in details all the previously mentioned methods (Bishop,

2003; Duda, Hart & Stork, 2001; Michaklski & Tecuci, 1994; Mitchell, 1997).

At the same time machine learning systems are capable of solving a number of problems related to pattern classification, data clustering, predicting purposes, and information retrieval. In traditional data mining, one can identify that machine learning is used for tackling four types of data mining problems: classification, clustering, association rules and prediction problems.

The task in classification problems is to assign classes to objects according to their characteristics (features). The central aim in designing a classifier is to train the classifier with patterns of known labels drawn out of the total number of available data, which usually are labeled as "positive examples" for samples that belong to a known class and "negative examples" for all those samples that do not belong to the known class. The classifier success is evaluated by the ability to generalize, that is, the ability to predict correctly the label of novel (unseen to the classifier) patterns that have been left out from the training process.

The training process uses an adjustment mechanism that iteratively adjusts the parameters of the classifier in order to get closer to learning the class of the training data. Evaluating classifier generalization one may have an estimation of the performance of the classifier in classifying and predicting labels of newly collected data. Since the class of each data example during the training phase is provided to the classifier, the type of learning is called "supervised learning," where the supervision takes place in adjustments of the classifier parameters so that a misclassified data example in an iteratively learning process is classified correctly.

Classification problems also deal with prediction, as the task in classification is to minimize the error of misclassified test data and therefore classifiers according to the quality of collected data and to the accuracy rate of performance of the classifier may predict classes. In this aspect, prediction is harder when instead of a discrete class one try to find the next value in the range of hypothesis after training the model with historical data, as for example, predicting the closing price of a security in stock-market based on historical financial data.

On the contrary, clustering is a method that uses a machine learning approach called "unsupervised learning," where no predefined classes exist and the task is to find groups of similar objects creating a cluster for each group. Therefore, in a cluster belong data that have similar features between them and at they same time they have dissimilar features with the rest of data. Association rules aims to find relationships and interesting patterns from large data sets.

Although the overwhelming majority of machine learning research is based on supervised and unsupervised learning models, there exist two more types of learning: reinforcement learning and multi-instance learning. Reinforcement learning tries to learn behavior through trial-and-error interactions with a dynamic environment. The difference from supervised learning is that correct classes are never presented, nor suboptimal actions explicitly corrected. In multi-instance learning (Dietterich, Lathrop & Lozano-Perez, 1997) the training data consists of "bags" each containing many instances, while in supervised learning the data set for training consists of positive and negative examples. A bag is labeled positively if it contains at least one positive instance. The task is to learn some concept from the training set of bags for predicting the label of unseen bags. Training bags have known labels, however, the instances have unknown labels so the training process comprises labeled data that are composed of unlabeled instances and the task is to predict the labels of unseen data.

Normally decision trees, and rule-based models are used to solve supervised learning problems; self-organizing maps (SOM), and clustering models are typically used in unsupervised learning problems; and genetic and evolutionary algorithms are typically used in reinforcement learning problems. The rest of machine learning methods are used in both supervised, unsupervised, reinforcement, and multi-instance learning problems.

MACHINE LEARNING APPLIED TO WEB MINING

Machine learning techniques can be very helpful when applied to the process of Web mining. Although there is a close relation between machine learning and Web mining one should denote that Web mining is not actually the application of machine learning techniques on the Web (Kosala & Blockeel, 2000). Other methods studying interesting patterns on the Web may be methods of statistical analysis (Gibson & Ward, 2000; Sharma & Woodward, 2001; Yannas & Lappas, 2005; Yannas & Lappas, 2006), or heuristics (Sutcliffe, 2001).

Primitive Web mining attempts to find patterns to explain various Web "phenomena" can be also found in qualitative Web research methods (Demertzis, Diamantaki, Gazi, & Sartzetakis, 2005; Gillani, 1998; Margolis, Resnick, & Tu, 1997; Maule, 1998; Li, 1998; Reeves, and Dehoney, 1998) that usually rely on observations, annotation of online and archived Web objects, interviews or surveys of Web administrators and users, textual analysis, focus groups, social experiments (Schneider & Foot, 2004) to analyze and explain various Web phenomena.

This approach is usually originated from social science and communication researchers, where the ability to apply more advanced computerized methods like machine learning is limited, however, the interesting of such methods is the expressive power to interpret and explain such phenomena. To the best of our knowledge the author has not identified any combined machine learning and qualitative studies. It could be very interesting to see how such studies will be empowered from the intelligence and automations of machine learning and from the interpretation ability of social sciences.

In comparison to data mining, Web mining may have a few common characteristics similar to machine learning methods and approaches. However, working with Web data is more difficult due to fact that Web data are formed dynamically, change frequently, and their structure cannot be stored in a fixed length database with known features and characteristics. On the contrary, most data mining systems are well structured and remain static over time. Moreover, Web data have many different data type such as text, tables, links, sidebars, layouts, images, audio, video, pdf files, word files, postscript files, executable files, animation files, and so forth, to name a few. Detection of such data types can be a hard problem that needs considerable effort to solve it as with table detection in Websites, where support vector machines and decision trees can be used for attacking this problem (Wang & Hu, 2002). Lastly, the Web is considerably larger than traditional databases in terms of magnitude due to the billions of existing Websites. Before going to the next section, the author presents an indicative research that relates machine learning with the three Web mining categories.

Machine Learning and Web Content Mining

Intelligent indexing text on the Web is the primary goal of search engines building their databases. Machine learning techniques and Web content mining are widely used in this task. Neural networks are commonly used for Web document classification. They are trained by existing Web date for learning to correctly classify patterns of Web documents. They produce high classification accuracy and are very popular among researchers for learning and classifying Web documents (Cirasa, Pilato, Sorbello, & Vassallo,

2000; Fukuda, Passos, Pacheco, Neto, Valerio, & Roberto, 2000; Pilato, Vitabile, Vassallo, Conti, & Sorbello, 2003).

Apart from neural networks classifiers, systems based on support vector machines for Web document classification are presented in Sun, Lim, and Ng (2002), and Yu, Han, and Chang (2002), whereas Esposito, Malerba, Di Pace, and Leo (1999) use three different classification models (decision trees, centroids and k-nearest-neighbor) for automated classification of Web pages; whereas hybrid systems like in Kuo, Liao, and Tu (2005) combine neural networks with genetic algorithms to analyse Web browsing paths for a recommendation system based on intelligent agents.

Also, Bayesian classifiers for text categorization in Syskill and Webert (Pazzani & Billsus, 1997) are used in a recommendation system to recommend Web pages, and in Mooney and Roy (2000) to produce content-based book recommendations. Semeraro, Basile, Degemmis, and Lops (2006) train a Bayes classifier that infers user profiles as binary text classifiers (likes and dislikes) in an application that acts like a conference participant advisor that suggests conference papers to be read and talks to be attended by a conference participant.

Similarly, reinforcement learning and Bayes networks are used as intelligent agents in Rennie and McCallum (1999) for learning and classifying efficiently Web documents. Stamatakis, Karkaletsis, Paliouras, Horlock, Grover, and Curran (2003) compare various machine learning approaches (decision trees, support vector machines, nearest neighbour classifier, naïve baies) for identifying domain-specific Websites.

Machine Learning and Web Usage Mining

Classifiers and clustering algorithms are usually used for analyzing hyperlinks in Web usage mining. Pierrakos, Paliouras, Papatheodorou, Karkaletsis, and Dikaiakos (2003) use clustering applied to Web usage mining for creating community specific directories to offer users a more personalized view of the Web according to their preferences. Hu and Meng (2005) present a system that combines the intelligent agent approach with collaborative filtering using neural networks and Bayes network in order to retrieve relevant information. Zhou, Jiang, and Li (2005) apply multi-instance learning on Web mining by using the browsing history of the user in the Web index recommendation problem for recommending unseen Web pages.

Yao, Hamilton, and Wang (2002) combine three different machine learning techniques: association rules, clustering and decision trees to help users navigate a Website by analysing and learning from Web usage mining and user behavior. A hybrid approach that uses self-organizing maps (SOM), (Kohonen, 1990) and a neuro-fuzzy model is applied on log files by Wang et al. (2005) for Web traffic mining in order to predict Web server traffic. Genetic algorithms are used in (Tug, Sakiroglu & Arslan, 2006) for the discovery of user sequential accesses from log files.

Machine Learning and Web Structure Mining

The most famous application of Web structure mining is the Google search engine based on Brin and Page's (1998) PageRang algorithm for ranking pages relevance. Mladenic and Grobelnik (1999) use the k-nearest-neighbor algorithm to train a system for predicting Web content from hyperlinks. Wu, Gordon, DeMaagd, and Fan (2006) use principal cluster analysis to identify a small number of major topics from millions of navigational data. Lu and Getoor (2003) apply classifiers for link-based object classification. Probabilistic models are used in Matsuo, Ohsawa, and Ishizuka (2001) for Web search and identifying Web communities; in Lempel and Moran, (2001) for Web search; Cohn and Chang, (2000), Getoor, Segal, Tasker, and Koller (2001), and

Richardson and Domingos, (2002) for Web page classification.

APPLICATIONS OF WEB MINING TO SOCIETAL BENEFIT AREAS

Web mining may benefit those organizations that want to utilize the Web as a knowledge base for supporting decision-making. Pattern discovery, analysis, and interpretation of mined patterns may lead to take better decisions for the organization and for the provided services. E-commerce and e-business are two fields that may be empowered by Web mining with lots of applications to increase sales, doing intelligence business or even used in crisis management as in Tango-Lowy and Lewis (2005), where Web mining and self organizing maps are used in crisis scenarios.

Lots of Web mining applications found in the literature describe the effectiveness of the application from the Web administration point of view. The target in these applications is taking advantage of the mined knowledge from the users to increase the benefits of the organization. In this chapter, the author focuses on social beneficial areas from Web mining, and hence the point of view is on Web mining applications that can help users or group of users. An obvious societal benefit is that Web mining research efforts lead to user (or group of users) satisfaction by providing accurate and relevant information retrieval; by providing customized information; by learning about user's demands so that services can target specific groups or even individual users; and by providing personalized services. The author identified research on the following areas, where Web mining offers societal benefits: Helpdesks and recommendation systems; digital libraries; security and crime investigation; e-learning; e-government services; and e-politics and e-democracy.

Helpdesks and Recommendation Systems

Recommendation systems are based on user modeling that are mainly derived from content-based learning or from collaborative filtering (Zukerman & Albrecht, 2001). Content–based learning uses a user's past usage behavior and acts as an indicator of his/her future behavior. Collaborative filtering is based on ratings of user favors, like rating music or movies, so that rating history of a user can be associated with similar preferences of other users. So a user is classified in a user model, where recommendations can be addressed to the user according to favors of other people from the specific classified user model. Hybrid recommendation systems that take benefits from both collaborative filtering and content-based learning have been also investigated in literature (Melville, Mooney, & Nagarajan, 2002; Sarwar, Karypis, Konstan, & Riedl, 2000).

Martin-Guerrero, Palomares, Balaguer-Ballester, Soria-Olivas, Gomez-Sanchis, and Soriano-Asensi (2006) propose a recommender model for predicting user preferences based on common clustering algorithms in a citizen Web portal. Clustering and collaboration filtering is used in Hayes, Avesani, and Veeramachaneni (2006) for a blog recommendation system. A blog is a journal-style Website usually written by a single user, where entries are presented in a reverse chronological order.

ReferralWeb (Kautz, Selman & Shah, 1997) is a project that mines social networks from the Web by using collaborative filtering for identifying experts that could answer questions asked by individuals. Nasraoui and Pavuluri (2004) using neural networks provide accurate Web recommendations based on a committee of predictors. Yao et al. (2002) created PagePrompter, an agent-based recommender that helps users navigate a Website by analysing and learning from Web usage mining and user behavior. The interesting part of Pageprompter is that it combines three different machine learning techniques: association rules, clustering, and decision trees for achieving its task.

Pierrakos et al. (2003) use clustering applied to Web usage mining for creating community specific directories to offer users a more personalized view of the Web according to their preferences and may be assisted by using these directories as starting points on their navigation. Garofalakis, Kappos, and Mourloukos (1999) studied Website optimization using Webpage popularity. Scheffer (2004) created an e-mail answering assistant by semisupervised text classification.

Fast and accurate Web services are practical implications from improved helpdesks and recommendation systems.

Digital Libraries

Digital libraries provide precious information distributed all around the world without necessarily having the need to be physically present in a traditional library building. In this context, Web mining research aiming to offer better services on digital libraries have been identified in literature. Adafre and Rijke (2005) use clustering for discovering missing hypertext links in Wikipedia, the largest encyclopedia on the Web that is created and modified by many volunteer authors. Web mining on Wikipedia is also investigated by Gleim, Mehler, and Dehmer (2006). Bhattacharya and Getoor (2004) use clustering for detecting group of entities, like authors, from links and resolving the coreference problem of multiple references to the same paper in autonomous citation indexing engines, like CiteSeer (Giles, Bollacker, & Lawrence, 1998).

CiteSeer is an important resource for computer scientists for searching electronic versions of papers. Cai, Shao, He, Yan, and Han (2005) work also is based on another well-known digital library in computer science community, the DBLP library at http://dblp.uni-trier.de. Their work is related to machine learning feature extraction algorithms in order to discover hidden communities in heterogeneous social networks. Graph analysis for discovering Web communities can be modeled by using Bayesian networks as demonstrated by Goldenberg and Moore (2005) for identifying coauthorship networks.

A content-based learning book recommendation system is proposed by Mooney and Roy (2000) based on Web pages of the Amazon online digital store. Large portals with news updated frequently per day consist of rich information and may be considered as part of digital libraries in the way that newspaper articles are indexed and available to readers in traditional libraries.

At the same time, news sites are large portal sites that increase their content on a daily basis. For such sites, the interpretation of Web content to meaningful content can be classified into semantic categories in order to make both information retrieval and presentation easier for individuals and group of users is very important (Eirinaki & Vazirgiannis, 2003). Liu, Yu, and Le (2005) use fuzzy clustering to identify meaningful news patterns from Web news stream data. However, the wide distribution of knowledge on the one side and the easiness of access to this knowledge on the other side from various groups of people like researchers, academics, students, pupils, professionals or independents are the most valuable practical implications of Web mining to this societal interest area.

E-Learning

Web mining may be used for enhancing the learning process in e-learning applications. Bellaachia, Vommina, and Berrada (2006), introduce a framework, where they use log files to analyze the navigational behavior and the performance of e-learners so that to personalize the learning content of an adaptive learning environment in order to make the learner reach his learning objective. Zaiane (2001) studies the use of machine learning techniques and Web usage mining to enhance Web-based learning environments for the educator to better evaluate the learning process and for the learners to help them in their learning task. Students' Web logs are investigated and analysed in Cohen and Nachmias (2006) in a Web-supported instruction model. Improved e-learning services that accommodate user needs are practical implications from Web mining and machine learning to the e-learning area.

Security and Crime Investigation

Web mining techniques may be used for identifying cyber-crime actions like Internet fraud and fraudulent Websites, illegal online gambling, hacking, virus spreading, child pornography distribution, and cyberterrorism. Chen, Qin, Reid, Chung, Zhou, and Xi (2004) note that clustering and classification techniques can reveal identities of cybercriminals, whereas neural networks, decision tress, genetic algorithms, and support vector machines can be used to crime patterns and network visualization. Chen et al. (2004) provide a detailed study on methods against terrorist groups on the Web for predictive modeling, terrorist network analysis, visualization of terrorists' activities, linkages and relationships.

Similarly, Wu et al. (2006) based on user's online activities use principal cluster analysis to identify a small number of major topics from millions of navigational data in an approach that can be useful in security against terrorism. Do, Chang, and Hui (2004) implemented a system that can benefit safe Internet browsing in school, home and workplace. The system monitors and filters Web access by applying Web mining for performing Web data classification in order to classify Web data in a "white list" of allowed pages or blacklist of blocked Web pages. Social Networks extracted from instant messaging by using clustering is investigated by Resig, Dawara, Homan, and Teredesai (2004). Enjoying a more secure environment having better online and offline protection are implications of Web mining and machine learning to this societal interest area.

E-Government Services

The processes through which government organizations interact with citizens for satisfying user (or group of users) preferences leads to better social services. The major characteristics of e-government systems are related to the use of technology to deliver services electronically focusing on citizens needs by providing adequate information and enhanced services to citizens to support political conduct of government. Empowered by Web mining methods e-government systems may provide customized services to citizens resulting to user satisfaction, quality of services, support in citizens decision making, and finally leads to social benefits. However, such social benefits mainly rely on the organization's willingness, knowledge, and ability to move on the level of using Web mining.

The e-government dimension of an institution is usually implemented gradually. E-government maturity models (Irani, Al-Sebie & Elliman, 2006; Lappas & Yannas, 2006) describe the online stages an organization goes through time, becoming more mature in using the Web for providing better services to citizens. The maturity stages start from the organization's first attempt to be online aiming at publishing useful citizens' information and move to higher maturity stages of being interactive, making transactions and finally transforming the functionality of the organization to operate their business and services electronically through the Web. But, maturity stages described in literature do not have a Web mining dimension, which the author considers that should be the climax in maturity stages.

Riedl (2003) states that by using interviews and Web mining the actual access to information by citizens should be tracked, analyzed and used for the redesign of e-government information services. E-Government literature reveals that only recently Web mining has attracted researchers in e-government applications. Fang and Sheng (2005) present a Web mining approach for designing better Web portal for e-government. Hong and Lee (2005) propose an intelligent Web information system of government based on Web usage mining to help disadvantaged users make good decisions-making for their profit improvements. In the health sector, Mayer, Karkaletsis, Stamatakis, Leis, Villarroel, and Thomeczek (2006) investigate improvements of health services by quality labelling of medical Web content in the recently announced MedIEQ project. Conclusively, e-government aim to improve government's services to citizens and any improvement to this direction lead to valuable implications of Web mining and machine learning to national and local societies.

E-Politics and E-Democracy

E-politics provides political information and politics *"on demand"* to the citizens by improving the political transparency and democracy, benefiting parties, candidates, citizens and the society. Election campaigners, parties, members of parliament, and members of local governments on the Web are part of e-politics. Despite the importance of e-politics in democracy there is limited Web mining methods to meet citizen needs. The author has identified in the literature research that only refers in mining political social networks on the Web. Link analysis has been used to estimate the size of political Web graphs (Ackland, 2005), to map political parties network on the Web (Ackland & Gibson, 2004) and to investigate the U.S. political Blogosphere (Ackland, 2005b). Political Web linking is also studied by Foot et al. (2003) during the U.S. congressional election campaign season on the Web. In this aspect, expanding e-democracy boarders will lead to more transparent and participating democracy, which are vital to the society.

FUTURE TRENDS

Nowadays the Web is a rich and huge information repository, where a number of methods and automatic systems have been created for identifying, locating, accessing, and retrieving information. The main open question in Web mining is how to provide information relevant to specific users' needs. Semantic Web (Berners-Lee et al., 2001) works toward this direction and is considered as the next Web generation. The current Web is based on the hypertext mark-up language (HTML), which specifies how to layout Web pages so that they can be readable to humans, thus it is human-centralized. The problem is the retrieval of relevant information by search engines because machines cannot understand Web content to retrieve relevant information. This is expected to change by semantic Web technologies as in semantic Web "information is given well-defined meaning better enabling computer and people to work in cooperation" (Berners-Lee et al., 2001).

Consequently, machine-learning techniques will continue to play the most important role in the semantic Web (Hess & Kushmerick, 2004) for information retrieval and knowledge discovery. Berendt et al. (2002) introduce "semantic Web mining" as the field where semantic Web meets Web mining. It is expected that machine learning techniques and semantic Web mining will be in the focus of research for the next years.

In this chapter the author has introduced areas of societal interest that may be benefited by Web mining and machine learning. The literature review revealed that most research in these areas has just recently been started. The future trend seems to be the convergence of Web mining and machine learning to practical solutions in the six areas of societal benefit: Helpdesks and recommendation systems, digital libraries, e-learning, security and crime investigation, e-government services, e-politics and e-democracy.

CONCLUSION

This chapter has provided a survey on Web mining and machine-learning methods focusing on current Web mining research in societal benefit areas identifying that most of this research has been recently developed. Therefore, one of the current trends of Web mining is toward the connection between intelligent Web services and applications of social benefits, which brings to work closer scientists from various disciplines. Furthermore, this integrating tendency benefits researchers from various fields.

Social studies on the Web may benefit from machine learning and Web mining methods for providing them with tools and methods to better collect, manage and analyze Web based-phenomena. Moreover, a social interpretation of the meaning of outcomes from computer science Web mining methods is the key question from social and communications studies (Thelwall, 2006). Finally, Web mining and machine learning community may benefit from social and communication expertise on the Web to better interpret their outcomes in the direction of why this happens; or whether mining patterns have meaningful or useful knowledge; or whether hidden knowledge found from Web mining creates a new view that needs further investigation and explanation.

REFERENCES

Ackland, R. & Gibson, R. (2004). Mapping political party networks on the WWW. In *Proceedings of the Australian Electronic Governance Conference*, Melbourne, Australia.

Ackland, R. (2005). *Estimating the size of political Web graphs*. Revised paper presented to ISA Research Committee on Logic and Methodology Conference. Retrieved April 10, 2008, from http://acsr.anu.edu.au/staff/ackland/papers/political_web_graphs.pdf

Ackland, R. (2005b). *Mapping the U.S. political blogosphere: Are conservative bloggers more prominent?* Paper presented to BlogTalk Downunder, Sydney. Retrieved April 10, 2008, from http://acsr.anu.edu.au/staff/ackland/papers/polblogs.pdf

Adafre, S. F. & Rijke, M. D. (2005). Discovering missing links in Wikipedia. In *Proceedings of the 3rd International Workshop on Link Discovery* (pp. 90-97). ACM Press.

Badia, A. & Kantardzik, M. (2005). Graph building as a mining activity: Finding links in the small. In *Proceedings of the 3rd International Workshop on Link Discovery* (pp. 17-24). ACM Press.

Baldi, P., Frasconi, P., & Smyth, P. (2003). *Modeling the Internet and the Web: Probabilistic methods and algorithms*. West Sussex, UK: John Wiley.

Bellaachia, A., Vommina, E., & Berrada, B. (2006). Minel: A framework for mining e-learning logs. In *Proceedings of the 5th IASTED International Conference on Web-based Education* (pp. 259-263). Puerto Vallarta, Mexico.

Berendt, B. (2002). Using site semantic to analyze, visualize and support navigation. *Data Mining and Knowledge Discovery, 6*, 37-59.

Berendt, B., Hotho, A., & Stumme, G. (2002). Towards semantic web mining. *Lecture Notes in Computer Science* (vol. 2342, pp. 264-278).

Berners-Lee, T., Hendler, J., & Lassila, O. (2001). The semantic Web. *Scientific American, 284*(5), 34-43.

Bhattacharya, I. & Getoor, L. (2004). Deduplication and group detection using links. In *Proceedings of the SIGKDD Workshop on LinkAnalysis and Group Detection,* Seattle, WA.

Bishop, C. M. (2003). *Neural networks for pattern recognition.* Oxford University Press.

Brin, S. & Page, L. (1998). The anatomy of a large-scale hypertextual Web search engine. In *Proceedings of the 7th International World Wide Web Conference, Elsevier Science* (pp. 107-117), New York.

Buchner, A. G., Mulvenna, M. D., Anand, S. S. & Hughes, J. G. (1999). Navigation pattern discovery from Internet data. In *Proceedings of the Web Usage Analysis and User Profiling Workshop* (pp. 25-30), San Diego, CA.

Cai, D., Shao, Z., He, X., Yan, X., & Han, J. (2005). Minning hidden community in heterogeneous social networks. In *Proceedings of the 3rd International Workshop on Link Discovery* (pp. 58-65). ACM Press.

Chakrabarti, S., (2000). Data mining for hypertext: A tutorial survey. *ACM SIGDDD Explorations, 1*(2), 1-11.

Chakrabarti, S. (2003). *Mining the Web: Discovering knowledge from hypertext data.* San Francisco: Morgan Kaufmann Publishers.

Chen, H. & Chau, M. (2004). Web mining: Machine learning for Web applications. *Annual Review of Information Science and Technology (ARIST), 38*, 289-329.

Chen, H., Chung, W., Xu, J. J., Wang, G., Qin, Y., & Chau, M. (2004). Crime data mining: A general framework and some examples. *Computer, 37*(4), 50-56.

Chen, H., Qin, J., Reid, E., Chung, W., Zhou, Y., Xi, W., Lai, G., Bonillas, A., & Sageman, M., (2004). The dark Web portal: Collecting and analyzing the presence of domestic and international terrorist groups on the Web. In *Proceedings of the 7th International Conference on Intelligent Transportation Systems (ITSC),* Washington D.C.

Cirasa, A., Pilato, G., Sorbello, F., & Vassallo, G. (2000). EaNet: A neural solution for Web pages classification. In *Proceedings of the 4th World Multiconference on Systemics, Cybernetics, and Informatics SCI2000*, Orlando, Florida.

Cohn D. & Chang, H. (2000). Learning to probabilistically identify authoritative documents. In *Proceedings of the17th International Conference on Machine Learning (ICML2000)* (pp. 167-174), Stanford, California.

Cooley, R., Mobasher, B., & Srivastava, J. (1997). Web mining: Information and pattern discovery on the World Wide Web. In *Procceding of the 9th International Conference on Tools with Artificial Intelligence(ICTAI '97)* (pp. 558-567), New Port Beach, CA: IEEE Computer Society.

Cohen, A. & Nachmias, R. (2006). A quantitative cost effectiveness model for Web-supported academic instruction. *The Internet and Higher Education, 9*(2), 81-90.

Dai, H. & Mobasher, B. (2003). A road map to more effective Web personalization; Integrating domain knowledge with Web usage mining. In *Proceedings of the International Conference on Internet Computing* (IC 2003), Las Vegas, Nevada.

Demertzis, N., Diamantaki, K., Gazi, A., & Sartzetakis, N. (2005). Greek political marketing on-line: An analysis of parliament members' Web sites. *Journal of Political Marketing, 4*(1), 51-74.

Desikan, P., Srivastava, J., Kumar, V., & Tan, P. N. (2002). *Hyperlink analysis: Techniques and applications* (Tech. Rep. TR 2002-0152). Army High Performance Computing Center.

Dietterich, T. G., Lathrop, R. H., & Lozano-Perez, T. (1997). Solving the multiple-instance problem with axis-parallel rectangles. *Artificial Intelligence, 89*(1-2), 31-71.

Do, T. D., Chang, K., & Hui, S. C. (2004). Web mining for cyber monitoring and filtering. In *Proceedings of the 2004 IEEE Conference on Cybernetics and Intelligent Systems Vol. 1* (pp. 399-404). Singapore.

Duda, R. O., Hart, P. E., & Stork, D. G. (2001). *Pattern classification*. New York: John Wiley.

Eirinaki, M. & Vazirgiannis, M. (2003). Web mining for Web personalization. *ACM Transactions on Internet Technology, 3*(1), 1-27.

Esposito, F., Malerba, D., Di Pace, L., & Leo, P. (1999). A learning intermediary for automated classification of Web pages. In *Proceedings of the 16th International Workshop on Machine Learning in Text Data Analysis (ICML1999)* (pp. 37-46).

Faca, F. M. & Lanzi, P. L. (2005). Mining interesting knowledge from Weblogs: A survey. *Data Knowledge Engineering, 53*(3), 225-241.

Fang, X., Sheng, O. R. L. (2005). Designing a better Web portal for digital government: A Web-mining based approach. In *Proceedings of the 2005 National Conference on Digital Government Research* (pp. 277-278), Atlanta, Greorgia.

Foot, K., Schneider, S., Dougherty, M., Xenos, M., & Larsen, E. (2003). Analyzing linking practices: Candidate sites in the 2002 U.S. electoral Web sphere. *Journal of Mediated Communication, 8*(4).

Fukuda, H., Passos, E., Pacheco, A. M., Neto, L. B., Valerio, J., Roberto, V. J. D., Antonio, E. R., & Chigener, L. (2000). Web text mining using a hybrid system. In *Proceedings of the 6th Brazilian Symposium on Neural Networks* (pp.131–136).

Garofalakis, J., Kappos, P., & Mourloukos, D. (1999). Website optimization using page popularity. *IEEE Internet Computing, 3*(4), 22-29.

Getoor, L., Segal, E., Tasker, B., & Koller, D. (2001). Probabilistic models of text and link structure for hypertext classification. In *Proceedings of the IJCAI Workshop on Text Learning: Beyond Supervision*, Seattle, Washington.

Getoor, L. & Diehl, C. P. (2005). Link mining: A survey. *ACM SIGKDD Explorations Newsletter, 7*(2), 3-12.

Gibson, R. K. & Ward, S. J. (2000). A proposed methodology for studying the functions and effectiveness of party and candidate Web-sites. *Social Science Computer Review, 18*(3), 301-319.

Giles, C. L., Bollacker, K., & Lawrence, S. (1998). CiteSeer: An automatic citation indexing system. In *Proceedings of the 3rd ACM Conference on Digital Libraries*, 89-98.

Gillani, B. (1998). The Web as a delivery mechanism to enhance instruction. *Educational Media International, 35*(3), 197-202.

Gleim, R., Mehler, A., & Dehmer, M. (2006). Web corpus mining by instance of Wikipedia. In *Proceedings of the EACL 2006 Workshop on Web as Corpus*, Trento, Italy.

Goldenberg, A. & Moore, A. W. (2005). Bayes net graphs to understand co-authorship networks? In *Proceedings of the 3rd International Workshop on Link Discovery* (pp. 1-8). ACM Press.

Hayes, C., Avesani, P., & Veeramachaneni, S. (2006). An analysis of bloggers and topics for a blog recommender system. In *Proceedings of the Workshop on Web Mining, 7th European Conference on Machine Learning and the 10th European Conference on Principles and Practice of Knowledge Discovery in Databases (ECML/PKDD)*, Berlin, Germany.

Herrera-Viedma, E. & Pasi, G. (2006). Soft approaches to information retrieval and information access on the Web: An introduction to the special topic section. *Journal of the American Society for Information Science and Technology, 57*(4), 511-514.

Hess, A. & Kushmerick, N. (2004). Machine learning for annotating semantic Web services. In *Proceedings of the AAAI Spring Symposium on Semantic Web Services*, Palo Alto, California.

Hong, G. H. & Lee, J. H. (2005). Designing an intelligent Web information system of government based on Web mining. *Lecture notes in computer science* (Vol. 3614, pp. 1071-1078).

Hu, W. & Meng, B. (2005). Design and implementation of Web mining system based on multi-agent. *Lecture notes on artificial intelligence* (Vol. 3584, pp.491-498).

Irani, Z., Al-Sebie, M., & Elliman, T. (2006). Transaction stage of e-government systems: Identification of its location & importance. In *Proceedings of the 39th Hawaii International Conference on System Sciences,* Hawai.

Kautz, H., Selman, B., & Shah, M. (1997). Referral Web: Combining social networks and collaborating filtering. *Communications of the ACM, 40*(3), 63-65.

Kohonen, T. (1990). The self-organizing maps. *Proceedings of the IEEE,* 78, 1464-1480.

Kosala, R. & Blockeel, H. (2000). Web mining research: A survey. *ACM, 2*(1), 1-15.

Kuo, R. J., Liao, J. L., & Tu, C. (2005). Integration of ART2 neural network and genetic k-means algorithm for analyzing Web browsing paths in electronic commerce. *Decision Support Systems, 40*, 355-374.

Lappas, G. & Yannas, P. (2006). A framework to evaluate political party Websites. In *Proceedings of the 4th International Conference on Politics and Information Systems: Technologies and Applications Vol. II* (pp. 226-231), Orlando, Florida.

Lawrence, S. & Giles, C. L. (1999). Accessibility of information on the Web. *Nature, 400,* 107-09.

Lempel, R. & Moran, S. (2001). SALSA: The stochastic approach for link-structure analysis. *ACM Transactions on Information Systems,* 19(2), 131-160.

Levene, M. & Loizou, G. (1999). *Computing the entropy of user navigation in the Web* (Tech. Rep. No. RN/99/42), University College London.

Li, X. (1998). Web page design and graphic use of three U.S. newspapers. *Journalism and Mass Communication Quarterly, 75*(2), 353-365.

Liu, J. W., Yu, S. J., & Le, J. J. (2005). Online mining dynamic Web news patterns using machine learn methods. *Lecture notes on artificial intelligence* (Vo. 3614, pp. 462-465).

Lu, Q. & Getoor, L. (2003). Link-based text classification. In *Proceedings of the 3rd International Workshop on Link Discovery* (pp. 1-8). ACM Press.

Margolis, M., Resnick, D., & Tu, C.-C. (1997). Campaigning on the Internet: Parties and candidates on the World Wide Web in the 1996 primary season. *Harvard International Journal of Press/Politics, 2*(1), 59-78.

Martin-Guerrero, J. D., Palomares, A., Balaguer-Ballester, E., Soria-Olivas, E., Gomez-Sanchis, J., & Soriano-Asensi, A. (2006). Studying the feasibility of a recommender in a citizen Web portal based on user modeling and clustering algorithms. *Expert Systems with Applications, 30*, 299-312.

Matsuo, Y., Ohsawa, Y., & Ishizuka, M. (2001). Average-clicks: A new measure of distance on the WWW. In *Proceedings of First Asia-Pacific Conference, Web Intelligence*, Japan.

Maule, R. W. (1998). Content design frameworks for Internet studies curricula and research. *Internet Research: Electronic Networking Applications and Policy, 8*(2), 174-184.

Mayer, M. A., Karkaletsis, V., Stamatakis, K., Leis, A., Villarroel, D., Thomeczek, C., Labsky, M., Lopez-Ostenero, F., & Honkela, T. (2006). MedIQ-Quality labelling of medical Web content using multilingual information extraction. *Studies in Health Technology and Informatics, 121*, 183-190.

Melville, P., Mooney, R. J., & Nagarajan, R. (2002). Content-boosted collaborative filtering for improved recommendations. In *Proceedings of the 18th National Conference on Artificial Intelligence* (pp. 187-192).

Michalski, R. S. & Tecuci, G. (1994). *Machine learning: A multistrategy approach* (Vol. IV). Morgan Kaufmann

Mitchell, T. (1997). *Machine learning*. McGraw Hill.

Mladenic, D. & Grobelnik, M. (1999). Predicting content from hyperlinks. In *Proceedings of the 16th International ICML99 Workshop on Machine Learning in Text Data Analysis* (pp. 109-113).

Mobasher, B., Jain, N., Han, E., & Srivastava, J. (1996). *Web Minning: Pattern discovery from WWW transaction* (Tech. Rep. TR-96050). Department of Computer Science, University of Minnesota, Minneapolis. Retrieved April 12, 2008, from http://citeseer.ist.psu.edu/mobasher-96web.html

Mobasher, B., Cooley, R., & Srivastava, J. (1999). Creating adaptive Web sites through usage based clustering of URLs. In *Proceedings of the IEEE Knowledge and Data Engineering Exchange Workshop* (KDEX99), Chicago, Illinois.

Mobasher, B., Dai, H., Luo, T., Sung, Y., & Zhu, J. (2000). Integrating Web usage and content mining for more effective Web personalization. In *Proceedings of the International Conference on E-Commerce and Web Technologies (ECWeb 2000)* (pp. 165-176). Greenwich, UK.

Mooney, R. J. & Roy, L. (2000). Content-based book recommending using learning for text categorization. In *Proceedings of the 5th ACM Conference on Digital Libraries* (pp. 195-204). ACM Press.

Nasraoui, O. & Pavuluri, M. (2004). Complete this puzzle : A connectionist approach to accurate Web recommendations based on a committee of predictors. In *Proceedings of the 6th WEBKDD Workshop*, Seattle, Washington.

Ngu, D. S. W. & Wu, X. (1997). Sitehelper: A localized agent that helps incremental exploration of the World Wide Web. *Computer Networks, 29*(8-13), 1249-1255.

Oberle, D., Berendt, B., Hotho, A., & Gonzalez, J. (2003). Conceptual user tracking. *Lecture notes on artificial intelligence* (Vol. 2663, pp. 155-164).

Pal, S., Talwar, V., & Mitra, P. (2002). Web mining in soft computing framework: Relevance, state of the art and future directions. *IEEE Transactions on Neural Networks, 13*(5), 1163-1177.

Park, H. W. (2003). Hyperlink network analysis: A new method for the study of social structure on the Web. *Connections, 25*(1), 49-61.

Pazzani, M. & Billsus, D. (1997). Learning and revising user profiles: The identification of interesting Web sites. *Machine Learning, 27*(3), 313-331.

Pierrakos, D., Paliouras, G., Papatheodorou, C., Karkaletsis, V., & Dikaiakos, M. (2003). Web community directories: A new approach to Web personalization. *Lecture notes on artificial intelligence* (Vol. 3209, pp. 113-129).

Pilato, G., Vitabile, S., Vassallo, G., Conti, V., & Sorbello, F. (2003). A concurrent neural classifier for HTML documents retrieval. *Lecture notes in computer science* (Vol. 2859, pp. 210-217).

Reeves, T. C. & Dehoney, J. (1998). Cognitive and social functions of courseWeb sites. In H. Maurer & R.G. Olson (Eds.), *Proceedings of WebNet World Conference 98—World Conference of the WWW, Internet & Intranet*. Orlando, FL: Association for the Advancement of Computing in Education.

Rennie, J. & McCallum, A. K. (1999). Using reinforcement learning to spider the Web efficiently. In *Proceedings of the 16th International ICML99 Workshop on Machine Learning in Text Data Analysis* (pp. 335-343).

Resig, J., Dawara, S., Homan, C. M., & Teredesai, A. (2004). Extracting social networks from instant messaging populations. In *Proceedings of LinkKDD'04*, Seattle, Washington.

Richardson, M. & Domingos, P. (2002). The intelligent surfer: Probabilistic combination of link and content information in PageRank. *Advances in Neural Information Processing Systems, 14*.

Riedl, R. (2003). Design principles for E-government services. In *Proceedings of eGov Day 2003*, Vienna, Austria.

Sarwar, B., Karypis, G., Konstan, J., & Riedl, J. (2000). Analysis of recommendation algorithms for e-commerce. In *Proceedings of the ACM Conference on Electronic Commerce* (pp. 158-162).

Schneider, S. & Foot, K. (2004). The Web as an object of study. *New Media & Society, 6*(1), 114-122.

Scime, A. (2005). *Web mining: Application and techniques.* Hershey, PA: Idea Group Inc.

Scheffer, T. (2004). Email answering assistance by semi-supervised text classification. *Intelligent Data Analysis, 8*(5), 2004.

Sharma, A. & Woodward, R. (2001). Political economy Websites: A researcher's guide. *New Political Economy, 6*(1), 119-130.

Sebastiani, F. (2002). Machine learning in automated text categorization. *ACM Computing Surveys, 34*(1), 1–47.

Semeraro, G., Basile, P., Degemmis, M., & Lops, P. (2006). Discovering user profiles from papers by using word sense disambiguation. In *Proceedings of the ECML/PKDD Workshop on Web Mining* (pp. 69-79), Berlin, Germany.

Spiliopoulou, M., Pohle, C., & Faulstich, L. (1999). Improving the effectiveness of a Web site with Web usage mining. In *Proceedings of WEBKDD99* (pp. 142-162), San Diego, CA.

Spiliopoulou, M. & Pohle, C. (2001). Data mining for measuring and improving the success of Web sites. *Data Mining and Knowledge Discover, 5*(1-2), 85-114.

Srivastava, J., Cooley, R., Deshpande, M., & Tan, P. (2000). Web usage mining: Discovery and applications of usage patterns from Web data. *SIGKDD Explorations, 1*, 12-23.

Stamatakis, K, Karkaletsis, V., Paliouras, G., Horlock, J., Grover, C., Curran, J. R. & Dingare, S. (2003). Domain-specific Web site identification: The CROSSMARC focused Web crawler. In *Proceedings of the Second International Workshop on Web Document Analysis (WDA 2003)* (pp. 75-78), Edinburgh, UK.

Sun, A., Lim, E. P. & Ng, W.K. (2002). Web classification using support vector machine. In *Proceedings of the Fourth ACM CIKM International Workshop on Web Information and Data Management (WIDM'02)*, McLean, Virginia.

Sutcliffe, A. (2001). Heuristic evaluation of Website attractiveness and Web usability. *Lecture notes in computer science* (Vol. 2220, pp. 183-198).

Tango-Lowy, R. & Lewis, L. (2005). Situation management in crisis scenarios based on self-organizing neural mapping technology. In *Proceedings of the IEEE Military Communications Conference* (pp. 1-7), Atlantic City, New Jersey.

Thelwall, M. (2006). Interpreting social science link analysis research: A theoretical framework. *Journal of the American Society for Information Science and Technology, 57*(1), 60-68.

Tug, E., Sakiroglu, M., & Arslan, A. (2006). Automatic discovery of the sequential accesses from Web log data files via a genetic algorithm. *Knowledge-Based Systems, 19*(3), 180-186.

Wang, X., Abraham, A., & Smith, K. (2005). Intelligent Web traffic mining and analysis. *Journal of Network and Computer Applications, 28*, 147-165.

Wang, Y. & Hu, J. (2002). A machine learning approach for table detection on the Web. In *Proceedings of the 11th International World Web Conference*, Honolulu, Hawaii.

Wasserman, S. & Faust, K. (1994). *Social network analysis: Methods and applications*. Cambridge University Press.

Wolfgang, G. & Lars, S. (2000). Mining Web navigation path fragments. In *Proceedings of the Workshop on Web Mining for E-Commerce (KDD2000)* (pp. 105-110). Boston, MA.

Wu, H., Gordon, M., DeMaagd, K., & Fan, W. (2006). Mining Web navigations for intelligence. *Decision Support Systems, 41*, 574-591.

Yannas, P. & Lappas, G. (2005). Web campaign in the 2002 Greek municipal elections. *Journal of Political Marketing, 4*(1), 33-50.

Yannas, P. & Lappas, G. (2006). Web candidates in the 2002 Greek prefecture elections. *Journal of E-Government, 3*(1), 53-67.

Yao, Y. Y., Hamilton, H. J. & Wang, X. (2002). PagePrompter: An intelligent agent for Web navigation created using data mining techniques. *Lecture notes in computer science* (Vol. 2475, pp. 506-513).

Yu, H., Han, J., & Chang, K. C. (2002). PEBL: Positive example based learning for Web page classification using SVM. In *Proceedings Of The International Conference On Knowledge Discovery In Databases (KDD02)* (pp. 239-248), New York.

Zaiane, O. R. (2001). Web usage mining for a better Web-based learning environment. In *Proceedings of Conference on Advanced Technology for Education* (pp. 60-64). Banff, Alberta, Canada.

Zhang, Y., Yu, J. X., & Hou, J. (2005). *Web communities: Analysis and construction*. Berlin: Springer.

Zhou, Z., Jiang, K., & Li, M. (2005). Multi-instance learning based Web mining. *Applied Intelligence, 22*(2), 135-147.

Zukerman, I. & Albrecht, D. (2001). Predictive statistical models for user modeling. *User Modelling and User Adapted Interaction, 11*, 5-18.

Chapter VI
The Importance of Data Within Contemporary CRM

Diana Luck
London Metropolitan University, UK

ABSTRACT

In recent times, customer relationship management (CRM) has been defined as relating to sales, marketing, and even services automation. Additionally, the concept is increasingly associated with cost savings and streamline processes as well as with the engendering, nurturing and tracking of relationships with customers. Much less associations appear to be attributed to the creation, storage and mining of data. Although successful CRM is in evidence based on a triad combination of technology, people and processes, the importance of data is unquestionable. Accordingly, this chapter seeks to illustrate how, although the product and service elements as well as organizational structure and strategies are central to CRM, data is the pivotal dimension around which the concept revolves in contemporary terms. Consequently, this chapter seeks to illustrate how the processes associated with data management, namely: data collection, data collation, data storage and data mining, are essential components of CRM in both theoretical and practical terms.

INTRODUCTION

Throughout the past decade, customer relationship management (CRM) has become such a buzzword that in contemporary terms the concept is used to reflect a number of differing perspectives. In fact, although in essence CRM pivots on the fundamental underpinnings of data mining, the concept has been defined as essentially relating to sales, marketing, and even services automation. Additionally, CRM is increasingly associated with cost savings and streamline processes as well as with the engendering, nurturing, and tracking of relationships with customers. Much less associations appear to be attributed to the creation, storage and mining of data; all essential components of CRM in both theoretical and practical terms.

In support of the close connection of CRM with data mining, it should be emphasized that in contemporary terms, the acronym CRM is used to refer to both customer relationship marketing and customer relationship management. Although customer relationship marketing and customer relationship management are indeed often regarded as specialised fields of study, within the discourse of this chapter it is argued that they are in fact inter-related. Subsequently, throughout this chapter, the scope of CRM is intended to span from the development and marketing of relationships between organizations and their customers to the day-to-day management of these relationships. The collation, storage and mining of data are by all means implicitly encompassed within the associated processes conducted as part of CRM.

Throughout the past decade, CRM has been associated with various objectives and differing perspectives. Accordingly, while it is at times referred to as being synonymous to a form of marketing such as database marketing (Khalil & Harcar, 1999), services marketing (Grönroos, 1994), and customer partnering (Kamdampully & Duddy, 1999) for instance, at other times it is specified in terms of more specific marketing objectives such as customer retention (Walters & Lancaster, 1999a), customer share (Rich, 2000), and customer loyalty (Reichheld & Schefter, 2000). In fact, as Lindgreen and Crawford (1999, p. 231) succinctly summarise, more often than not the concept seems to be "described with respect to its purposes as opposed to its instruments or defining characteristics". Meanwhile, the exact nature of the CRM approach remains persistently elusive while the realm of CRM remains unquestionably complex. This blurred outlook is poignantly emphasised in the definition that:

Essentially CRM relates to sales, marketing, services automation, but it is increasingly embracing enterprise-resource planning applications in order to deliver cost savings and more streamlined services within organizations, as well as tracking the relationships organizations have with their customers, and indeed, their suppliers. (Key Note, 2002a, p. 1)

Notwithstanding such complexity, it simply cannot be denied that CRM is intricately connected with data mining.

In line with the wide latitude afforded by its complexity, various themes have been discussed under the title CRM in both trade and academic literature. However, in spite of being extensive, as a whole this coverage still seems to lack coherence. Although in recent times, CRM has been described as a triad combination of technology, people, and processes (Chen & Popovich, 2003; Galbreath & Rogers, 1999); the importance of data is unquestionable. Accordingly, this chapter seeks to illustrate how, although the product and service elements as well as organizational structure and strategies are central to CRM, data is the pivotal dimension around which the concept revolves in contemporary terms in practice as well as in theory.

The technologies associated with data management, namely: data collection, data collation, data storage and data mining, have undoubtedly influenced the evolution and implementation of information systems within companies. In fact, the central role of databases and data mining within the context of current CRM practices is so evident that it could even be argued that the concept quintessentially revolves around the collection and usage of data. Accordingly, current and emerging technologies have been associated and are expected to continue to be associated with databases. Database technologies indeed appear to have significantly contributed to the evolution of CRM. However, regardless how theoretically valid the relevance of data may be to concept of CRM, unless it is adequately implemented within operations, that is to say unless its significance can be translated to operations, its benefits are unlikely to be fully reaped. Thus it is crucial that all the processes, which complement the process of data mining, are also focused upon.

In an attempt to illustrate the significance of data mining applications within the context of CRM, this chapter has opted to focus on one industry: the hotel industry. Hence, throughout this chapter, the actions of hotel companies have been used to consolidate the arguments and explicate how businesses within the hotel industry are trying to optimize the range of opportunities, which the adequate use of data affords. Accordingly, the types of managerial practices, strategies and tactics, being deployed by hotel chains in their attempts to facilitate CRM at the organizational and individuals levels have been reviewed.

As a synthesis, the aim of this chapter is two-fold. While on one level, it is intended to set a platform for briefly exploring how databases are subject to a range of important influences with respect to the underlying connections with customers; on another level it expects to consolidate how ultimately data or the search or even the exploitation of data needs to be aligned with the individual capabilities and strategies of companies, and indeed with the reality as well as aspirations of organizations. Although the examples used within this chapter focused on the hotel industry, it is suggested that similar opportunities can be extended to other companies. However, the opportunities and benefits will be aligned to the dynamics of the said industries and the forces operating in the specific market within which these companies operate.

PROCESSES: THE KEY TO UNLOCKING THE SECRETS OF DATA

Until fairly recently, efficient facilities, standardized products, and lower costs have arguably been sufficient for companies to be able to satisfy customer needs (Chen & Popovich, 2003). However, with increased competition, mass marketing appears to have lost its glitter. Instead, relationship marketing and CRM have been hailed by organizations and academics as the solution to this change of consumer expectations. Notwithstanding, several academics including Palmer (1996) and Murphy (2001) have argued that if companies intend to optimally embrace CRM, they will need to realign their business offerings.

Furthermore, developments in information technology have dramatically enhanced the scope for the collection, analysis, and exploitation of information on customers (Long et al., 1999). However, these technological developments have also highly likely led to an important trend, which has in evidence centered itself on database marketing. As a concept, database marketing revolves around the implications that organizations can acquire and maintain extensive files of information on past and current customers as well as on prospects. Although database marketing may be regarded as being traditionally inherent to the specialised field of direct marketing, numerous authors including Moncrief and Cravens (1999) and Long, Hogg, Hartley, and Angold (1999) have acknowledged how its functions are being increasingly applied to enhance and refine relationships with customers in other areas of marketing too. Consequently, the database, the fundamental tool of traditional direct marketing, has become a pivotal instrument within such areas as the CRM arena, not only as far as interaction and the exchange of information between an organization and its customers are concerned, but also in the facilitation of processes such as the segmentation and targeting of these customers. Consequently, when companies engage in CRM, they also clearly have to engage data collection, data storage and data mining processes. Therefore, strategic and even tactical CRM centers on data mining.

Supporters of the important role of technology within the CRM and the general business arena are numerous. Consequently, databases and information systems have been increasingly favored. Fraser, Fraser, and McDonald (2000) even advocate that it is only when companies ensure that their organizational and systems changes remain one step ahead of their competitors' that they can be said to be making the most appropriate use of technology. By contending that technology can set companies ahead of their competitors, Fraser et al. quintessentially appear to equate technology to a competitive advantage. However failure is expected if companies believe that CRM is only a technology solution. CRM will only be successful if companies learn how to disseminate and exploit the information, which they have collected on their customers, on their databases. In other words, unless data mining is effectively conducted, CRM is highly unlikely be utilized optimally and beneficially. In an attempt to explicate how these distinct yet complementary processes can be integrated to operate from a level platform, Overell (2004, p.1) testifies that although within the business environment, companies have instead tended to follow a flawed contingency and expected information technology "to solve management problems," they should learn to "rethink functionally fragmented processes from the customers' viewpoint."

A consortium of academics, including Joplin (2001) and Nitsche (2002), consider that CRM is not a technological solution but a strategy. In fact, according to Joplin, CRM is the most important strategy, which any company must adopt and develop if it wishes to remain competitive. The evolving properties of CRM as a strategic solution are emphasised by Nitsche when he argues that "technology is not a panacea" (2002, p. 208) and that the people and markets, around which CRM revolves, are "changing just as the competition is" (2002:207). To be able to embrace the fast occurring changes, Nitsche advises that companies must reorganize themselves. Thus companies arguably need to restructure their strategies and tactics in line with emerging new market forces in order to capture the inherent and changing opportunities afforded CRM and by databases. This reorganization and restructuring can indeed only be achieved through a review of functional and business processes. Accordingly, CRM should implicitly linked to the capture, collation, storage, and mining of data. Additionally, it implies that the means and processes through which companies acquire, mine and use data also need to be continuously and consistently monitored.

In this context, the changes of the market environment have a direct impact on relationship marketing. According to Prabhaker (2001, p. 113), within the business environment, "two specific evolving forces have led to organizations having to rethink their business models." These are: the power of customers and the changes in technology. The effect of the dyadic synergy created by these two evolving forces is said to have been two-fold. On one hand, companies appear to have in general attempted to keep up with and adapt to these changes, and on the other hand more proactive companies appear to have learnt how to additionally leverage the advances in technology and computer-integrated control systems to significantly improve their own initial strategic capabilities. This latter contention is arguably aligned to Zahra et al. (1999) explanation about how technology can impact a company's internal and external capabilities. Indeed, advances in database technologies have influenced the way in which companies have used databases and data within their operations as well as the processes they have followed to capture and to mine the said data.

Within the hotel industry, a tiered adaptation to the changing market forces associated with databases seems to prevail. Advances in technology have enabled operational benefits in terms of automation in back-office functions such as reservations processes and check-in processes being generally reaped by hotel chains. Some hotel companies have in evidence additionally attempted to benefit from other opportunities. For example in June 2002, when Travelodge started to develop a new database of online customers as part of its strategy to double the 5000 reservations which the company took ever week, the company's website was also redesigned in line with up-to-date technologies in order to streamline its reservation process (Key Note, 2002b). During 2003, Corus & Regal Hotels Plc recategorized the profiles of the customers on its existing database in an attempt to engage in precise targeting. As part of their strategy to increase return on investment, all bookings made for any of the hotels within the group was redirected via the central reservations office or to their new marketing database. This new process was implemented in order to update existing records consistently and continuously and automatically create the profiles of new customers (Key Note, 2003). Through the centralization of its customer contacts, this hotel chain arguably also put itself in a better position to offer a more controlled view of the company to its existing and prospective customers. Perhaps even more importantly, such an integrated system arguably enables the hotel company to enhance its data mining opportunities.

CRM implies a detailed examination of the guest (Davies, 2000). As databases are essentially associated with the ability provide exactly that, it is evident how CRM and databases are intricately linked. CRM systems may include functions relating to customer retention, customer profitability, customer response to marketing campaigns and even more mundane details such as whether customers prefer still to sparkling water. However, in order to achieve such objectives, companies have to adhere to some specific processes. These processes pivot around the processes associated with the capturing, storing, and mining of data on customers, as well as around the company's use of the mined data. It is indeed argued that not only are data acquisition and data mining quintessential in the success of CRM, but also are the ways companies use the mined data with regards to their strategies and even tactics.

THE DATABASE: THE PIVOTAL TOOL OF CRM

The Pareto principle states that 80% of company's income comes from 20% of its customers. According to Bentley (2005), the ongoing challenge for hotel companies is to determine which specific customers represent that 20%. In an attempt to identify the profitable customers, hotel companies are increasingly investing in database infrastructure. Meanwhile, technological developments have highly likely led to an important trend, which is evidently centred itself on database marketing.

As a concept, database marketing revolves around organizations acquiring and maintaining extensive files of information on past and current customers as well as on prospects. Although the objective of databases is to enable a better portrait of customers and their buying habits, ultimately they are intended to not only enable hotel companies to market their products, services and even special offers more effectively, but to also provide an improved personalised service to customers (Bentley, 2005). Although database marketing is traditionally associated with the specialised field of direct marketing, numerous authors including Moncrief and Cravens (1999) and Long et al. (1999) have acknowledged how its functions are being increasingly applied to enhance and refine relationships with customers. Consequently, the database has become a pivotal instrument within the CRM arena, not only as far as interaction and the exchange of information between an organization and its customers are concerned, but also in the facilitation of processes such as the segmentation and targeting of customers. Furthermore, as a result of the mining of the data captured on customers, precise targeting can be achieved.

According to Bradbury (2005), a database is a structured collection of information, which is not only set as indexes but also searchable. In general, databases are used for business applications such as the storage of customers' data. Thus, previous hotel reservations and even restaurant reservations may be held in a hotel chain's database. In layman's terms, databases may be compared to an electronic library, which receives fresh data, stores information and make the latter accessible to an organization; thereby helping maintain a continuous learning loop (McDonald, 1998). In more implicit marketing terms, databases can be extended to form an extensive and multilevel process (Tapp, 2001). Within the CRM arena, it could be argued that databases are used not only to promote and facilitate interaction between an organization and its customers from the time of an initial response, but also to help with the measurement and analysis of such interactions. Simply put, the ongoing relationship between an organization and a customer can be systematically recorded in databases. In this aspect, a sophisticated database cannot only store data on active, dormant or lapsed customers but it may even have the potential to identify prospects (McDonald, 1998; Tapp, 2001). Subsequently, the increasingly integral role, which databases have come to play in CRM campaigns appears well founded.

As stated earlier, developments in information technology have dramatically enhanced the scope for the collection, analysis and exploitation of information on customers (Long et al., 1999) and for these purposes, data warehouses have been increasingly created by businesses. A data warehouse is essentially a giant database, which takes the raw information from the various systems within a hotel, such as central reservations and room service, and converts the data collated from all the sources into one easily accessible and ideally user-friendly set of data (Davies, 2000). When used effectively, data warehouses cannot only gather data on a continuous basis but they can also allow the precise segmentation of information about customers. Subsequently, profitable interaction with customers can be increased and operations such as targeting and even customer service can be improved. As succinctly summarised by Davies (2000), the ultimate aim of data warehouses is by all means to help create customer retention. Large hotel chains have in evidence been acquiring and storing customer data in a combined attempt to achieve competitive edge and improve the experience of customers. It even appears that hotel chains have realised the associated benefits of databases. For example, as a consequence of investing in customer relationship management software, Marriott International registered improvements in other areas, such as cross-selling and yield management (Caterer and Hotelkeeper, 2004).

The capability of databases to help track actual purchases of customers and enable inferences to predict future behaviour patterns may undoubtedly encourage the assumption that database marketing is routine within the embracing of CRM. Moncrief and Cravens' (1999, p. 330) contention that "customer service levels increase when customer information becomes so easy to obtain and disperse," and could by all means imply that databases are being efficiently and effectively used to acquire and maintain information on existing and prospective customers. Abbott (2001, p. 182) even advocates that refinements in technology has provided companies with increasing opportunities and well-structured channels to not only collect abundant amount of data but also to manipulate this data in various ways so as to unravel any unforeseen areas of knowledge. However, several academics have reservations on how databases are not being optimally used.

CONVERTING DATA INTO COMPETITIVE EDGE

CRM is arguably a progression from data warehousing. At present, one of the principle functions of CRM systems is to collect as much data about each customer as is possible. As discussed earlier, this information is then stored, and to be used at a later stage to give guests as much of a personalised service as possible when they return

(Davies, 2001). According to Cindy Green, the Senior Vice-President of Pegasus Business Intelligence, this will not only lead to a change in the sales and marketing arena but even more importantly this will imply that hotel companies will need to become as advanced in the management of their customer relationships as technology will enable them to be (Davies, 2001). This change of perspective is arguably expected to engender a transition from the management of data about the customers to the management of interactive relationships. Accordingly the data which hotel chains have compiled over the years about their customers, would need to be used intelligently in order to enable predictions about consumer behavior as well as the anticipation of needs or even problems. Such data can be used precisely for target marketing campaigns. Indeed, as succinctly summarized by Green, CRM is in actual fact simply about a hotel company being willing and flexible enough to change its behavior in line with what customers are saying and what the data collated reveal about them.

CRM concept has grown out of companies' attempts to offer a better service to their customers than their competitors are offering (Gledhill, 2002). Within the hotel industry, as identified earlier in this chapter, one of the major elements appears to be the pursuit to streamline back-office processes in order to achieve greater operational efficiencies. Technology has revolutionized operations within the hotel industry as applications have already managed to smoothly link front-office processes such as check-in, with back-office functionality such as reservation details. Additionally, in order to enhance their engagement in CRM, many hotel chains have invested in customized systems. Notwithstanding, as is succinctly reminded by Chen and Popovich (2003, p. 682), despite the crucial role that technology and people play within the CRM arena the philosophical bases of CRM: relationship marketing, customer profitability, lifetime value, retention and satisfaction, are in fact created through the business process management.

According to Cindy Estis Green, from Driving Revenue, a consultancy that aims to help hotel companies add value to the data they collect from and about the customer, the management of a database involves three crucial stages (Goymour, 2001). Firstly, when all the data collected about a guest is consolidated into a usable set of information, the automated cleaning of data must be conducted. Secondly, the analysis of the information about the guests must undergone segmentation in order for the hotel company to be able to precisely target the most attractive prospects and discard those suspects who do not meet the profiling criteria. Thirdly, the results of the targeting of specific guests must be tracked in order to determine which guests responded to the campaigns. This step will not only identify the profitable customers, but also will ultimately also indicate which promotions are successful. Subsequently, the adequacy of campaigns can be evaluated.

The general consensus is that an integrated and centralized database will enable a complete view of the customers within a hotel chain. Such a database is expected to collect ongoing information from all relevant sources and outlets, such as reservations and other point of sale systems located within the various hotels. Information from customer satisfaction questionnaires, surveys or even e-mails can also be fed into the database. The database would ideally be compiled so as to produce an integrated set of information in order to create a unified profile about each customer (Bentley, 2005).

According to Jane Waterworth, the marketing director at Shire Hotels, the standardizing of data is a process, which hotel companies should take seriously, as it is vital to ascertain that they in fact are inputting the right data in their CRM system. According to Steve Clarke, the account director at marketing database company CDMS, companies which are serious about CRM must consolidate their data. Otherwise customers may end up receiving the same information from various sources, thereby diluting marketing initiatives, and more specifically for the company, no full view of a customer's behavior would be achievable. Furthermore, as emphasised by Bentley (2005), without all the relevant information about a customer, any attempt to use data in a meaningful and precise way to enhance loyalty schemes or even marketing campaigns will be essentially flawed.

A central data warehouse can by all means combine information from many sources and help consolidate a comprehensive and reliable picture of a hotel's clients. Although data warehouses can be clear and immediately accessible, Velibor Korolija, the operations director with software specialist of the Bromley Group, argue that for business and marketing analysts, data warehouses are by no means enough. In fact, it is data mining, a process which involves the analysis of the data in an attempt to seek meaningful relationships not previously known, which Korolija advocates to be of utmost importance (Davies, 2000).

Data mining refers to the process of retrieving data from a data warehouse for analysis purposes. Data mining tools and technologies have been accredited by such academics as Nemati and Barko (2003, p. 282) with having the potential to enhance the decision-making process by transforming data into valuable and actionable knowledge to gain a competitive advantage.

Although many databases may by all means be deemed to be appropriate data warehouses, it has been argued that the data mining process associated with many of these has been consistently flawed. In fact, in spite of several academics acknowledging the technological trend to rely on database marketing to acquire and maintain extensive information on existing and potential customers (Krol, 1999; Long et al., 1999; Moncrief & Cravens, 1999), such academics as Dyer (1998), Rich (2000), Joplin (2001) and Overell (2004) provide evidence to confirm that companies are not adequately using the information at their disposal to build and strengthen relationships with customers.

Moreover, according to Dyer (1998), many practitioners are failing to make optimum use of their client databases because not only their information is being updated, but also the available data is not even being analysed adequately so as to produce pertinent qualitative and quantitative information, from which future strategies and tactics could be taken. Yet, Murphy (2001) advocates that not only does personalized data have to exist and be correct, but also this data should be correctly updated and be made available to the rest of the organization. Here, the general consensus is that this process should be rigidly adhered to whichever channel of communication the customer uses to interact with an organization (Key Note, 2002b). Although this step may not already be adhered to within the hotel industry, there is an indication that some hotel chains have integrated this process in their systems. For instance, from 2003 all bookings made for any of the hotels within the Corus & Regal hotel chain have been redirected via the central reservations office or to their new marketing database so that the information on the database can be continuously updated. Accordingly, the records about existing customers are consistently updated while the profiles of new customers are automatically created (Key Note, 2003).

Highlighting a different shortcoming, Rich (2000) argues that companies are failing to use the information stored in their databases to build relationships with their customers even though the latter could prove vital for marketers in their attempts to outperform their competitors in terms of providing a better service to customers. According to Overell (2004), marketers and companies are not even attempting to adequately analyze the data to an accepted level of depth. In spite of such contentions, Michael Gadbury, the vice-president of Aremissoft, a CRM software company, advocates that while two years ago, only ten percent of hotel companies showed interest in making use of the data, which they had collected about their customers, this percentage has risen to almost ninety percent in contemporary terms (Davies, 2001). It is anticipated that in recent years, even more companies have shown interest in adequately mining their customer database.

Although the integrated process of capturing, sifting, and interrogating data about customers may have been somehow flawed within some companies; companies have been so eager to capturing data about their customers that according to Overell (2004:1), "many organisations are sitting on mushrooming stockpiles of data." This over zealous attitude towards the collection of data seems to have gripped the hotel companies too. Indeed, as is advocated by Geoffrey Breeze, the Vice-President of marketing and alliance development at Hilton International, "hotels have far more information about their guests than they can actually use" (Caterer and Hotelkeeper, 2000, p. 14). However, Overell (2004) advocates that the general consensus among database experts is that companies do not have much more understanding of customers than they did prior to their embracing of CRM.

Nemati and Barko (2003, p. 282) offer a plausible explanation for the limited benefits reaped from data mining when they explain that although "management factors affecting the implementation of IT projects have been widely studied," "there is little empirical research investigating the implementation of organizational data-mining projects." Furthermore, in pointing to a plausible differential level of expertise between the collection of data and the actual mining and usage of this data, they also shed light on the inadequacy of training for the people at the various stages of the data mining process. For instance, it is notable that within the hotel industry, technical systems tend not to be developed in-house (Luck & Lancaster, 2003) but commissioned through expert agencies. While CRM systems are developed by experts in line with specifications requested by a hotel company, once unfolded within an organization, such systems tend to be monitored in-house. Luck suggests that internal employees may not have the adequate level of expertise that some of the filtering processes may call for. Furthermore, she also suggests that the high financial, human and technological resources needed to keep a data-mining system up to date may also place too high demands on some companies. Arguably in attempts to curtail limitations and perhaps to enhance their CRM opportunities, hotel companies have increasingly entered in partnerships with specialist agencies. While De Vere Group Plc enlisted the GB group to help create more targeted and cost effective database campaign; Thistle Hotels Ltd worked closely with Arnold Interactive to design, develop and handle its online strategy to increase its database from 50,000 to 500,000 profiles by the end of 2003 and its series of e-marketing campaigns (Key Note, 2003).

As identified by Bradbury (2005), CRM is meant to not only to help companies collect information about guests, but also, and even more importantly it is meant to help companies use the information collected about its customers more effectively. One of the ultimate steps within the data mining process is undeniably to cluster customers into segments, which are not only meaningful but also reachable by CRM campaigns. According to Korolija, it is by all means possible to cluster a hotel's guests into very specific demographic groups (Davies, 2000). In serving a number of closely-related purposes, customer segmentation has been portrayed as a means of predicting behavior (Clemons & Row, 2000), a method of detecting, evaluating, and selecting homogeneous groups (Reichheld & Schefter, 2000) and a way of identifying a target market for which a competitive strategy can be formulated (Gulati & Garino, 2000). In more general terms, customer segmentation is accredited by enabling the identification of key consumer groups, thereby favoring the effective targeting of such strategic tools as CRM programmes. It could also be plausible to posit that customer segmentation enables precision targeting. Some hotel chains in evidence appreciate the opportunities afforded by customer segmentation. For instance, in an attempt to precisely and cost effectively target its guests, De Vere Group Plc restructured its customer database in 2003 into a range of customer categories such as debutantes and devoted stayers. This strategy was also intended to enhance cross-selling across the various brands to existing customers. In the same year, Corus & Regal Hotels Plc divided its database, which consisted of 68,000 profiles, into categories. These spanned from cold prospects to loyal customers (Key Note, 2003).

The varied outcomes of customer segmentation have been well documented. Benefits such as added protection against substitution, differentiation, and pricing stability have been quoted by several authors including Walters and Lancaster (1999b) and Sinha (2000). Moreover, Ivor Tyndall, the head of customer intelligence at Le Meridien advocates that as the company segments their consumer base, they can precisely target different sectors or segments with different offers (Bentley, 2005). Although Botschen, Thelen, and Pieters (1999) support the importance of segmenting customers on the benefit-level, Long and Schiffman (2000) offer evidence to suggest that different segments of consumers may perceive benefits differently and consequently have differing degrees of affinity and commitment to CRM programmes and other benefits on offer.

The popularity of databases is increasing and as is highlighted by Abbott (2001, p.182), "vast databases holding terabytes of data are becoming commonplace." However, if companies do not follow the correct processes to tap into this valuable data they have in their databases, new knowledge about customers will be largely uncovered (Rich, 2000). Indeed, it is likely that the assiduous collection of information about customers will be largely wasted. Consequently, although in theory borrowing from the arena of direct marketing seems pertinent to CRM strategy, transferring the theoretical advantages into practice appears to be an altogether different scenario. Meanwhile, according to Felix Laboy, the chief executive officer of E-Site Marketing, when hotels are able to access more information about a guest and then be able to offer the latter the individual service the guest needs, loyalty will be encouraged (Edlington, 2003). Moreover while such authors as Davies (2001a) and Bentley (2005) advocate that when the data is correctly structured and hotel companies can target their marketing more effectively, it is expected that loyalty schemes will become more effective. To strengthen these arguments, Cindy Estis Green from Driving Revenue advocates that when a company shows that it cares about its guests through its offering of benefits, it can strongly influence the creation of customer loyalty (Goymour, 2001).

RETAINING OLD CUSTOMERS AND REACHING NEW ONES

It is well documented that retaining customers is more profitable than building new relationships. While Reichheld and Schefter (2000) discuss how the dynamics of customer retention are less costly that initiatives focusing on customer acquisition, Kandampully and Duddy (1999) even state that attracting new customers is five times more costly than retaining an existing customer. Consequently, the retention of existing customers has become a priority for businesses to survive and prosper. In view of its inherent long-term perspective, databases and CRM explicably appears to be ideal platforms for the achievement of this ongoing objective. As pertinently summarized by Chen and Popovich (2003), CRM strategy (and databases) can help to attract new customers, but even more importantly, helps develop and maintain existing ones.

Efforts to retain customers have led to the refining of processes such as target marketing and segmentation within hotel operations. Furthermore, with direct marketing and database marketing having been repeatedly identified as two of the immediate forebears of relationship marketing in consumer markets (Long et al., 1999), companies have adopted processes initially associated with the specialised field of direct marketing to facilitate their CRM objectives. As such, precise targeting, described by Lester (2004, p.4) as: "the ability to deliver accurate and exact marketing messages to people at a narrow customer segment level", is almost expected to be routinely applied as part of CRM programs. In fact, such is the commonality between direct marketing and CRM that despite the criticisms highlighted earlier in this chapter about the current poor level of data-mining, it would seem almost impetuous for hotel chains embracing CRM not to focus on this process within in their operations.

Through a combination of technology and business processes, which attempts to find out and understand who customers are, what they do and to identify their likes and dislikes (Couldwell, 1998), CRM and its databases can and do facilitate the understanding of the customer. In turn, an understanding of customers in line with the dynamics of organizations is expected not only to help design systems, which meet customers' needs more effectively, but also this balance is highly likely to lead to stronger customer loyalty and lasting relationships. As the achievement of loyalty appears to be sought within the hotel industry (Palmer, McMahon-Beattie & Beggs, 2000; Tepeci, 1999), several processes have been integrated into hotel operations to work alongside databases in order to find out more about customers' needs and wants. Accordingly, customer satisfaction surveys and customer service questionnaires are routinely distributed to guests in an attempt to improve operations and understand customers better, and indeed to gather more information for the databases.

Meanwhile, in describing CRM as "an enterprise-wide customer centric business model that must be built around the customer," Chen and Popovich (2003, p. 682) arguably imply that when hotel chains embrace CRM, rather than keep customers at the end of the value chain (Jobber, 1998), hotel chains should instead put customers at the start of operations. This reversal of the direction of the traditional value chain means that companies will have to shift their CRM endeavours from a mass marketing perspective where customers are sought for products or services, to develop products and services, which are actually tailored to fit the needs of the company's targeted customers. In this perspective, CRM appears to call for a reversal of some traditional processes and the integration of data into all levels of an organization. Thus, it is crucial that hotel companies strive to maintain and enhance the data they collect about their targeted guests. Only when they ensure that processes are being correctly implemented within their operations, would hotel companies be able to truly be in a position to assess what their customers actually seek. Only then, would a unified and optimum CRM offering be possible. Consequently, the database is indeed the pivotal tool around which CRM revolves in contemporary terms.

FUTURE TRENDS

Databases are subject to a range of important influences in addition to technological advances. Although internal capabilities is by all means a determining factor in how a company uses databases and data, the way in which these are embraced is going to differ from company to company and perhaps even more importantly from industry to industry. It is argued that although there is a common platform from which data can be used, the means and usage is going to be dictated by the dynamics of the industry and product or service. As such future research is recommended into the dynamics of specific industries in order to determine the relevancy of databases and data to that specific industry. Indeed, this chapter set out to illustrate the applicability of data and databases to the hotel industry, and not only are individual capabilities and strategies of companies are relevant, but also the reality of organizations within a micro and macro business environment concern. These unquestionably demand intensive research and experimentation through proper feedback and monitoring tools.

CONCLUSION

CRM has been hailed as a powerful tool in the quest for strengthening relationships with customers. A triad combination of technology, people and processes can arguably enable hotel chains to not only implement CRM within their operations, but also to reap the opportunities, which the concept can provide. However, in recent times it appears that most attention has been focused on technology rather than on the capturing and mining of data.

Technology has greatly enhanced the processes associated with the implementation, evaluation and monitoring of CRM. Database technologies have by all means driven CRM into a new era not only in terms of storing and mining information to help make sales, but also to access customers, gather data and even target campaigns. The importance of the database within CRM is in fact so unquestionable that it can be said that database is now the central tool of CRM.

Although it is not denied that technology is crucial in the facilitation of CRM and as such attracted much investment, it is emphasised that the optimization of CRM also requires the organization of business processes. Although data mining processes associated with the hotel industry have been somehow flawed, and depending on databases, hotel chains of all sizes appear to increasingly be developing and implementing database technologies. However, it is argued that the acquisition of a sophisticated database is by no means sufficient to reap the benefits of CRM. Indeed, the effectiveness of data mining procedures is crucial if successful CRM is to be achieved. Subsequently, companies are expected not only to continuously view their organizations from the customers' perspective but also as importantly, gear operations to actively involve customer feedback and market as well as technological changes.

When these processes are consistently and continuously integrated, applied, and monitored, it is expected that companies would be able to gather and disseminate the right type of data to optimally achieve their CRM objectives. Indeed, successful CRM does not just emerge or simply exist. Thus, it is advocated that the creation and establishing of successful customer relationships confront companies with a complex range of relationship and network management tasks above the ones, which is inherent to their traditional operations and structures. The principal processes of data management, namely data acquisition, collation and mining, are indeed integral to these business functions.

REFERENCES

Abbott, J. (2001). Data data everywhere – and not a byte of use? *Qualitative Market Research: An International Journal, 4*(3), 182-192.

Bentley, R. (2005, August 25). Data with destiny. *Caterer & Hotelkeeper, 38.*

Bradbury, D. (2005, August 31). Technology Jargon Buster. *Caterer & Hotelkeeper,*

Botschen, G., Thelen, E. M. & Pieters, R. (1999). Using means-end structures for benefit segmentation: An application to services. *European Journal of Marketing, 33* (1/2).

Caterer & Hotelkeeper (2000, September 7). Hotel groups deny they're missing Web opportunities, 14.

Caterer & Hotelkeeper (2004, 24 June). Do the knowledge, 34.

Chen, I. J. & Popovich, K. (2003). Understanding customer relationship management (CRM); People, process and technology. *Business Process Management Journal, 9*(5), 672-688.

Clemons, E. & Row, M. (2000, November 13). Behaviour is key to web retailing strategy. *Financial Times.*

Couldwell, C. (1998, May 21). A data day battle. *Computing*, 64-66.

Davies, A. (2000, 29 June). Data's the way to do it, *Caterer & Hotelkeeper*, 31-32.

Davies, A. (2001, 26 July). On-line, on course. *Caterer & Hotelkeeper*, 37-39.

Dyer, N. A. (1998). What's in a relationship (other than relations)? *Insurance Brokers Monthly & Insurance Adviser, 48*(7), 16-17.

Edlington, S. (2003, January 20). Future perfect? *Caterer & Hotelkeeper*, 26.

Fraser, J., Fraser, N., & McDonald, F. (2000). The strategic challenge of electronic commerce. *Supply Chain Management: An International Journal, 5*(1), 7-14

Galbreath, J. & Rogers, T. (1999). Customer relationship leadership: A leadership and motivation model for the twenty-first century business. *The TQM Magazine, 11*(3), 161-171.

Gledhill, B. (2002, February 28). Learning from history. *Caterer & Hotelkeeper*, 33.

Goymour, A. (2001, 26 July). Host in the machine. *Caterer & Hotelkeeper*, 43-45.

Grönroos, C. (1994). From scientific management to service management: A management perspective for the age of service competition. *International Journal of Service Management, 5*(1), 5-20.

Gulati, R. & Garino, J. (2000, May-June). Get the right mix of bricks and mortar. *Harvard Business Review*, 107-114.

Jobber, D. (1998). *Principles of marketing* (2nd ed.). McGraw-Hill

Joplin, B. (2001, March/April). Are we in danger of becoming CRM lemmings? *Customer Management*, 81- 85

Kandampully, J. & Duddy, R. (1999). Relationship marketing: a concept beyond primary relationship. *Marketing Intelligence &Planning, 17*(7), 315-323.

Key Note (2002a), Customer Relationship Management

Key Note (2002b), Hotels

Key Note (2003), Hotels

Khalil, O. E. M. & Harcar, T. D. (1999). Relationship marketing and data quality management. *SAM Advanced Management Journal, 64* (2).

Krol, C. (1999, May). A new age: It's all about relationships. *Advertising Age, 70*(21), S1-S4.

Lester, T. (2004, March 31). Pitfalls of precision bombing. *FT Management*, 4.

Lindgreen, A. & Crawford, I. (1999). Implementing, monitoring and measuring a programme of relationship marketing. *Marketing Intelligence & Planning, 17*(5), 231-239.

Long, G., Hogg, M. K., Hartley, M. & Angold, S. J. (1999). Relationship marketing and privacy: Exploring the thresholds. *Journal of Marketing Practice: Applied Marketing Science, 5*(1), 4-20.

Long, M. M. & Schiffman, L. G. (2000). Consumption values and relationships: Segmenting the market for frequency programs. *Journal of Consumer Marketing, 17*(3).

Luck, D. & Lancaster, G. (2003). E-CRM: Customer relationship marketing in the hotel industry. *Managerial Auditing Journal – Accountability and the Internet, 18*(3), 213-232.

McDonald, W. J. (1998). *Direct marketing: An integrated approach.* McGraw-Hill International Editions.

Moncrief, W. C. & Cravens, D. (1999). Technology and the changing marketing world. *Marketing Intelligence and Planning, 17*(7), 329-332.

Murphy, J. M. (2001, March-April). Customer excellence: From the top down. *Customer Management*, 36-41.

Nemati, H. R. & Barko, C. D. (2003). Key factors for achieving organizational data-mining success. *Industrial Management & Data Systems, 103*(4), 282-292.

Nitsche, M. (2002, January-March). Developing a truly customer-centric CRM system: Part One – Strategic and artchitectural implementation. *Interactive Marketing, 3*(3), 207-217.

Overell, S. (2004, March 31). Customers are not there to be hunted. *FT Management*, 2.

Palmer, A. (1996). Relationship marketing: A universal paradigm or management fad? *The Learning Organisation, 3*(3), 18-25.

Palmer, A., McMahon-Beattie, U. & Beggs, R. (2000). A structural analysis of hotel sector loyalty programmes. *International Journal of Contemporary Hospitality Management, 12*(1), 54-60.

Prabhaker, P. (2001). Integrated marketing-manufacturing strategies. *Journal of Business & Industrial Marketing, 16*(2), 113-128.

Reichheld, F. & Schefter, P. (2000, July/ August). E-loyalty. *Harvard Business Review*, 105-113.

Rich, M. K. (2000). The direction of marketing relationships. *The Journal of Business & Industrial Marketing, 15*(2/3), 170-179.

Sinha, I. (2000, March/ April). Cost transparency: The Net's real threat to prices and brands. *Harvard Business Review*, 43-55.

Tapp, A. (2001). *Principles of direct marketing* (2nd ed). Prentice Hall.

Tepeci, M. (1999). Increasing brand loyalty in the hospitality industry. *International Journal of Contemporary Hospitality Management, 11*(5).

Walters, D. & Lancaster, G. (1999a). Value and information – Concepts and issues for management. *Management Decision, 37*(8), 643-656.

Walters, D. & Lancaster, G. (1999b). Using the Internet as a channel for commerce. *Management Decision, 37*(10), 800-816.

Zahra, S., Sisodia, R., & Matherne, B. (1999, April). Exploiting the dynamic links between competitive and technology strategies. *European Management Journal, 17*(2), 188-201.

Chapter VII
Mining Allocating Patterns in Investment Portfolios

Yanbo J. Wang
University of Liverpool, UK

Xinwei Zheng
University of Durham, UK

Frans Coenen
University of Liverpool, UK

ABSTRACT

An association rule (AR) is a common type of mined knowledge in data mining that describes an implicative co-occurring relationship between two sets of binary-valued transaction-database attributes, expressed in the form of an ⟨*antecedent*⟩ ⇒ ⟨*consequent*⟩ *rule. A variation of ARs is the (WARs), which addresses the weighting issue in ARs. In this chapter, the authors introduce the concept of "one-sum" WAR and name such WARs as allocating patterns (ALPs). An algorithm is proposed to extract hidden and interesting ALPs from data. The authors further indicate that ALPs can be applied in portfolio management. Firstly by modelling a collection of investment portfolios as a one-sum weighted transaction-database that contains hidden ALPs. Secondly the authors show that ALPs, mined from the given portfolio-data, can be applied to guide future investment activities. The experimental results show good performance that demonstrates the effectiveness of using ALPs in the proposed application.*

INTRODUCTION

Investments (Bodie, Kane, Marcus, & Ryan, 2003; Cuthbertson & Nitzsche, 2001) are one of the major schools in financial research that parallels corporate finance (Damodaran, 2001), personal financial planning (Ho & Robinson, 2001), financial engineering (Neftci, 2004), and so forth. Portfolio management, aiming to minimize the overall risk while maximizing the total expected return for an investment activity, is perhaps one of the most indispensable tools available in investments. It diversely "allocates" a given amount of assets/funds in a variety of investment-items (i.e., bonds, funds, options, stocks, etc.). In Ho and Robinson (2001) diversification (Farrell, 2006) was introduced as a principle of investments. There are three dimensions in diversification (Ho & Robinson, 2001): (1) diversity across items/assets within the same investment-security, (2) diversity across different securities of investments, and (3) diversity internationally. When addressing diversification in portfolio management, choosing to invest a portfolio that consists of a set of uncorrelated investment-items or negatively correlated investment-item pairs, noted as the correlation coefficient based portfolio theory (Ho & Robinson, 2001), is recommended.

Data mining (Bramer, 2007; Han & Kamber, 2001; Han & Kamber, 2006; Hand, Mannila & Smyth, 2001; Thuraisingham, 1999) is a promising area of current research and development in computer science, which is attracting more and more attention from a wide range of different groups of people. It aims to extract various types of hidden, interesting, previously unknown and potentially useful knowledge (i.e., rules, patterns, regularities, customs, trends, etc.) from databases, where the volume of a collected database can be measured in GBytes. In data mining common types of mined knowledge include: association rules (Agrawal & Srikant, 1994), classification rules (Quinlan, 1993), prediction rules (Han & Kamber, 2001), classification association rules (Ali, Manganaris & Srikant, 1997), clustering rules (Mirkin & Mirkin, 2005), emerging patterns (Dong & Li, 1999), sequential patterns (Wang & Yang, 2005), and so forth. In the past decade, data mining techniques have been widely applied in, for example, bioinformatics (Wang, Zaki, Toivonen & Shasha, 2005), e-commerce (Raghavan, 2005), geography (Miller & Han, 2001), marketing and sales studies (Berry & Linoff, 1997). Kovalerchuk and Vityaev (2000) systematically discussed, in the scope of computational finance, the necessities and/or possibilities of employing data mining technologies and/or methodologies in financial research.

Portfolio management, in a general prospect, refers to the overall process of creating appropriate portfolio strategies that will ensure/almost-ensure profits in future investment activities. The portfolio management process has been analysed in many literatures, but a unique scheme has not yet been agreed upon. The stages of the portfolio management process usually include:

1. **Investment-item selection:** Where a number of investment-items/assets that will be comprised in a "potential" portfolio are selected.
2. **Investment-item return prediction:** Where the expected return of each asset, selected in stage 1, is predicted.
3. **Investment-item weight determination:** Where a candidate portfolio is generated by assigning a suitable weight to each asset, based on the result of stage 2.
4. **Portfolio selection:** Where the "best" portfolio strategy is selected from a number of alternative candidate portfolios that are generated by iteratively processing stages 1, 2 and 3. Best in this case is defined according to the return and risk of the candidate portfolio.

In the past decade, research in portfolio management has demonstrated an interest in some data mining and/or machine learning concepts (Hung, Liang & Liu, 1996; Lazo, Maria, Vellasco, Aurelio & Pacheco, 2000; Tseng, 2004; Wang & Weigend, 2004; Zhang & Zhou, 2004). A number of approaches in such research are summarized as follows:

- John, Miller, and Kerber (1996) developed a rule induction based stock selection system, namely Recon. This system marks "stocks with returns in the top 20% in a given quarter as exceptional and the rest as unexceptional" (pp. 52-53); and analyses "a historical database and produce rules that would classify present stocks as exceptional or unexceptional future performers" (p. 53).

- A hybrid approach that generates candidate portfolios by integrating the well-known APT (arbitrage pricing theory) model (Ross, 1976) with neural networks was introduced by Hung, Liang, and Liu (1996). In this approach, "an APT model can be used to determine prices, and then a neural network predicts the trend of each risk factor in the future" (Zhang & Zhou, 2004, p. 517). A portfolio selection mechanism was further developed in Hung, Liang, and Liu (1996) to select the optimal/best portfolio(s) by computing a performance score for each generated candidate portfolio.
- Kohara, Ishikawa, Fukuhara, and Nakamura (1997) incorporated prior knowledge with artificial neural networks "to improve the performance of stock market prediction" (Yu, Wang & Lai, 2005, p. 336).
- Quah and Srinivasan (1999) proposed an artificial neural network stock selection system "to select stocks that are top performers from the market and to avoid selecting under performers" (Yu, Wang & Lai, 2005, p. 336).
- Lazo, Maria, Vellasco, Aurelio, and Pacheco (2000) describes "a hybrid model for portfolio selection and management, which comprised three modules: a genetic algorithm for the selection of the assets that are going to form the investment portfolio, a neural net for the prediction of the returns on the assets in the portfolio, and a genetic algorithm for the determination of the optimal weights for the assets" (Zhang & Zhou, 2004, p. 517).
- Enke and Thawornwong (2005) proposed an approach that utilizes data mining and/or machine learning techniques to forecast stock market returns. In this approach, "an information gain technique used in machine learning for data mining" (p. 927) is introduced to evaluate "the predictive relationships of numerous financial and economic variables" (p. 927); and "neural network models for level estimation and classification are then examined for their ability to provide an effective forecast of future values" (p. 927).

Contribution

In this chapter, the authors introduce a novel type of mined knowledge in data mining, namely allocating patterns (ALPs), which can be recognized as a variation of the traditional association rules (ARs) in a special weighted setting. An ALP is a "one-sum" weighted AR (WAR), where each item involved in an AR is associated with a weighting score between 0 and 1, and the sum of all AR item weights is 1. An ALP can not only indicate the implicative co-occurring relationship between two sets of binary-valued transaction-database attributes (items) in a weighting setting, but also inform the allocating relationship among AR items (e.g. ⟨allocating weight/quota a to item X⟩ ⇒ ⟨the allocation of both quotas b and c to items Y and Z⟩, where $0 < a, b, c < 1$, and $a + b + c = 1$). An algorithm is proposed to extract all hidden and interesting ALPs from a one-sum weighted transaction-database (the well-established transaction-database in a one-sum weighting fashion). With regard to portfolio management, the authors model a collection of investment portfolios as a one-sum weighted transaction-database that contains hidden ALPs; and suggest that a set of ALPs, mined from the given portfolio-data, can be treated as the candidate portfolios that can be further applied to guide future investment activities. It is believed that ALPs will prove be useful in several different areas as well. The experiments are conducted using two sets of possible investment portfolios generated from the CSMAR (China Stock Market & Accounting Research) China Stock Trade and Quote Research Database (CSTQR·Database). The experimental results show good performance regarding both the rate of obtaining "qualified" candidate portfolios and the monthly average return of the obtained candidate portfolios (as used in Hung, Liang & Liu, 1996). The results evidence the effectiveness of addressing ALPs in the proposed portfolio management application.

Chapter Organization

The following section describes the related data mining aspects in association rule mining (ARM) and weighted association rule mining (WARM). In the third section the concept of ALP is introduced, based on describing the one-sum weighted transaction-database, one-sum weighted itemsets and such WARs. An algorithm is proposed in the fourth section that identifies all hidden and interesting ALPs in a given one-sum weighted transaction-database. In the fifth section, the authors further suggest an application of mining a set of ALPs in a collection of investment portfolios. Experiments are presented in the sixth section that demonstrates the effectiveness of using ALPs in the proposed application. Finally the conclusions and a number of open issues for future research are discussed at the end of this chapter.

RELATED WORK

Association Rule Mining

Association rule mining (ARM) aims to extract a set of ARs from a given transaction-database D_T, first introduced in Agrawal, Imielinski, and Swami (1993). Let $I = \{a_1, a_2, \ldots, a_{n-1}, a_n\}$ be a set of items (database attributes), and $\mathcal{T} = \{T_1, T_2, \ldots, T_{m-1}, T_m\}$ be a set of transactions (database records), D_T is described by $\mathcal{T}$, where each $T_j \in \mathcal{T}$ comprises a set of items $I' \subseteq I$. In ARM, two threshold values are usually used to determine the significance of an AR:

- **Support:** The frequency that the items occur or co-occur in $\mathcal{T}$. A support threshold σ, defined by the user, is used to distinguish frequent items from the infrequent ones. A set of items S is called an itemset, where $S \subseteq I$ and $\forall a_i \in S$ co-occur at least once in $\mathcal{T}$. If the frequency of S in $\mathcal{T}$ exceeds σ, S is defined as a Frequent Itemset (FI).
- **Confidence:** Represents how "strongly" an itemset X implies another itemset Y, where $X, Y \subseteq I$ and $X \cap Y = \{\varnothing\}$. A confidence threshold α, supplied by the user, is used to distinguish high confidence ARs from low confidence ARs.

An AR $X\langle\text{antecedent}\rangle \Rightarrow Y\langle\text{consequent}\rangle$ is said to be valid when the support for the co-occurrence of X and Y exceeds σ, and the confidence of this AR exceeds α. The computation of support is: $(X \cup Y) / |\mathcal{T}|$, where $|\mathcal{T}|$ is the size function of the set $\mathcal{T}$. The computation of confidence is: $support(X \cup Y) / support(X)$. Informally, $X \Rightarrow Y$ can be interpreted as: if X exists, it is likely that Y also exists. With regards to the history of ARM investigation, three major categories of ARM algorithms can be identified: (1) mining ARs from all possible FIs, (2) mining ARs from maximal frequent itemsets (MFIs), and (3) mining ARs from frequent closed itemsets (FCIs).

Mining ARs from FIs

In the past decade, many algorithms have been introduced to mine ARs from identified FIs. These algorithms can be further grouped into different "families", such as pure-apriori like, semi-apriori like, set enumeration tree like, and so forth.

- Pure-apriori like where FIs are generated based on the generate-prune level by level iteration that was first promulgated in the apriori algorithm (Agrawal & Srikant, 1994). In this family archetypal algorithms include: Apriori, AprioriTid and AprioriHybrid (Agrawal & Srikant, 1994), partition (Savasere, Omiecinski & Navathe, 1995), DHP (direct hashing and pruning) (Park, Chen & Yu, 1995), sampling (Toivonen, 1996), DIC (dynamic itemset counting) (Brin, Motwani, Ullman & Tsur, 1997), CARMA (continuous association rule mining algorithm) (Hidber, 1999), and so forth. It can be remarked that the well-established apriori algorithm has been the basis of many subsequent ARM and/or ARM-related algorithms. The apriori algorithm is sketched as follows (see Algorithm 1).
- Semi-apriori like where FIs are generated by enumerating candidate itemsets but do not apply the apriori generate-prune iterative approach founded on (1) the join procedure, and (2) the prune procedure that employs the "closure property" of itemsets — if an itemset is frequent then all its subsets will also be frequent; if an itemset is infrequent then all its supersets will also be infrequent. In this family typical algorithms include: AIS (Agrawal·Imielinski·Swami) (Agrawal, Imielinski & Swami, 1993), OCD (off-line candidate determination) (Mannila, Toivonen & Verkamo, 1994), SETM (SET oriented mining) (Houtsma & Swami, 1995), and so forth.
- Set enumeration tree like where FIs are generated through constructing a set enumeration tree structure (Rymon, 1992) from D_T, which avoids the need to enumerate a large number of candidate itemsets. In this family a number of approaches can be further divided into two main streams: (1) Apriori-TFP (apriori-total·from·partial) based (i.e. Coenen & Leng, 2001; Coenen & Leng,

Algorithm 1. The apriori algorithm

Input: (a) A transaction-database D_T;
(b) A support threshold σ;
Output: A set of frequent itemsets *SFI*;
Begin Algorithm:
(1) $k := 1$;
(2) *SFI* := **prepare** an empty set for holding the identified frequent itemsets;
(3) **generate** all candidate 1-itemsets from D_T;
(4) **while** (candidate *k*-itemsets exist) **do**
(5) **determine** support for candidate *k*-itemsets from D_T;
(6) **add** frequent *k*-itemsets into *SFI*;
(7) **remove** all candidate *k*-itemsets that are not sufficiently supported to give frequent *k*-itemsets;
(8) **generate** candidate $(k + 1)$-itemsets from frequent *k*-itemsets using "closure property" (see **semi-apriori like**);
(9) $k \leftarrow k + 1$;
(10) **end while**
(11) **return** (*SFI*);
End Algorithm

2002; Coenen, Goulbourne & Leng, 2001; Coenen, Leng & Ahmed, 2004; Coenen, Leng & Goulbourne, 2004; etc.), and (2) FP-tree (Frequent-Pattern-tree) based (i.e., El-Hajj & Zaiane, 2003; Han, Pei & Yin, 2000; Liu, Pan, Wang & Han, 2002; etc.).

Mining ARs from MFIs

It is apparent that the size of a complete set of FIs can be very large. The concept of MFI (Roberto & Bayardo, 1998) was proposed to find several "long" (super) FIs in D_T, which avoids the redundant work required to identify "short" FI. The concept of vertical mining has also been effectively promoted in this category (Zaki, Parthasarathy Ogihara, & Li, 1997). Vertical mining, first mentioned in Holsheimer, Kersten, Mannila, and Toivonen (1995), deals with a vertical transaction database D_{TV}, where each database record represents an item that is associated with a list of its relative transactions (the transactions in which it is present). MFI algorithms include: MaxEclat/Eclat (Zaki, Parthasarathy, Ogihara & Li, 1997), MaxClique/Clique (Zaki, Parthasarathy, Ogihara & Li, 1997), Max-Miner (Roberto & Bayardo, 1998), Pincer-Search (Lin & Kedem, 1998), MAFIA (MAximal Frequent Itemset Algorithm) (Burdick, Calimlim & Gehrke, 2001), Genmax (Gouda & Zaki, 2001), and so forth.

Mining ARs from FCIs

Algorithms belonging to this category extract ARs through generating a set of FCIs from D_T. In fact the support of some subitemsets of an MFI might be hard to identified resulting in a further difficulty in the computation of confidence. The concept of FCI (Pei, Han & Mao, 2000) is proposed to improve this property of MFI, which avoids the difficulty of identifying the support of any sub-itemsets of a relatively long FI. A FCI f is an itemset $S \in D_T$, where f is frequent, and $\neg\exists$ itemset $f' \supset f$ and f' shares a common support with f. The relationship between FI, MFI and FCI is that MFI $\subseteq$ FCI $\subseteq$ FI (Burdick, Calimlim & Gehrke, 2001). In this category algorithms include: CLOSET (mining CLOsed itemSETs) (Pei, Han & Mao, 2000), CLOSET+ (Wang, Han & Pei, 2003), CHARM (closed association rule mining; the 'H' is gratuitous) (Zaki & Hsiao, 2002), MAFIA (Burdick, Calimlim & Gehrke, 2001), and so forth.

Weighted Association Rule Mining

Weighted association rule mining (WARM), first introduced in (Cai, Fu, Cheng & Kwong, 1998), aims to address the weighting issue in ARM investigation and extract WARs from a weighted transaction-database. In the past decade, a number of alternative approaches have been subsequently described in WARM (i.e., Lu, Hu & Li, 2001; Tao, Murtagh & Farid, 2003; Wang, Yang, & Yu 2000; etc.). Broadly WARM approaches can be categorized into three groups: (1) mining horizontal WARs, (2) mining vertical WARs, and (3) mining mixed WARs.

Mining Horizontal WARs

The Traditional Approach

Cai, Fu, Cheng, and Kwong (1998) introduced the concept of weighted item based on a "real-life" marketing experience—not all goods share the same importance in a market. With regard to a retailing business, mining from weighted items/goods enables the generation of such ARs with more emphasis on some particular goods (e.g., goods that are under promotion, goods that always make significant profits) and less emphasis on other goods. The idea of mining ARs in a special transaction-database, where each item is assigned with a weighting score, directly depicts the problem of mining horizontal WARs. Let $I^W = \{a^W_1, a^W_2, \ldots, a^W_{n-1}, a^W_n\}$ be a set of weighted items, where each $a^W_i \in I^W$ is an item $a_i \in I$ labelling with a user-defined weighting score w_i ($0 \leq w_i \leq 1$). Let

$\mathcal{T} = \{T_1, T_2, \ldots, T_{m-1}, T_m\}$ be a set of transactions. A horizontal weighted transaction-database $D^W{}_T$ is described by $\mathcal{T}$, where each $T_j \in \mathcal{T}$ comprises a set of weighted items $I^{W\prime} \subseteq I^W$.

To measure the significance of a horizontal WAR, the "weighted-support—weighted-confidence" approach, as an extension of the well-established "support—confidence" framework, was introduced in Cai, Fu, Cheng, and Kwong (1998). A horizontal weighted support threshold σ^W is supplied by the user that distinguishes frequent horizontal weighted itemsets from the infrequent ones. A horizontal weighted itemset $X^W \cup Y^W$ is considered to be frequent if $(\Sigma\{a^W{}_i \in (X^W \cup Y^W)\}\ w_i) * support(X^W \cup Y^W) \geq \sigma^W$, where X^W, $Y^W \subseteq I^W$ and $X^W \cap Y^W = \{\varnothing\}$. Having a set of frequent horizontal weighted itemsets generated from $D^W{}_T$, a set of horizontal WARs can be further obtained. A horizontal WAR $X^W \Rightarrow Y^W$ is said to be valid when $X^W \cup Y^W$ is frequent, and $((\Sigma\{a^W{}_i \in (X^W \cup Y^W)\}\ w_i) * support(X^W \cup Y^W)) / ((\Sigma\{a^W{}_i \in X^W\}\ w_i) * support(X^W)) \geq \alpha^W$, where α^W is a user-defined horizontal weighted confidence threshold.

The Variation Approach

Wang, Yang, and Yu (2000) proposed an alternative approach of mining horizontal WARs by introducing a variational horizontal weighted transaction-database $D^W{}_T{}^*$. With regards to the real-life marketing, the newly mined horizontal WARs "can not only improve the confidence in the rules, but also provide a mechanism to do more effective target marketing by identifying or segmenting customers based on their potential degree of loyalty or volume of purchases" (p. 270). In Table 1 several points in terms of item weighting score properties that differentiate $D^W{}_T{}^*$ from $D^W{}_T$ are listed.

In a marketing context, a typical horizontal WAR mined from $D^W{}_T{}^*$ can be exemplified as ⟨bread[9, 13]⟩ ⇒ ⟨milk[1, 3]⟩, which can be interpreted as: when bread is purchased in the quantity between 9 and 13, it is likely that the milk in the quantity between 1 and 3 is also purchased. In Wang, Yang, and Yu (2000) the proposed WAR generation approach comprises two phases: (1) generating a set of frequent itemsets from $D^W{}_T{}^*$ regardless the weighting issue; and (2) extract-

Table 1. The difference between $D^W{}_T$ and $D^W{}_T{}^*$

Properties of Item Weighting Scores	$D^W{}_T$	$D^W{}_T{}^*$
Single-value like vs. **Interval-value like**	The weighting score of an item in $D^W{}_T$ is given as a single value v. The weighting score is defined as *single-value like*.	The weighting score of an item in $D^W{}_T{}^*$ is given as an interval of two values $[v_1, v_2]$, where $v_1 < v_2$. The weighting score is defined as *interval-value like*.
Percentage like vs. **Positive-integer like**	The value of the weighting score for an item in $D^W{}_T$ is given as $0 \leq v \leq 1$. The weighting score is defined as *percentage like*.	Both lower and upper values of the weighting score interval for an item in $D^W{}_T{}^*$ are given as $v_1, v_2 \geq 1$ and $v_1, v_2 \in Z$ (both v_1, v_2 are positive integers). The weighting score is defined as *positive-integer like*.
Static like vs. **Dynamic like**	The weighting score of an item in $D^W{}_T$ is given as a fixed value in all transactions. The weighting score is defined as *static like*.	The weighting score of an item in $D^W{}_T{}^*$ can be valued differently in different transactions. The weighting score is defined as *dynamic like*.

ing hidden and interesting horizontal WARs based on (1). In (2) a set of candidate rules can be enumerated from the result of (1), where the consequent of each candidate rule "*only contains one weighted item for the sake of simplicity*" (p. 271). A number of "qualified" horizontal WARs can be further identified in the set of candidate rules regarding the user-specified threshold values of support, confidence and density. Since the proposed approach shows an interest in producing maximum rules only, a set of maximum horizontal WARs — "*a qualified WAR* $X \Rightarrow Y$ *is a maximum WAR if for any generalization* X' *of* X *and* Y' *of* Y *where* $X' \neq X$ *and* $Y' \neq Y$, *neither of* $X' \Rightarrow Y$, $X \Rightarrow Y'$, *nor* $X' \Rightarrow Y'$ *is a qualified WAR*" (p. 271)—is finally obtained. Tao, Murtagh, and Farid (2003) classified the process of mining horizontal WARs from $D^W{}_T{}^*$, proposed in (Wang, Yang, & Yu, 2000), as a technique of post-processing or maintaining ARs.

The Improved Approach

Tao, Murtagh, and Farid (2003) identified the main challenge of mining horizontal WARs: the downward closure property of itemsets is invalid in the generation of significant/frequent horizontal weighted itemsets. To solve this problem, an improved approach of mining horizontal WARs was proposed in Tao, Murtagh, and Farid (2003), which takes an alternative horizontal weighted transaction-database $D^W{}_T{}^+$ as the input. The only difference between $D^W{}_T{}^+$ and $D^W{}_T$ is that the item weighting scores in $D^W{}_T{}^+$ can be valued as any real number. This improved approach automatically assigns a weighting score w_t_j to each transaction T_j in $D^W{}_T{}^+$, where the computation of w_t_j is: $(\sum\{a^W{}_i \in T_j\}\ w_i) / |T_j|$. Based on the assigned transaction scores, a set of frequent horizontal weighted itemsets SFI^W can be generated. A horizontal weighted itemset $X^W \cup Y^W$ is considered to be frequent if $(\sum\{j = 1 \ldots |\mathcal{T}|\ \&\ (X^W \cup Y^W) \subseteq T_j\}\ w\_t_j) / (\sum\{j = 1 \ldots |\mathcal{T}|\}\ w\_t_j) \geq \sigma^W$, where $X^W, Y^W \subseteq I^W$, $X^W \cap Y^W = \{\varnothing\}$, and σ^W is a user-supplied horizontal weighted support threshold. In the generation of frequent horizontal weighted itemsets, the downward closure property can be proved works properly. With respect to the idea presented in Agrawal and Srikant (1994), all horizontal WARs can be further mined from SFI^W. In this improved approach of mining horizontal WARs, automatically assigning a weighting score to each transaction (in a vertical fashion) signifies the approach of mining vertical WARs.

Mining Vertical WARs

Lu, Hu, and Li (2001) extended the traditional approach of mining horizontal WARs in a vertical manner by introducing the vertical weighted transaction-database $D^W{}_{TV}$. With regards to the real-life marketing, it can be indicated that not all transactions share the same importance in a market. For example, transactions that have been dealt ages ago may be less important than current transactions; transactions that are processed in a particular region may be more interesting than other transactions; and so forth. Thus assigning non-identical weighting scores to different transactions is suggested.

In Lu, Hu, and Li (2001) the concept of transaction interval was introduced that allows a number of adjacent transactions share a common weighting score. In this vertical WARM approach, items are treated as uniformity. Let $I = \{a_1, a_2, \ldots, a_{n-1}, a_n\}$ be a set of items, $\mathcal{T} = \{T_1, T_2, \ldots, T_{m-1}, T_m\}$ be a set of m-many transactions, and $\mathcal{T}I = \{TT_1, TT_2, \ldots, TT_{M-1}, TT_M\}$ be a set of M-many transaction intervals that covers all transactions in $\mathcal{T}$ in a non-overlapping manner, where $M \leq m$. A vertical weighted transaction-database $D^W{}_{TV}$ is described by $\mathcal{T}I$, where each $TT_l \in \mathcal{T}I$ contains a number of T_j, and each $T_j \in \mathcal{T}$ comprises a set of items $I' \subseteq I$. In $D^W{}_{TV}$ a vertical weighting score w_v_l is assigned to each $TT_l \in \mathcal{T}I$, where $0 \leq w\_v_l \leq 1$.

The process of mining vertical WARs, described in Lu, Hu, and Li (2001), consists of two stages: (1) generating a set of large vertical weighted itemsets from $D^W{}_{TV}$; and (2) extracting vertical

WARs based on (1). In (1) a vertical weighted support threshold σ^{W}_{V} is supplied by the user that distinguishes large vertical weighted itemsets from the small ones. The weighted support of an vertical weighted itemset $X \cup Y$ is calculated as: $(\sum\{l = 1 \ldots M\} (w\_v_l * count((X \cup Y)_l))) / (N_v)$, where $X, Y \subseteq I, X \cap Y = \{\varnothing\}$, $count((X \cup Y)_l)$ is the number of transactions that contain $X \cup Y$ in the transaction interval TT_l, and N_v is the weighted transaction number. The calculation of N_v is: $\sum\{l = 1 \ldots M\} (w\_v_l * N_l)$, where N_l is the number of transactions that are found in the transaction interval TT_l. It can be proved that the closure property works properly in this stage. In (2) a vertical weighted WAR generation approach is applied, which is similar to the rule-generation approach provided in Agrawal and Srikant (1994).

Mining Mixed WARs

A further extension in mining WARs was presented in Lu, Hu, and Li (2001), which combines both approaches of mining horizontal and vertical WARs. This hybrid WARM approach takes a mixed weighted transaction-database D^{W}_{TM} as the input. Let $I^{W} = \{a^{W}_1, a^{W}_2, \ldots, a^{W}_{n-1}, a^{W}_n\}$ be a set of weighted items, where each $a^{W}_i \in I^{W}$ is an item $a_i \in I$ labelling with a user-defined weighting score w_i $(0 \leq w_i \leq 1)$. Let $T = \{T_1, T_2, \ldots, T_{m-1}, T_m\}$ be a set of m-many transactions, and $TI = \{TT_1, TT_2, \ldots, TT_{M-1}, TT_M\}$ be a set of M-many transaction intervals that covers all transactions in T in a non-overlapping manner, where $M \leq m$. A mixed weighted transaction-database D^{W}_{TM} is described by TI, where each $TT_l \in TI$ contains a number of T_j, each $T_j \in T$ comprises a set of weighted items $I^{W'} \subseteq I^{W}$, and the weighted items in each transaction are ordered in an ascending manner based on their item weights. In D^{W}_{TM} a vertical weighting score w_v_l is assigned to each $TT_l \in TI$, where $0 \leq w\_v_l \leq 1$.

The process of mining mixed WARs (Lu, Hu & Li, 2001) is similar to the process of mining vertical WARs. In the stage of generating large mixed weighted itemsets, a weighted support threshold σ^{W}_{M} is specified by the user that distinguishes large mixed weighted itemsets from the small ones. The weighted support of a mixed weighted itemset $X^{W} \cup Y^{W}$ is calculated as: $(1/k) * (\sum\{a^{W}_i \in (X^{W} \cup Y^{W})\} w_i) * ((\sum\{l = 1 \ldots M\} (w\_v_l * count((X^{W} \cup Y^{W})_l))) / (N_v))$, where X^{W}, $Y^{W} \subseteq I^{W}$, $X^{W} \cap Y^{W} = \{\varnothing\}$, $count((X^{W} \cup Y^{W})_l)$ is the number of transactions that contain $X^{W} \cup Y^{W}$ in the transaction interval TT_l, N_v is the vertical weighted transaction number (see the previous subsection), $(1/k) * (\sum\{a^{W}_i \in (X^{W} \cup Y^{W})\} w_i)$ is the horizontal weight of $X^{W} \cup Y^{W}$, and k is the size of $X^{W} \cup Y^{W}$. In this stage, the closure property works by checking the lower bound of the vertical weighted support for each candidate itemset, where the calculation of such lower bound for a mixed weighted k-itemset $X^{W} \cup Y^{W}$ is: $(k * \sigma^{W}_{M}) / (\sum\{i = n - k \ldots n\} (w_i))$. In the stage of generating mixed WARs, an approach that is similar to the rule-generation approach provided in Agrawal and Srikant (1994) is employed.

ALLOCATING PATTERNS

A new type of horizontal WAR, namely allocating pattern (ALP), is described in this section. It can not only indicate the implicative co-occurring relationship between two sets of items in a weighing setting, but also inform the allocating relationship among AR items. In a marketing context, an archetypal ALP can be exemplified as ⟨bread[0.25] ham[0.35]⟩ ⇒ ⟨milk[0.40]⟩, which can be interpreted as: when people spend 25% and 35% of their money to purchase bread and ham together, it is likely that people also spend 40% of the money to purchase milk. The approach of mining ALPs requires a special horizontal weighted transaction-database $D^{W}_{T\text{-}OS}$ as the input.

One-Sum Weighted Transaction Database

In Table 1 three sets of item score properties are defined to analyse different horizontal weighted transaction-databases. These properties are "single-value like vs. interval-value like," "percentage like vs. positive-integer like," and "static like vs. dynamic like." In $D^W_{T\text{-}OS}$ item weighing scores show an additional property ("one-sum" like) that distinguishes $D^W_{T\text{-}OS}$ from other horizontal weighted transaction-databases — the sum of all item scores in each transaction is 1. Hence $D^W_{T\text{-}OS}$ can be named as one-sum weighted transaction-database.

Let $I^{OSW} = \{a^{OSW}_1, a^{OSW}_2, \ldots, a^{OSW}_{n-1}, a^{OSW}_n\}$ be a set of one-sum weighted items, and $\mathcal{T} = \{T_1, T_2, \ldots, T_{m-1}, T_m\}$ be a set of transactions. Each $a^{OSW}_i \in I^{OSW}$ represents an item $a_i \in I$ assigning with a set of weighting scores $\theta_i = \{wi_1, wi_2, \ldots, wi_{m-1}, wi_m\}$, where $0 \leq wi_j \leq 1$ and $|\theta_i| = |\mathcal{T}|$ that means: for different transactions $T_j \in \mathcal{T}$, different scores $wi_j \in \theta_i$ can be assigned to a particular item $a^{OSW}_i \in I^{OSW}$. A one-sum weighted transaction-database $D^W_{T\text{-}OS}$ is described by $\mathcal{T}$, where each $T_j \in \mathcal{T}$ comprises a set of one-sum weighted items $I^{OSW'} \subseteq I^{OSW}$, and $\sum\{i = 1 \ldots |I^{OSW'}| \text{ or } |T_j|\}\ wi_j = 1$. An overall comparison, in terms of item weighting score properties, of four different horizontal weighted transaction-databases is provided in Table 2.

One-Sum Weighted Itemsets

An itemset can be recognized in a transaction-database D_T if this particular set of items appears as a subset of at least one transaction T_j in D_T. A one-sum weighted itemset can be treated as an itemset that is presented in a particular weighting frame, where the item scores are assigned in a one-sum percentage manner. For example, $\{I_1[0.1], I_2[0.3], I_3[0.3], I_5[0.3]\}$ and $\{I_1[0.1], I_2[0.3], I_3[0.5], I_5[0.1]\}$ are two different weighting frames for the itemset $\{I_1, I_2, I_3, I_5\}$. An itemset can produce as many as infinity possible weighting frames. If an itemset weighting frame *IWF* appears as a subset of at least one transaction T_j in a one-sum weighted transaction-database $D^W_{T\text{-}OS}$, this *IWF* can be identified as a one-sum weighted itemset in $D^W_{T\text{-}OS}$.

The Score Transformation Procedure

To determine whether an *IWF* is a subset of a particular T_j in $D^W_{T\text{-}OS}$ or not, the actual weighting

Table 2. The comparison of D^W_T, $D^W_T{}^*$, $D^W_T{}^+$, and $D^W_{T\text{-}OS}$

Properties of Item Weighting Scores	D^W_T	$D^W_T{}^*$	$D^W_T{}^+$	$D^W_{T\text{-}OS}$
Single-value like vs. Interval-value like	*Single-value like*	*Interval-value like*	*Single-value like*	*Single-value like*
Percentage like vs. Positive-integer / Positive-real like	*Percentage like*	*Positive-integer like*	*Positive-real like*	*Percentage like*
Static like vs. Dynamic like	*Static like*	*Dynamic like*	*Static like*	*Dynamic like*
One-sum like	*No*	*No*	*No*	*Yes*

score wj_i that is assigned to each item $a^{OSW}_i \in T_j$ where $a^{OSW}_i \in IWF$ needs to be transformed as $(wj_i) / (\Sigma\{q = 1\ldots|T_j| \ \& \ (a^{OSW}_q \in IWF)\} \ wj_q \in T_j)$. The transformed scores clarify the actual allocating relationship among these *IWF*-related items in T_j. An *IWF* is defined as a subset of T_j if the score of each item involved in *IWF* matches the relative item score transformed in T_j. For example, an *IWF* can be given as $\{I_1[0.2], I_2[0.4], I_3[0.4]\}$ while a transaction T_j may be $\{I_1[0.1], I_2[0.2], I_3[0.2], I_4[0.25], I_5[0.25]\}$; the weighing scores for items I_1, I_2 and I_3 are concentrated since the item intersection $IWF \cap T_j = \{I_1, I_2, I_3\}$; although the actual scores of I_1, I_2 and I_3 are presented differently in *IWF* (as "0.2", "0.4" and "0.4") and T_j (as "0.1," "0.2" and "0.2"), *IWF* is still noticed as a subset of T_j because the transformed scores of I_1, I_2 and $I_3 \in T_j$ are computed as "0.1 / (0.1 + 0.2 + 0.2) = 0.2", "0.2 / (0.1 + 0.2 + 0.2) = 0.4" and "0.2 / (0.1 + 0.2 + 0.2) = 0.4," that match the scores given in *IWF*. The transformation of transaction item scores enables the one-sum weighted property to be lasted from transactions (a special case of itemsets) to the extracted weighted itemsets.

Frequent One-Sum Weighted Itemsets

A one-sum weighted itemset is considered to be frequent if it can be found as a subset of more than $(\sigma^W_{OS} * |\mathcal{T}|)$-many transactions in $D^W_{T\text{-}OS}$, where σ^W_{OS} is a user-supplied one-sum weighted support threshold. It should be noted that the well-known closure property of itemsets can also be found in one-sum weighted itemsets, so that: (1) if a one-sum weighted itemset is frequent then all its subsets will also be frequent; and (2) if a one-sum weighted itemset is infrequent then all its supersets will also be infrequent.

One-Sum Weighted Association Rules

A frequent one-sum weighted itemset is presented as $X^{OSW} \cup Y^{OSW}$, where $X^{OSW}, Y^{OSW} \subseteq I^{OSW}$ and $X^{OSW} \cap Y^{OSW} = \{\emptyset\}$. A one-sum WAR in the form of $X^{OSW} \Rightarrow Y^{OSW}$ can be further produced by a rule formalisation procedure, namely rule-formalisation (see Algorithm 2). In rule-formalisation, $w(a^{OSW}_i) \in (X^{OSW} \cup Y^{OSW})$ represents the corre-

Algorithm 2. The rule-formalisation procedure

Input: A frequent one-sum weighted itemset in terms of (X^{OSW}, Y^{OSW});
Output: A formalized one-sum weighted association rule *p* (as $X^{OSW} \Rightarrow Y^{OSW}$);
Begin Algorithm:
(1) **prepare** *p* to be a formalized one-sum weighted association rule;
(2) **formalize** "⟨" as the first part of *p*;
(3) **for each** $a^{OSW}_i \in X^{OSW}$ **do**
(4) **update** *p* iteratively by formalising "a^{OSW}_i '[' $w(a^{OSW}_i) \in (X^{OSW} \cup Y^{OSW})$ ']'" as its second part;
(5) **end for**
(6) **update** *p* by formalising "⟩ ⇒ ⟨" as its third part;
(7) **for each** $a^{OSW}_i \in Y^{OSW}$ **do**
(8) **update** *p* iteratively by formalising "a^{OSW}_i '[' $w(a^{OSW}_i) \in (X^{OSW} \cup Y^{OSW})$ ']'" as its fourth part;
(9) **end for**
(10) **update** *p* by formalising "⟩" as its last part;
(11) **return** (*p*);
End Algorithm

sponding (actual) weighting score for the item a^{OSW}_i in $X^{OSW} \cup Y^{OSW}$.

A one-sum WAR $X^{OSW} \Rightarrow Y^{OSW}$ is said to be valid when $count((X^{OSW} \cup Y^{OSW}) \subseteq (T_j \in \mathcal{T})) / count(X^{OSW} \subseteq (T_j \in \mathcal{T})) \geq \alpha^W_{OS}$, where α^W_{OS} is a user-supplied one-sum weighted confidence threshold, $count(J)$ is the count function that returns the number of occurences of an object J, and the previously described score transformation procedure is employed to verify the "$\subseteq$" relationship.

ALLOCATION PATTERN MINING

In this section, an allocating pattern mining (ALPM) approach is proposed to extract all hidden and interesting ALPs from a one-sum weighted transaction-database $D^W_{T\text{-}OS}$. With respect to the traditional ARM approach presented in (Agrawal & Srikant, 1994), the proposed ALPM approach consists of two phases: (1) generating a set of frequent one-sum weighted itemsets from $D^W_{T\text{-}OS}$; and (2) mining one-sum WARs (noted as ALPs) based on (1).

Generating Frequent One-Sum Weighted Itemsets

An algorithm, namely apriori-ALP, is proposed to generate a set of frequent one-sum weighted itemsets from $D^W_{T\text{-}OS}$, which takes the apriori algorithm (see Algorithm 1) as its basis. A one-sum weighted

Algorithm 3. The apriori-ALP algorithm

Input: (a) A one-sum weighted transaction-database $D^W_{T\text{-}OS}$;
(b) A one-sum weighted support threshold σ^W_{OS};
Output: A set of frequent one-sum weighted itemsets SFI^W_{OS};
Begin Algorithm:
(1) $k := 1$;
(2) SFI^W_{OS} := **prepare** an empty set for holding the identified frequent one-sum weighted itemsets;
(3) C_k: = **generate** the set of candidate k-itemsets from $D^W_{T\text{-}OS}$;
(4) **while** ($C_k \neq \{\varnothing\}$) **do**
(5) **for each** element $e_i \in C_k$ **do**
(6) **generate** all itemset weighting frames for e_i through scanning all transactions in $D^W_{T\text{-}OS}$;
(7) **initialize** a Boolean variable *frequentFlag* as false;
(8) **for each** itemset weighting frame $f_j \in e_i$ **do**
(9) *support* := **count**($f_j \subseteq$ transactions in $D^W_{T\text{-}OS}$); *// the previously described score transformation procedure is employed to verify the "$\subseteq$" relationship*
(10) **if** (($support$ / $|D^W_{T\text{-}OS}|$) $\geq \sigma^W_{OS}$) **then**
(11) **add** f_j into SFI^W_{OS}; *// f_j is stored with its actual support value*
(12) **set** *frequentFlag* to be true;
(13) **end for**
(14) **if** ($\neg$*frequentFlag*) **then**
(15) **remove** e_i from C_k;
(16) **end for**
(17) $k \leftarrow k + 1$;
(18) $C_k \leftarrow$ **apriori-gen**(C_{k-1}); *// the apriori-gen function is introduced in (Agrawal & Srikant, 1994)*
(19) **end while**
(20) **return** (SFI^W_{OS});
End Algorithm

support threshold, as a parameter of apriori-ALP, is taken from the user. The apriori-ALP algorithm is described as follows (see Algorithm 3).

Mining One-Sum WARs (ALPs)

Given a set of frequent one-sum weighted itemsets SFI^W_{OS} that is generated from apriori-ALP, an algorithm, namely ALP-generation, is further proposed to extract ALPs from SFI^W_{OS}. A one-sum weighted confidence threshold, as a parameter of ALP-generation, is taken from the user. According to the closure property of one-sum weighted itemsets, all subsets of a frequent one-sum weighted itemset f_i are included in SFI^W_{OS}, where $\lfloor f_i \rfloor \geq 2$. Hence the process of ALP-Generation can be designed as follows (see Algorithm 4).

APPLYING ALPS IN PORTFOLIO MANAGEMENT

Portfolio management as the core study in investments research aims to determine the best portfolio strategy, in terms of total expected return and overall risk, that guides how individuals concurrently invest a number of investment-items. With respect to the real-life financial market, an investment-item can be any type of securities, that is, bonds, cash equivalents, funds, futures, options, stocks, and so forth. The primary goal of a portfolio strategy is "to choose a set of risk assets to create a portfolio in order to maximize the return under certain risk or to minimize the risk for obtaining a specific return" (Zhang & Zhou, 2004, p. 517).

Modeling a Collection of Portfolios

Traditional Transaction-Database Model

A number of "popular" investment-items (e.g., the stocks issued by Microsoft, Royal Bank of Scotland, Wal-Mart, etc.) can always be easily listed. This list of items/assets already depicts an investment portfolio in a particular weighted setting, where items are weighted identically—spending/

Algorithm 4. The ALP-generation algorithm

Input: (a) A set of frequent one-sum weighted itemsets SFI^W_{OS};
(b) A one-sum weighted confidence threshold α^W_{OS};
Output: A set of allocating patterns *SALP*;
Begin Algorithm:
SALP := **prepare** an empty set for holding the identified allocating patterns;
for each frequent one-sum weighted itemset $f_i \in SFI^W_{OS}$ **do**
  for each frequent one-sum weighted itemset $f_j \in SFI^W_{OS}$ **do**
    if ($f_j \subset f_i$) **then** *// the previously described score transformation procedure is employed to verify the "⊂" relationship*
      confidence := f_i.*support* / f_j.*support*;
      if (*confidence* $\geq \alpha^W_{OS}$) **then**
        allocating pattern *p* := **Rule-Formalisation**(f_j, $f_i - f_j$);
        add *p* into *SALP*;
  end for
end for
return (*SALP*);
End Algorithm

allocating the same amount of assets/funds to each portfolio-item. A collection of such portfolios can be modelled as a traditional transaction-database *P*. Mining a set of ARs from *P*, each AR illustrates an implicative co-occurring relationship between two sets of investment-items. For example an AR may be mined as ⟨stock_no.1 stock_no.2 bond_no.1⟩ ⇒ ⟨stock_no.3 fund_no.1⟩, which can be interpreted as: when stock_no.1, stock_no.2 and bond_no.1 are invested together, it is likely that both stock_no.3 and fund_no.1 are also invested. In real-life investment activities, ARs are not generally applicable because an amount of assets/funds is usually allocated to each portfolio-asset in a non-identical manner.

One-Sum Weighted Transaction-Database Model

In this subsection, the authors model a collection of investment portfolios as a one-sum weighted transaction-database *P**, where each attribute in a database record represents an investment-item assigning with a weighting score between 0 and 1 (i.e., the ratio – the amount of assets/funds spent on this portfolio-item to the total amount of assets/funds spent on this portfolio), and the sum of all investment-item scores in a portfolio (database record) is 1. A number of ALPs can be identified in *P** that illustrate such implicative allocating relationships between two sets of investment-items. An ALP mined from *P** can be exemplified as ⟨stock_no.1[0.3] stock_no.2[0.15] bond_no.1[0.2]⟩ ⇒ ⟨stock_no.3[0.05] fund_no.1[0.3]⟩, which can be interpreted as: when people invest 30%, 15% and 20% of their total assets/funds to stock_no.1, stock_no.2 and bond_no.1 together, it is likely that people also invest 5% and 30% of the total assets/funds to stock_no.3 and fund_no.1. Based on the additional information of one-sum item/asset weights, ALPs can be generally applied in real-life investment activities.

Collecting "Meaningful" Portfolios

A collection of portfolios is assumed "meaningful" if each portfolio is collected under some subscriptive conditions. An aspect of the conditions can be described as: each collected portfolio must be "successful" — the realized return of a portfolio exceeds a user-defined return threshold δ, where the lower bound of δ is defined as the return of investing the same amount of assets/funds to a risk free investment-item in the same period of time. Other conditions that may be considered/specified by the user include: (1) each collected portfolio must be invested in a particular time interval; (2) each collected portfolio must contain a particular number of investment-items; (3) each collected portfolio must be produced by a particular portfolio selection technique; (4) each collected portfolio must be produced by a particular financial institution; (5) the risk of each collected portfolio must be less than a user-supplied risk threshold; (6) the investment-items contained in each collected portfolio must be traded in a particular stock exchange; and so forth.

Guiding Future Investment Activities

It can be prospected that ALPs will prove be useful in a range of applications. With respect to portfolio management, mining ALPs from a set of meaningful portfolios can be applied to guide future investment activities. Given a collection of successful portfolios P^*_S that is invested in a particular time interval τ_1 (e.g., all portfolios are purchased on day_1 and sold on day_{90}), where the return threshold δ is suggested to be a percentage ψ times the average realized return of all investment-items involved in P^*_S in τ_1 (noted that in this chapter, ψ is simply determined as 50%), a set of mined ALPs can be treated as a number of candidate portfolios that will be further applied in future investment activities. The quality of each obtained candidate portfolio can be evaluated using a quality threshold μ, where μ is always chosen to be the yearly return of a risk free investment-item.

In Ye, Liu, Yao, Wang, Zhou, and Lu (2002) the yearly returns of two risk free investment-items are determined as: (1) 1.5%—deposit to bank or money market, and (2) 3%—bonds. In this chapter μ is chosen to be 5% in a conservative fashion. Hence a candidate portfolio is "qualified" if its realized return in a "test" time interval τ_2 (noted that the beginning of τ_2 is later than the end of τ_1) is greater than μ in τ_2 (i.e., if the range of τ_2 is three months, μ should be calculated as 5% / 12 * 3 = 1.25%). The overall performance of the proposed application can be measured by the rate of obtaining qualified candidate portfolios, which is: *count* (qualified candidate portfolios) / *count* (all generated candidate portfolios). On the other hand, the overall performance of the obtained candidate portfolios can be further evaluated by their monthly average return, as suggested in Hung, Liang, and Liu (1996).

EXPERIMENTAL RESULTS

In this section, the authors aim to evaluate the effectiveness of the proposed ALP application in portfolio management—a set of ALPs mined from a collection of successful investment portfolios can be treated as a number of candidate portfolios used to guide future investment activities. The evaluation is performed regarding both the rate of obtaining qualified candidate portfolios and the monthly average return of the generated candidate portfolios. All evaluations are obtained using the proposed apriori-ALP and ALP-generation algorithms. Experiments are run on a 1.20 GHz Intel Celeron CPU with 256 MByte of RAM running under Windows Command Processor.

The CSMAR-CSTQR·Database

The CSMAR Data

The experiments are conducted using two sets of possible investment portfolios generated from the CSMAR (China Stock Market & Accounting Research) China Stock Trade and Quote Research Database (CSTQR·Database).[1] In CSMAR eight databases and one system are included, they are China Stock Market Trading Database, China Stock Market Financial Database, China Securities Investment Fund Research Database, China Stock Market Information Disclosure System, China IPO Research Database, China Listed Firm Corporate Governance Research Database, China Listed Firm's Financial Ratios Research Database, China Stock Market Quarterly Report Database and the CSTQR·Database. The CSTQR·Database covers all details of every transaction and related information within every working day, providing data by bid and ask record.

Data Related Stock Exchanges

There are two stock markets in China: Shanghai Stock Exchange (SHSE) and Shenzhen Stock Exchange (SZSE). SHSE was opened in December 1990, whereas SZSE was established in July 1991. The CSTQR·Database involves two types of shares: A-shares and B-shares. The A-shares are domestic investment shares that are issued by Chinese companies, and listed on SHSE and SZSE. The A-shares are denominated in the Chinese money, that is, the RMB. Foreign individuals or institutions are not allowed to directly buy and sell these shares. On the other hand, the B-shares were issued to, and traded by overseas investors only; domestic investors were not able to purchase B-shares before 2001. Since 2001 the B-shares have been opened to domestic investors. The denomination currencies for the B-shares are the US dollar used on SHSE, and the Hong Kong dollar used on SZSE (Chan, Menkveld & Yang, 2003).

Two Simulated Portfolio Collections

Due to the sake of simplicity, for each of SHSE and SZSE only the first 50 listed stocks for A-shares are taken in the period between January 2003 and June 2003 from the CSTQR·Database. It should be noted that the stocks are listed in CSTQR·Database according to their stock IDs (i.e., "600000" represents Shanghai Pudong Development Bank Co., Ltd., "600001" represents Handan Iron & Steel Co., Ltd., "600002" represents Qilu Petroche Mical Company Ltd., etc.), where the stock IDs are issued to each stock without a specific order. For each of SHSE and SZSE, 5,000 successful portfolios are randomly created based on the 50 taken stocks, where each portfolio is limited to contain at least 7 and at most 15 stocks. To decide which stocks should be included in a simulated portfolio, a random procedure is applied. In Algorithm 5, the random

Algorithm 5. The portfolio-simulation procedure

```
Input:  (a) The number of stocks num; // 'num' is decided to be 50
        (b) The min size of a portfolio s; // 's' is decided to be 7
        (c) The max size of a portfolio t; // 't' is decided to be 15
Output: A simulated portfolio Φ;
Begin Algorithm:
(1) Φ := prepare an empty set for holding the selected SHSE/SZSE stocks;
(2) k := 1;
(3) c := 1;
(4) while ((k ≤ num) and (c ≤ t)) do
(5)    r1 := generate a random integer under 5; // including 0 and 5
(6)    r2 := 0;
(7)    if (r1 = 0) then
(8)      r2 ← generate a random integer under 1;
(9)    else if (r1 = 1) then
(10)     r2 ← generate a random integer under 2;
(11)   else if (r1 = 2) then
(12)     r2 ← generate a random integer under 3;
(13)   else if (r1 = 3) then
(14)     r2 ← generate a random integer under 5;
(15)   else if (r1 = 4) then
(16)     r2 ← generate a random integer under 8;
(17)   else
(18)     r2 ← generate a random integer under 13;
(19)   if (r2 = 0) then
(20)     Φ ← Φ ∪ k;
(21)     c ← c + 1;
(22)   k ← k + 1;
(23) end while
(24) if (c < s) then
(25)   return Portfolio-Simulation(num, s, t); // recursive procedure
(26) return (Φ);
End Algorithm
```

procedure that forms a portfolio structure, namely portfolio-simulation, is described.

In portfolio-simulation, the ranges of the random integer variable r_2 are designed to be increased in a Fibonacci pattern (i.e., 0, 1, 1, 2, 3, 5, 8, 13…). It should be noted that the Fibonacci pattern can be substituted by any other patterns. Having an overall structure of the simulated portfolio collection generated by iteratively processing portfolio-simulation, the one-sum weighting score is then assigned to each portfolio (transaction) item. Firstly, an integer ϖ_i is assigned to each item a_i in a transaction T_j, where ϖ_i is randomly chosen from {1, 2, 3, 4, 5}. Secondly, the one-sum weighing score w_i for a_i is then calculated as: $\varpi_i / (\Sigma\{k = 1 \ldots |T_j|\} \varpi_k)$. Two simulated portfolio collections (one for SHSE and another one for SZSE) are named as "shse.D50.N5000.Ifibonacci.W5" and "szse.D50.N5000. Ifibonacci.W5," where "shse"/"szse" specifies the stock exchange, "D" represents the number of stocks taken from a stock exchange, "N" denotes the number of simulated portfolios, "I" indicates the pattern applied in random integer generation in portfolio-simulation, and "W" signifies the size of the random integer set in the process of item weighting.

In both "shse.D50.N5000.Ifibonacci.W5" and "szse.D50.N5000.Ifibonacci.W5," only the successful portfolios are comprised. It is assumed that all portfolios are invested in the time interval between the first trade in January 2003 and the first trade in April 2003. Hence the return of each stock/item taken from the CSTQR-Database can be calculated as $(p_1 - p_0) / p_0$, where p_0 represents the purchasing price (the price of the first trade in January 2003), and p_1 indicates the selling price (the price of the first trade in April 2003). The overall return of a simulated portfolio (transaction) T_j is then calculated as: $\Sigma\{i = 1 \ldots |T_j|\} (w_i * ((p_1 - p_0) / p_0)_i)$. In the described time interval, the average return of the 50 taken stocks is that 13.423% on SHSE and 11.926% on SZSE. Thus the return threshold δ (used to measure whether a simulated portfolio is successful or not) is calculated as: (1) for SHSE, 50% * 13.423% = 6.7115%; and (2) for SZSE, 50% * 11.926% = 5.963% (as commenced in the previous section — ψ = 50%). Table 3 and Table 4 illustrate the first 5 simulated portfolios/transactions in "shse.D50.N5000.Ifibonacci.W5" and "szse.D50.N5000.Ifibonacci.W5". Noted that the integers listed before the square brackets are the stock IDs (i.e., "1" represents "600000" in SHSE and "000001" in SZSE), and the real numbers shown in the square brackets are the stock weights.

Mining ALPs from Simulated Portfolios

The evaluation undertaken used a one-sum weighted support threshold value of 1% and a one-sum weighted confidence threshold value of 75%, as used in (Coenen & Leng, 2004) to generate an extension of ARs, that is, the classification association rules (CARs), which parallels ALPs. The proposed apriori-ALP and ALP-generation algorithms are implemented and run on both simulated portfolio sets. There are 18 ALPs generated from "shse.D50.N5000.Ifibonacci.W5", while 16 ALPs are mined from "szse.D50.N5000.Ifibonacci.W5". In Table 5 and Table 6 the generated ALPs are listed for "shse.D50.N5000.Ifibonacci.W5" and "szse.D50.N5000.Ifibonacci.W5".

Evaluation of the Mined ALPs

ALPs from "shse.D50.N5000.Ifibonacci.W5"

The 18 ALPs mined from "shse.D50.N5000.Ifibonacci.W5" are treated as the candidate portfolios, and tested by investing them in the test time interval: purchasing at the first trade in May 2003 and selling at the first trade in June 2003. In Table 7, the return of each candidate portfolio for the test time interval is shown.

Table 3. The first five transactions in "shse.D50.N5000.Ifibonacci.W5"

T1	1[0.1111111111111111] 3[0.027777777777777776] 13[0.08333333333333333] 14[0.05555555555555555] 15[0.027777777777777776] 17[0.1111111111111111] 22[0.05555555555555555] 23[0.08333333333333333] 28[0.05555555555555555] 31[0.08333333333333333] 32[0.027777777777777776] 33[0.08333333333333333] 38[0.027777777777777776] 39[0.1111111111111111] 44[0.05555555555555555]
T2	1[0.06666666666666667] 2[0.044444444444444446] 7[0.08888888888888889] 11[0.06666666666666667] 12[0.022222222222222223] 13[0.08888888888888889] 15[0.06666666666666667] 16[0.06666666666666667] 8[0.022222222222222223] 19[0.022222222222222223] 22[0.08888888888888889] 23[0.08888888888888889] 24[0.08888888888888889] 25[0.08888888888888889] 27[0.08888888888888889]
T3	10[0.08823529411764706] 11[0.029411764705882353] 2[0.029411764705882353] 13[0.08823529411764706] 19[0.058823529411764705] 5[0.029411764705882353] 26[0.11764705882352941] 27[0.029411764705882353] 28[0.11764705882352941] 30[0.029411764705882353] 32[0.11764705882352941] 35[0.11764705882352941] 38[0.058823529411764705] 40[0.029411764705882353] 41[0.058823529411764705]
T4	6[0.05263157894736842] 9[0.10526315789473684] 11[0.05263157894736842] 12[0.10526315789473684] 20[0.07894736842105263] 21[0.07894736842105263] 23[0.02631578947368421] 25[0.05263157894736842] 26[0.07894736842105263] 27[0.10526315789473684] 28[0.05263157894736842] 29[0.05263157894736842] 32[0.02631578947368421] 33[0.10526315789473684] 34[0.02631578947368421]
T5	1[0.14285714285714285] 2[0.10714285714285714] 3[0.03571428571428571] 5[0.03571428571428571] 6[0.03571428571428571] 7[0.03571428571428571] 10[0.03571428571428571] 11[0.03571428571428571] 12[0.07142857142857142] 20[0.10714285714285714] 24[0.07142857142857142] 27[0.03571428571428571] 28[0.03571428571428571] 32[0.07142857142857142] 34[0.14285714285714285]

Table 4. The first five transactions in "szse.D50.N5000.Ifibonacci.W5"

T1	1[0.0975609756097561] 2[0.07317073170731707] 3[0.04878048780487805] 4[0.0975609756097561] 7[0.04878048780487805] 8[0.0975609756097561] 11[0.0975609756097561] 12[0.024390243902439025] 14[0.0975609756097561] 15[0.04878048780487805] 16[0.04878048780487805] 17[0.07317073170731707] 19[0.07317073170731707] 26[0.024390243902439025] 27[0.04878048780487805]
T2	1[0.05] 3[0.075] 4[0.1] 6[0.025] 7[0.025] 10[0.025] 16[0.075] 17[0.1] 18[0.075] 19[0.075] 20[0.1] 23[0.075] 24[0.1] 25[0.05] 34[0.05]
T3	1[0.08695652173913043] 5[0.06521739130434782] 6[0.06521739130434782] 7[0.043478260869565216] 15[0.043478260869565216] 18[0.08695652173913043] 20[0.06521739130434782] 21[0.06521739130434782] 24[0.043478260869565216] 26[0.06521739130434782] 27[0.08695652173913043] 30[0.08695652173913043] 31[0.08695652173913043] 32[0.021739130434782608] 33[0.08695652173913043]
T4	2[0.10256410256410256] 5[0.05128205128205128] 9[0.02564102564102564] 14[0.10256410256410256] 15[0.07692307692307693] 16[0.05128205128205128] 17[0.07692307692307693] 19[0.05128205128205128] 22[0.10256410256410256] 23[0.10256410256410256] 27[0.07692307692307693] 28[0.02564102564102564] 29[0.05128205128205128] 31[0.05128205128205128] 32[0.05128205128205128]
T5	13[0.11428571428571428] 14[0.08571428571428572] 16[0.08571428571428572] 17[0.05714285714285714] 23[0.08571428571428572] 24[0.05714285714285714] 30[0.08571428571428572] 34[0.11428571428571428] 38[0.02857142857142857] 40[0.11428571428571428] 42[0.08571428571428572] 48[0.08571428571428572]

Table 5. The 18 ALPs mined from the "shse.D50.N5000.Ifibonacci.W5"

ALP No. 1	⟨7[0.333333] 27[0.500003]⟩ ⇒ ⟨4[0.166663]⟩	conf = 0.865954
ALP No. 2	⟨8[0.333331] 20[0.222222]⟩ ⇒ ⟨4[0.444445]⟩	conf = 0.846154
ALP No. 3	⟨8[0.333331] 20[0.222222]⟩ ⇒ ⟨6[0.444445]⟩	conf = 0.813187
ALP No. 4	⟨17[0.333334] 18[0.22222]⟩ ⇒ ⟨23[0.444444]⟩	conf = 0.782609
ALP No. 5	⟨7[0.222222] 21[0.333333]⟩ ⇒ ⟨20[0.444444]⟩	conf = 0.77907
ALP No. 6	⟨11[0.444444] 14[0.333333]⟩ ⇒ ⟨7[0.222222]⟩	conf = 0.772277
ALP No. 7	⟨17[0.333331] 18[0.500002]⟩ ⇒ ⟨3[0.166665]⟩	conf = 0.771739
ALP No. 8	⟨26[0.333333] 31[0.444444]⟩ ⇒ ⟨23[0.222222]⟩	conf = 0.769231
ALP No. 9	⟨9[0.444445] 21[0.333333]⟩ ⇒ ⟨14[0.22222]⟩	conf = 0.767677
ALP No. 10	⟨7[0.222222] 21[0.333333]⟩ ⇒ ⟨8[0.444444]⟩	conf = 0.767442
ALP No. 11	⟨10[0.500002] 12[0.166663]⟩ ⇒ ⟨19[0.333333]⟩	conf = 0.761062
ALP No. 12	⟨17[0.500002] 18[0.333331]⟩ ⇒ ⟨14[0.166665]⟩	conf = 0.76087
ALP No. 13	⟨9[0.444445] 26[0.333333]⟩ ⇒ ⟨11[0.22222]⟩	conf = 0.758621
ALP No. 14	⟨8[0.333331] 20[0.222222]⟩ ⇒ ⟨23[0.444445]⟩	conf = 0.758242
ALP No. 15	⟨2[0.333333] 18[0.5]⟩ ⇒ ⟨10[0.166666]⟩	conf = 0.757282
ALP No. 16	⟨10[0.22222] 25[0.333333]⟩ ⇒ ⟨8[0.444445]⟩	conf = 0.755319
ALP No. 17	⟨18[0.500002] 25[0.333331]⟩ ⇒ ⟨3[0.166665]⟩	conf = 0.75
ALP No. 18	⟨17[0.333334] 18[0.22222]⟩ ⇒ ⟨6[0.444444]⟩	conf = 0.75

Table 6. The 16 ALPs mined from the "szse.D50.N5000.Ifibonacci.W5"

ALP No. 1	⟨3[0.2] 13[0.4] 16[0.3]⟩ ⇒ ⟨19[0.1]⟩	conf = 0.981132
ALP No. 2	⟨3[0.2] 13[0.099998] 17[0.299999]⟩ ⇒ ⟨15[0.400001]⟩	conf = 0.877193
ALP No. 3	⟨7[0.199999] 10[0.400002] 12[0.299998]⟩ ⇒ ⟨11[0.099999]⟩	conf = 0.847458
ALP No. 4	⟨1[0.444445] 28[0.333333]⟩ ⇒ ⟨4[0.22222]⟩	conf = 0.831169
ALP No. 5	⟨6[0.222222] 17[0.333331]⟩ ⇒ ⟨15[0.444445]⟩	conf = 0.821053
ALP No. 6	⟨17[0.333333] 18[0.5]⟩ ⇒ ⟨6[0.166666]⟩	conf = 0.792079
ALP No. 7	⟨5[0.333333] 8[0.444444]⟩ ⇒ ⟨6[0.222222]⟩	conf = 0.785047
ALP No. 8	⟨3[0.333333] 28[0.444444]⟩ ⇒ ⟨16[0.222222]⟩	conf = 0.78481
ALP No. 9	⟨19[0.444444] 23[0.333333]⟩ ⇒ ⟨10[0.222222]⟩	conf = 0.78125
ALP No. 10	⟨13[0.500002] 21[0.333331]⟩ ⇒ ⟨23[0.166665]⟩	conf = 0.77551
ALP No. 11	⟨1[0.333333] 10[0.444444]⟩ ⇒ ⟨6[0.222222]⟩	conf = 0.770833
ALP No. 12	⟨8[0.444444] 15[0.333334]⟩ ⇒ ⟨5[0.22222]⟩	conf = 0.770833
ALP No. 13	⟨4[0.222222] 8[0.333333]⟩ ⇒ ⟨10[0.444444]⟩	conf = 0.769231
ALP No. 14	⟨5[0.22222] 9[0.333334]⟩ ⇒ ⟨8[0.444444]⟩	conf = 0.765306
ALP No. 15	⟨9[0.333333] 10[0.444444]⟩ ⇒ ⟨2[0.222222]⟩	conf = 0.764706
ALP No. 16	⟨1[0.333333] 8[0.444444]⟩ ⇒ ⟨12[0.222222]⟩	conf = 0.761468

Table 7. The returns of the 18 candidate portfolios

No.	Candidate Portfolios	Return (%)
1	⟨7[0.333333] 27[0.500003]⟩ ⇒ ⟨4[0.166663]⟩	4.9111
2	⟨8[0.333331] 20[0.222222]⟩ ⇒ ⟨4[0.444445]⟩	4.5220
3	⟨8[0.333331] 20[0.222222]⟩ ⇒ ⟨6[0.444445]⟩	1.5153
4	⟨17[0.333334] 18[0.22222]⟩ ⇒ ⟨23[0.444444]⟩	1.9325
5	⟨7[0.222222] 21[0.333333]⟩ ⇒ ⟨20[0.444444]⟩	0.8196
6	⟨11[0.444444] 14[0.333333]⟩ ⇒ ⟨7[0.222222]⟩	2.2271
7	⟨17[0.333331] 18[0.500002]⟩ ⇒ ⟨3[0.166665]⟩	1.5136
8	⟨26[0.333333] 31[0.444444]⟩ ⇒ ⟨23[0.222222]⟩	6.0828
9	⟨9[0.444445] 21[0.333333]⟩ ⇒ ⟨14[0.22222]⟩	1.8606
10	⟨7[0.222222] 21[0.333333]⟩ ⇒ ⟨8[0.444444]⟩	2.6530
11	⟨10[0.500002] 12[0.166663]⟩ ⇒ ⟨19[0.333333]⟩	3.2090
12	⟨17[0.500002] 18[0.333331]⟩ ⇒ ⟨14[0.166665]⟩	0.7036
13	⟨9[0.444445] 26[0.333333]⟩ ⇒ ⟨11[0.22222]⟩	3.1105
14	⟨8[0.333331] 20[0.222222]⟩ ⇒ ⟨23[0.444445]⟩	3.0224
15	⟨2[0.333333] 18[0.5]⟩ ⇒ ⟨10[0.166666]⟩	-0.7885
16	⟨10[0.22222] 25[0.333333]⟩ ⇒ ⟨8[0.444445]⟩	2.5576
17	⟨18[0.500002] 25[0.333331]⟩ ⇒ ⟨3[0.166665]⟩	1.3371
18	⟨17[0.333334] 18[0.22222]⟩ ⇒ ⟨6[0.444444]⟩	0.4271
Average		2.6451

The quality threshold μ (used to determine whether a candidate portfolio is qualified or not) has been previously commenced as 5% (yearly). Since the range of the test time interval is one month, the quality threshold μ should be converted as 5% / 12 = 0.4167%. From Table 7 it can be seen that only the candidate portfolio no. 15 (highlighted) shows a return < μ. Thus the overall rate of obtaining qualified candidate portfolios herein is calculated as 17/18 = 94.44%. This very high rate of obtaining qualified candidate portfolios evidences that a set of mined ALPs can be used to guide future investment activities. It is worth giving further consideration to the following: for this test time interval of investment (one month), the average return of the 18 candidate portfolios (each comprises 3 stocks only; together involves 21 SHSE stocks only) is 2.6451%, while the average return of the 50 SHSE stocks is 2.785%; the average return of all candidate portfolios can be realized as high as 94.98% of the average return of the 50 stocks taken from SHSE. Hence the experimental result can be further interpreted as: almost all ALP-based portfolio strategies (candidate portfolios) can produce a return that is greater than the return of a risk free investment-item, and the average return of these strategies is considered high.

ALPs from "szse.D50.N5000.Ifibonacci.W5"

The 16 ALPs mined from "szse.D50.N5000.Ifibonacci.W5" are treated as the candidate portfolios

Table 8. The returns of the 16 candidate portfolios

No.	Candidate Portfolios	Return (%)
1	⟨3[0.2] 13[0.4] 16[0.3]⟩ ⇒ ⟨19[0.1]⟩	4.9060
2	⟨3[0.2] 13[0.099998] 17[0.299999]⟩ ⇒ ⟨15[0.400001]⟩	1.8930
3	⟨7[0.199999] 10[0.400002] 12[0.299998]⟩ ⇒ ⟨11[0.099999]⟩	7.1890
4	⟨1[0.444445] 28[0.333333]⟩ ⇒ ⟨4[0.22222]⟩	6.1988
5	⟨6[0.222222] 17[0.333331]⟩ ⇒ ⟨15[0.444445]⟩	1.4378
6	⟨17[0.333333] 18[0.5]⟩ ⇒ ⟨6[0.166666]⟩	2.1480
7	⟨5[0.333333] 8[0.444444]⟩ ⇒ ⟨6[0.222222]⟩	2.8499
8	⟨3[0.333333] 28[0.444444]⟩ ⇒ ⟨16[0.222222]⟩	5.8392
9	⟨19[0.444444] 23[0.333333]⟩ ⇒ ⟨10[0.222222]⟩	27.9634
10	⟨13[0.500002] 21[0.333331]⟩ ⇒ ⟨23[0.166665]⟩	8.1319
11	⟨1[0.333333] 10[0.444444]⟩ ⇒ ⟨6[0.222222]⟩	2.9985
12	⟨8[0.444444] 15[0.333334]⟩ ⇒ ⟨5[0.22222]⟩	2.4376
13	⟨4[0.222222] 8[0.333333]⟩ ⇒ ⟨10[0.444444]⟩	4.8600
14	⟨5[0.22222] 9[0.333334]⟩ ⇒ ⟨8[0.444444]⟩	1.7432
15	⟨9[0.333333] 10[0.444444]⟩ ⇒ ⟨2[0.222222]⟩	-8.7930
16	⟨1[0.333333] 8[0.444444]⟩ ⇒ ⟨12[0.222222]⟩	5.5649
Average		4.8355

as well, and tested by investing them in the same test time interval as described in the previous subsection. In Table 8, the return of each candidate portfolio for the test time interval is shown.

The value of μ is still taken as 0.4167% that is lasted from the previous subsection. From Table 8 it can be identified that the candidate portfolio no. 15 (highlighted) is the only non-qualified candidate portfolio (return < μ). The overall rate of obtaining qualified candidate portfolios herein is then calculated as 15/16 = 93.75%. This very high rate of obtaining qualified candidate portfolios evidences that a set of mined ALPs can be used to guide future investment activities. It can be further concerned: for this test time interval of investment (one month), the average return of the 16 candidate portfolios (each comprises 3 or 4 stocks only; together involves 21 SZSE stocks only) is 4.8355%, while the average return of the 50 SZSE stocks is 6.3024%; the average return of all candidate portfolios can be found as high as 76.72% of the average return of the 50 stocks taken from SZSE. Hence the experimental result can be further interpreted as: almost all ALP-based portfolio strategies (candidate portfolios) can produce a return that is greater than the return of a risk free investment-item, and the average return of these strategies is considered relatively high.

CONCLUSION AND FUTURE RESEARCH

This chapter is concerned with an investigation of applying a new type of data mining knowledge in portfolio management. This new type of knowledge can be recognized as an extension of the well-established ARs in a one-sum weighting setting. An overview of existing ARM and WARM approaches and/or algorithms was provided in the second section, where three categorizes of ARM algorithms and three categories of WARM approaches were reviewed. A new type of horizontal WARs was proposed in the third section, namely allocating pattern (ALP), which shows a one-sum percentage like item weighting property. A novel algorithm, separated in apriori-ALP and ALP-generation, was presented in the fourth section that effectively extracts ALPs from data. In the fifth section, the introduced ALPs were addressed in portfolio management, where the authors described the possibility of utilising ALPs to guide future investment activities. The experiments were performed in the sixth section, where two sets of simulated portfolios generated from CSMAR-CSTQR Database were taken to mine ALPs. The mined ALPs were then treated as the candidate portfolios, and tested by investing them in a test (later) time interval. The experimental result shows: a very high percentage of the mined ALPs (94.44% for the first set, and 93.75% for the second set) can produce a return that is greater than the return of a risk-free investment-item. With respect to the good evaluation performance, the effectiveness of the proposed ALP application in portfolio management can be demonstrated. In a further consideration, the average return of all candidate portfolios (each comprises 3 or 4 stocks only), for the test time interval of investment (one month), can be realized as a relatively high percentage (94.98% for the first set, and 76.72% for the second set) of the average return of all stock-items (taken from the CSTQR Database) comprised in each simulated portfolio set. Therefore the overall experimental result can be further interpreted as: it seems that almost all ALP-based portfolio strategies make a relatively high profit.

Future research is suggested to run more experiments on a wide range of stock market data, and conclude whether ALPs can be widely applied to guide future investment activities or not. Other obvious directions for future research include: finding other types of data mining knowledge based on ARM/WARM; investigating the improved algorithms of mining ALPs from data; applying ALPs in other areas; and so forth.

ACKNOWLEDGMENT

The authors would like to thank Professor Paul Leng, Dr. Robert Sanderson, and Dr. Mark Roberts of the Department of Computer Science at the University of Liverpool for their support with respect to the work described here.

REFERENCES

Agrawal, R., Imielinski, T., & Swami, A. (1993). Mining association rules between sets of items in large databases. In P. Buneman & S. Jajdia (Eds.), *Proceedings of the 1993 ACM SIGMOD International Conference on Management of Data* (pp. 207-216). New York: ACM Press.

Agrawal, R. & Srikant, R. (1994). Fast algorithms for mining association rules. In J. B. Bocca, M. Jarke & C. Zaniolo (Eds.), *Proceedings of the 20th International Conference on Very Large Data Bases* (pp. 487-499). San Francisco: Morgan Kaufmann Publishers.

Ali, K., Manganaris, S., & Srikant, R. (1997). Partial classification using association rules. In D. Heckerman, H. Mannila, D. Pregibon & R. Uthurusamy (Eds.), *Proceedings of the Third International Conference on Knowledge Discovery and Data Mining* (pp. 115-118). Menlo Park, CA: AAAI Press.

Berry, M. J. A. & Linoff, G. (1997). *Data mining techniques for marketing, sales, and customer support*. New York: John Wiley & Sons, Inc.

Bodie, Z., Kane, A., Marcus, A. J., & Ryan, P. J. (2003). *Investments (Fourth Canadian Edition)*. Toronto, ON (Canada): McGraw-Hill Ryerson Limited.

Bramer, M. (2007). *Principles of data mining—Undergraduate topics in computer science*. London, UK: Springer-Verlag.

Brin, S., Motwani, R., Ullman, J. D., & Tsur, S. (1997). Dynamic itemset counting and implication rules for market basket data. In J. Peckham (Ed.), *Proceedings of the 1997 ACM SIGMOD International Conference on Management of Data* (pp. 255-264). New York: ACM Press.

Burdick, D., Calimlim, M., & Gehrke, J. (2001). MAFIA: A maximal frequent itemset algorithm for transactional databases. In *Proceedings of the 17th International Conference on Data Engineering* (pp. 443-452). Los Alamitos, CA: IEEE Computer Society Publications.

Cai, C. H., Fu, A. W. C., Cheng, C. H., & Kwong, W. W. (1998). Mining association rules with weighted items. In B. Eaglestone, B. C. Desai & J. Shao (Eds.), *Proceedings of the 1998 International Database Engineering and Application Symposium* (pp. 68-77). Los Alamitos, CA: IEEE Computer Society Publications.

Chan, K. A., Menkveld, A. J., & Yang, Z. (2003). *Evidence on the foreign share discount puzzle in China: Liquidity or information asymmetry?* (Working Paper). Hong Kong, China: University of Science and Technology (HKUST).

Coenen, F. & Leng, P. (2001). Optimising association rule algorithms using itemset ordering. In M. Bramer, F. Coenen & A. Preece (Eds.), *Research and Development in Intelligent Systems XVIII–Proceedings of the Twenty-first SGES International Conference on Knowledge Based Systems and Applied Artificial Intelligence* (pp. 53-66). London, UK: Springer-Verlag.

Coenen, F. & Leng, P. (2002). Finding association rules with some very frequent attributes. In T. Elmaa, H. Mannila & H. Toivonen (Eds.), *Principles of Data Mining and Knowledge Discovery – Proceedings of the 6th European Conference on Principles and Practice of Knowledge Discovery in Databases* (pp. 99-111). Berlin Heidelberg, Germany: Springer-Verlag.

Coenen, F. & Leng, P. (2004). An evaluation of approaches to classification rule selection. In *Proceedings of the 4th IEEE International Conference on Data Mining* (pp. 359-362). Los Alamitos, CA: IEEE Computer Society Publications.

Coenen, F., Goulbourne, G., & Leng, P. (2001). Computing association rules using partial totals. In L. D. Raedt & A. Siebes (Eds.), *Principles of Data Mining and Knowledge Discovery – Proceedings of the 5th European Conference on Principles and Practice of Knowledge Discovery in Databases* (pp. 54-66). Berlin Heidelberg, Germany: Springer-Verlag.

Coenen, F., Leng, P., & Ahmed, S. (2004). Data structure for association rule mining: T-tree and p-tree. *IEEE Transactions on Knowledge and Data Engineering, 16*(6), 774-778.

Coenen, F., Leng, P., & Goulbourne, G. (2004). Tree structures for mining association rules. *Journal of Data Mining and Knowledge Discovery, 8*(1), 25-51.

Cuthbertson, K. & Nitzsche, D. (2001). *Investments: Spot and derivatives markets*. Chichester, West Sussex, UK: John Wiley & Sons, Ltd.

Damodaran, A. (2001). *Corporate finance theory and practice* (2nd ed.). New York: John Wiley & Sons, Inc.

Dong, G. & Li, J. (1999). Efficient mining of emerging patterns: Discovering trends and differences. In *Proceedings of the Fifth ACM SIGKDD International Conference on Knowledge Discovery and Data Mining* (pp. 43-52). New York: ACM Press.

El-Hajj, M. & Zaiane, O. R. (2003). Inverted matrix: Efficient discovery of frequent items in large datasets in the context of interactive mining.

In L. Getoor, T. E. Senator, P. Domingos & C. Faloutsos (Eds.), *Proceedings of the Ninth ACM SIGKDD International Conference on Knowledge Discovery and Data Mining* (pp. 109-118). New York: ACM Press.

Enke, D. & Thawornwong, S. (2005). The use of data mining and neural networks for forecasting stock market returns. *Expert Systems with Applications, 29*(2005), 927-940.

Farrell, M. (2006). Create a diversified portfolio. *©2006 Path to Investing—Leading the way to financial knowledge®*. New York: Lightbulb Press, Inc.

Gouda, K. & Zaki, M. J. (2001). Efficiently mining maximal frequent itemsets. In N. Cercone, T. Y. Lin & X. Wu (Eds.), *Proceedings of the 2001 IEEE International Conference on Data Mining* (pp. 163-170). Los Alamitos, CA: IEEE Computer Society Publications.

Han, J. & Kamber, M. (2001). *Data mining concepts and techniques*. San Francisco: Morgan Kaufmann Publishers.

Han, J. & Kamber, M. (2006). *Data mining concepts and techniques* (2nd ed.). San Francisco: Morgan Kaufmann Publishers.

Han, J., Pei, J., & Yin, Y. (2000). Mining frequent patterns without candidate generation. In W. Chen, J. F. Naughton & P. A. Bernstein (Eds.), *Proceedings of the 2000 ACM SIGMOD International Conference on Management of Data* (pp. 1-12). New York: ACM Press.

Hand, D., Mannila, H., & Smyth, P. (2001). *Principles of data mining*. Cambridge: MIT Press.

Hidber, C. (1999). Online association rule mining. In A. Delis, C. Faloutsos & S. Ghandeharizadeh (Eds.), *Proceedings of the 1999 ACM SIGMOD International Conference on Management of Data* (pp. 145-156). New York: ACM Press.

Ho, K. & Robinson, C. (2001). *Personal financial planning* (3rd ed.). North York, ON (Canada): Captus Press Inc.

Holsheimer, M., Kersten, M. L., Mannila, H., & Toivonen, H. (1995). A perspective on databases and data mining. In U. M. Fayyad & R. Uthurusamy (Eds.), *Proceedings of the First International Conference on Knowledge Discovery and Data Mining* (pp. 150-155). Menlo Park, CA: AAAI Press.

Houtsma, M. & Swami, A. (1995). Set-oriented mining of association rules in relational databases. In P. S. Yu & A. L. Chen (Eds.), *Proceedings of the Eleventh International Conference on Data Engineering* (pp. 25-33). Los Alamitos, CA: IEEE Computer Society Publications.

Hung, S.-Y., Liang, T.-P., & Liu, V. W.-C. (1996). Integrating arbitrage pricing theory and artificial neural networks to support portfolio management. *Decision Support Systems, 18*(1996), 301-316.

John, G. H., Miller, P., & Kerber, R. (1996). Stock selection using rule induction. *IEEE Expert, 11*(5), 52-58.

Kohara, K., Ishikawa, T., Fukuhara, Y., & Nakamura, Y. (1997). Stock price prediction using prior knowledge and neural networks. *International Journal of Intelligent Systems in Accounting, Finance and Management, 6*(1), 11-22.

Kovalerchuk, B. & Vityaev, E. (2000). *Data mining in finance: Advances in relational and hybrid Methods*. New York: Kluwer Academic Publisher.

Lazo, J. G., Maria, M., Vellasco, R., Aurelio, M., & Pacheco, C. (2000). A hybrid genetic-neural system for portfolio selection and management. In *Proceedings of the 7th International Conference on Engineering Applications of Neural Networks*. Kingston Upon Thames, UK: Kingston University.

Lin, D.-I., & Kedem, Z. M. (1998). Pincer search: A new algorithm for discovering the maximum frequent set. In H.-J. Schek, F. Saltor, I. Ramos & G. Alonso (Eds.), *Advances in Database Technology – Proceedings of the 6th International Conference on Extending Database Technology* (pp. 105-119). Berlin Heidelberg, Germany: Springer-Verlag.

Liu, J., Fan, Y., Wang, K., & Han, J. (2002). Mining frequent item sets by opportunistic projection. In *Proceedings of the Eighth ACM SIGKDD International Conference on Knowledge Discovery and Data Mining* (pp. 229-238). New York: ACM Press.

Lu, S., Hu, H., & Li, F. (2001). Mining weighted association rules. *Intelligent Data Analysis, 5*(2001), 211-255.

Mannila, H., Toivonen, H., & Verkamo, A. I. (1994). Efficient algorithms for discovering association rules. In U. M. Fayyad & R. Uthurusamy (Eds.), *Knowledge Discovery in Databases: Papers from the 1994 AAAI Workshop* (pp. 181-192). Menlo Park, CA: AAAI Press.

Miller, H. J. & Han, J. (2001). *Geographic data mining and knowledge discovery.* Bristol, PA: Taylor & Francis, Inc.

Mirkin, B. & Mirkin, B. G. (2005). *Clustering for data mining: A data recovery approach.* Virginia Beach, VA: Chapman & Hall / CRC.

Neftci, S. N. (2004). *Principles of financial engineering.* Burlington, MA: Elsevier Academic Press.

Park, J. S., Chen, M.-S., & Yu, P. S. (1995). An effective hash based algorithm for mining association rules. In M. J. Carey & D. A. Schneider (Eds.), *Proceedings of the 1995 ACM SIGMOD International Conference on Management of Data* (pp. 175-186). New York: ACM Press.

Pei, J., Han, J., & Mao, R. (2000, May). CLOSET: An efficient algorithm for mining frequent closed itemsets. In D. Gunopulos & R. Rastogi (Eds.), *Proceedings of the 2000 ACM SIGMOD Workshop on Research Issues in Data Mining and Knowledge Discovery* (pp. 21-30), Dallas, TX.

Quah, T. S. & Srinivasan, B. (1999). Improving returns on stock investment through neural network selection. *Expert Systems with Applications, 17*(4), 295-301.

Quinlan, J. R. (1993). *C4.5: Programs for machine learning.* San Francisco: Morgan Kaufmann Publishers.

Raghavan, S. N. R. (2005). Data mining in e-commerce: A survey. *Sadhana, 30*(2&3), 275-289.

Roberto, J. & Bayardo, Jr. (1998). Efficiently mining long patterns from databases. In L. M. Hass, & A. Tiwary (Eds.), *Proceedings of the 1998 ACM SIGMOD International Conference on Management of Data* (pp. 85-93). New York: ACM Press.

Ross, S. (1976). The arbitrage theory of capital asset pricing. *Journal of Economic Theory, 13*, 341-360.

Rymon, R. (1992). Search through systematic set enumeration. In B. Nebel, C. Rich & W. R. Swartout (Eds.), *Proceedings of the 3rd International Conference on Principles of Knowledge Representation and Reasoning* (pp. 539-550). San Francisco: Morgan Kaufmann Publishers.

Savasere, A., Omiecinski, E., & Navathe S. (1995). An efficient algorithm for mining association rules in large databases. In U. Dayal, P. M. D. Gray & S. Nishio (Eds.), *Proceedings of the 21st International Conference on Very Large Data Bases* (pp. 432-444). San Francisco: Morgan Kaufmann Publishers.

Tao, F., Murtagh, F., & Farid, M. (2003). Weighted association rule mining using weighted support and significance framework. In L. Getoor, T. E. Senator, P. Domingos & C. Faloutsos (Eds.), *Proceedings of the Ninth ACM SIGKDD International Conference on Knowledge Discovery and Data*

Mining (pp. 661-666). New York: ACM Press.

Thuraisingham, B. (1999). *Data mining: Technologies, techniques, tools, and trends*. Boca Raton, FL: CRC Press LLC.

Toivonen, H. (1996). Sampling large databases for association rules. In T. M. Vijayaraman, A. P. Buchmann, C. Mohan & N. L. Sarda (Eds.), *Proceedings of the 22nd International Conference on Very Large Data Bases* (pp. 134-145). San Francisco: Morgan Kaufmann Publishers.

Tseng, C.-C. (2004). Portfolio management using hybrid recommendation system. In *Proceedings of the 2004 IEEE International Conference on e-Technology, e-Commerce, and e-Services* (pp. 202-206). Los Alamitos, CA: IEEE Computer Society Publications.

Wang, H. & Weigend, A. S. (2004). Data mining for financial decision making. *Decision Support Systems, 37*(2004), 457-460.

Wang, W. & Yang, J. (2005). *Mining sequential patterns from large data sets*. Secaucus, NJ: Springer-Verlag New York, Inc.

Wang, W., Yang, J., & Yu, P. (2000). Efficient mining of weighted association rules (WAR). In *Proceedings of the Sixth ACM SIGKDD International Conference on Knowledge Discovery and Data Mining* (pp. 270-274). New York: ACM Press.

Wang, J., Han, J., & Pei, J. (2003). CLOSET+: Searching for the best strategies for mining frequent closed itemsets. In: L. Getoor, T. E. Senator, P. Domingos & C. Faloutsos (Eds.), *Proceedings of the Ninth ACM SIGKDD International Conference on Knowledge Discovery and Data Mining* (pp. 236-245). New York: ACM Press.

Wang, J. T. L., Zaki, M. J., Toivonen, H. T. T., & Shasha, D. E. (2005). *Data mining in bioinformatics*. London, UK: Springer-Verlag.

Ye, Z., Liu, X., Yao, Y., Wang, J., Zhou, X., Lu, P., & Yao, J. (2002). An intelligent system for personal and family financial service. In L. Wang, J. C. Rajapakse, K. Fukushima, S.-Y. Lee & X. Yao (Eds.), *Proceedings of the 9th International Conference on Neural Information Processing* (Vol. 5, pp. 2325-2327). Los Alamitos, CA: IEEE Computer Society Publications.

Yu, L., Wang, S., & Lai, K. K. (2005). Mining stock market tendency using GA-based support vector machines. In X. Deng & Y. Ye (Eds.), *Proceedings of the First International Workshop on Internet and Network Economics* (pp. 336-345). Berlin Heidelberg, Germany: Springer-Verlag.

Zaki, M. J. & Hsiao, C.-J. (2002) CHARM: An efficient algorithm for closed itemset mining. In R. L. Grossman, J. Han, V. Kumar, H. Mannila & R. Motwani (Eds.), *Proceedings of the Second SIAM International Conference on Data Mining* (Part IX No. 1). Philadelphia, PA: SIAM.

Zaki, M. J., Parthasarathy, S., Ogihara, M., & Li, W. (1997). New algorithms for fast discovery of association rules. In D. Heckerman, H. Mannila, & D. Pregibon (Eds.), *Proceedings of the Third International Conference on Knowledge Discovery and Data Mining* (pp. 283-286). Menlo Park, CA: AAAI Press.

Zhang, D. & Zhou, L. (2004). Discovery golden nuggets: Data mining in financial application. *IEEE Transactions on Systems, Man, and Cybernetics – Part C: Applications and Reviews, 34*(4), 513 –522.

ENDNOTE

[1] CSMAR-CSTQR·Database is provided by GuoTaiAn Information Technology Company, Shenzhen, China. (http://www.chinagtait.com)

Chapter VIII
Application of Data Mining Algorithms for Measuring Performance Impact of Social Development Activities

Hakikur Rahman
Sustainable Development Networking Foundation, Bangladesh

ABSTRACT

Social development activities are flourishing in diversified branches of society endeavor, despite numerous hurdles inflicting on their ways that are truly cross-sectoral. They vary from providing basic human services, as such education, health, and entrepreneurship to advance maneuvers depending on the demand at the outset. However, while talking about discovering true success cases around the globe, recapitulating their thoroughfares to accumulate knowledge; and foremost, utilizing newly emerged information technology methods to archive and disseminate model cases, not many stand on their own. This has happened due for many reasons, and a few of them are; improper program design, inaccurate site selection, incorrect breakeven analysis, insufficient supply of funding, unbalanced manpower selection, inappropriate budget allocation, inadequate feedback and monitoring. Apart from them, there are many hidden parameters that are not even visible. Furthermore, these visible parameters (including the invisible) are intricately intermingled to one another in such a way that lagging of one derailed the whole project and eventually the program fail. Not surprisingly, all of these parameters depend on data and information on implemented programs or projects of which they mostly lack. Thus, lack of data and information related to their appropriateness (or inappropriateness), made them failure projects, despite devoted efforts by the implementers, in most cases. This chapter has tried to focus on data mining applications and their utilizations in formulating performance-measuring tools for social development activities. In this context, this chapter has provided justifications to include data mining algorithm to establish monitoring and evaluation tools for various social development applications. Specifically, this chapter gave in-depth analytical observations to establish knowledge centers with various approaches and finally it put forward a few research issues and challenges to transform the contemporary human society into a knowledge society.

INTRODUCTION

All information pertaining to a successful organization is truly its asset. Information, such as client lists, vendor lists, product details, employee information, and corporate strategy, is invaluable. Without appropriate feeding of information, a business cannot operate properly (Utimaco, 2005). This is potentially true for any sort of ventures that may vary from providing services to the scientific community or academics or civil society or individuals. However, to take an intelligent decision, the information needs to be processed and compiled.

Data mining is a method of collecting and processing of data and eventually assisting to take knowledgeable decision. In today's modern information based environment, data mining is day by day coming at the front and beginning to acquire more and more attention. Because data mining is all about acquisition, assessment and analysis, and by automatic or semiautomatic means huge or small, all quantities of data can help to uncover meaningful patterns and rules. These patterns and schemes help enterprises improve their marketing, sales and customer support operations to better understand their end users. Over the years, corporate houses have accumulated very large databases from applications such as enterprise resource planning (ERP), client relationship management (CRM), or other operational systems. People believe that there are untapped values hidden inside these data, and data mining techniques can help these patterns out of this data.[1]

Currently data are being collected and accumulated across a wide variety of fields at an exaggerated pace. Data are no more a rigid matter for an entrepreneurship, or an organization, but have became an intrinsic part of any management process and most dynamic in nature. For these reasons, data mining algorithms are imperative to researches in the aspect of making intelligent decisions through data mining. To cope up with this new arena of research, there is an urgent need for a new generation of computational theories and tools to assist humans in extracting useful information (knowledge) from the rapidly growing volumes of digital data.

At the same time, data mining and knowledge discovery in databases have been attracting a significant amount of research, industry, and media attention (Boulicaut, Esposito, Giannotti & Pedreschi, 2004; Bramer, 1999; Fayyad, Piatetsky-Shapiro & Smyth, 1996; Freitas, 2002; Kargupta & Chen, 2001; Kloesgen & Zythkow, 2002; Larose, 2004; Miller & Han, 2001). This chapter provides a brief overview of this emerging field, clarifying how data mining and knowledge discovery in databases are related to each other, and especially focused on application of data mining algorithms in establishing social development management systems. In this aspect, this chapter intends to illustrate a few real-world applications, but specifically focused to data mining algorithms; challenges involved in those applications of knowledge discovery, including contemporary and future research directions in the arena of establishing knowledge centers to assist the society for taking intelligent decision.

Along the way, this chapter tries to provide a few hints on data mining algorithms and put forward a few illustrations with which data mining algorithms may be applied for making decision support systems. Furthermore, this chapter has endeavored to justify on several models on establishment of knowledge centers. The author finds that knowledge centers (information center, kiosks, community information centers) are being established in many countries during the last decade with aspirations for assisting the grass roots communities. However, until now, not many researches are being conducted to measure their impacts in the society, or any cost benefit analyses have carried out.

In recent years, many countries have seen evolution of telecenters in various forms, ranging from kiosks, information centers, community information centers, village information centers to multipurpose village information centers, knowledge centers, and the like. However, due to proper implementation, management, and monitoring, most of them failed in many countries and donors have withdrawn supporting them despite enormous demands exist in various parts of the world. Varying from sub-Saharan countries with minimum information access to South Asian countries with lack of information management framework, many effort of establishing these information centers remain exorbitant. Furthermore, most of the telecenters did not maintain any records on their clients, or their habits, nor the reasons for their failures, or any analytical studies. Given these perspectives, this chapter tries to devise a few algorithms to formulate the measuring criteria of knowledge centers, utilizing data mining. Finally it discusses a few challenges with some hints on future research directives before concluding.

BACKGROUND

In contrast to heuristics (which contain general recommendations based on statistical evidence or theoretical reasoning), algorithms are comprised of completely defined, finite sets of steps, operations, or procedures to produce a particular outcome. For example, with a few certain exceptions, all computer programs, mathematical formulas, and (ideally) health prescriptions and food recipes are algorithms.[2] Algorithms are based on finite patterns and occurrences in any incidents, and the outcome could be quantified using mathematical formulations (Abbass, Sarker & Newton, 2002; Adamo, 2001; Kantardzic, 2002;Yoon & Kerschberg, 1993).

Historically, the concept of finding useful patterns in data has been given a variety of names, including data mining, knowledge extraction, information discovery, information harvesting, data archeology, data warehousing, data repository, or data pattern processing. Furthermore, the term data mining has been mainly used by statisticians, data analysts, and management information system (MIS) communities. Though it has also gained popularity in the database field (Chakrabarti, 2002; Fayyad, Piatetsky-Shapiro & Smyth, 1996; Hand, Mannila & Smyth, 2001; Liu & Motoda, 1998a, 1998b; Pal & Mitra, 2004; Perner & Petrou, 1999; Pyle, 1999), but development partners and researchers in the field of implementing numerous development projects remain aloof of utilizing data mining techniques to preserve their data or content, and as well as utilizing data mining algorithms to derive their project outcomes. Data remain as critical means of project evaluation essence and data processing possesses as a simple means of conversion of raw data into tables or charts. The hidden pattern within the data remains hidden and transformation of those data into knowledge element could not gain concrete momentum until now.

Furthermore, there has not been any mathematical formulation derived that can take care the transformation of data into knowledge and at the same time, measure their impact in the society, or quantify the impact of data transformation. The traditional method of turning data into knowledge relies on manual analysis and interpretation. For example, in the health-care industry, it is common for physicians or specialists to periodically analyze current trends and changes in health-care data. The specialists then provide a report detailing the analysis to the authority; and ultimately this report becomes the basis for future decision making and planning for health-care management. In a totally different category of application, planetary geologists sift through remotely sensed images of planets and asteroids by carefully locating and cataloging such geologic objects of interest as impact craters.

Perhaps it can be a village information center, established at a very remote corner of a geographically dispersed region. There has not been evolved many readymade formulas, algorithms, hypothesis, or any measuring criteria to recognize their pattern of growth and implementation, nature of operation, sustainability of their existence, or replication of success cases in applicable states or stages.

Be it science, research, marketing, finance, health care, retail shop, community center, or any other field, the classical approach to data analysis relies fundamentally on one or more analysts becoming intimately familiar with the data ad serving as an interface between the data and the users and end products (Berthold & Hand, 1999; Fayyad, Piatetsky-Shapiro & Smyth, 1996; Maimon & Last, 2000; Mattison, 1997). Nevertheless, in recent years many entrepreneurs are formulating measuring criteria that include marketing, finance (especially investment), fraud detection, data access, data cleaning, manufacturing, telecommunications, and Internet agents.

In this chapter, a few data mining algorithms based on rough set theory (RS) (Cox, 2004; Curotto & Ebecken, 2005; Kantardzic, 2002; Myatt, 2006; Nanopoulos, Katsaros & Manolopoulos, 2003; Thuraisingham, 1999; Zhou, Li, Meng & Meng, 2004) are included which are used to extract decision-making rules from dataset. Rough set theory provides a neat methodology to formalize and calculate the results for data mining problems. In the early 1980's Z. Pawlak, in cooperation with other researchers developed the rough set data analysis (RSDA) (Pawlak, 1982). As recommended by its main adage "let the data speak for themselves", RSDA tried to distinguish internal characteristics of a data set, such as categorization, dependency, and association rules, without invoking external metrics and judgment (Drewry et al., 2002).

MAIN THRUST

The output of a data mining algorithm is typically a pattern or a set of patterns that are valid in the given data. A pattern is defined as a statement (expression) in a given language, that describes (relationships among) the facts in a subset of the given data, and is in some sense simpler than the enumeration of all the facts in the subset. (Drewry et al, 2002, p. 2)

A given data mining algorithm usually depends on a built-in class of patterns, and the particular language of patterns considered depends on the characteristics of given data (the attributes and their values).

This section constitutes the main thrust of the chapter and includes a few models/patterns of data mining algorithms that would be used to deduce possible measuring criteria of social development processes. However, to remain within the context of the chapter, specifically, the algorithms related

Table 1. Microcredit loan seekers' information

Loan-seekers ID	Debt level	Income level	Employment status	Credit risk	Remarks
1	High	High	Self-Employed	Bad	
2	High	High	Salaried	Bad	
3	High	Low	Self-Employed	Bad	
4	High	Low	Salaried	Bad	
5	Low	High	Self-Employed	Bad	Accepted
6	Low	High	Salaried	Bad	Accepted
7	Low	Low	Self-Employed	Bad	
8	Low	Low	Salaried	Bad	
9	High	High	Self-Employed	Good	Accepted
10	High	High	Salaried	Good	Accepted
11	High	Low	Self-Employed	Good	
12	High	Low	Salaried	Good	
13	Low	High	Self-Employed	Good	Accepted
14	Low	High	Salaried	Good	Accepted
15	Low	Low	Self-Employed	Good	
16	Low	Low	Salaried	Good	Accepted

to establishment of knowledge centers have been elaborated, with apparent hints to a few other types of development activities.

Data mining for association rules is an useful method for analyzing data that describe transactions, lists of items, unique phrases (in text mining), and so forth. Generally, association rules that take the form *If Body then Head*, where body and head stand for simple codes, text values, items, consumer choices, phrases, and so forth, or the conjunction of codes and text values and the like. (e.g., if (debt=high and age<35 and repayment=high) then (risk=high and insurance=high); here the logical conjunction before the then would be the body, and the logical conjunction following the then would be the head of the association rule). Based on some user-defined "threshold" values for rule, the apriori algorithm (Agrawal, Imielinski & Swami, 1993; Agrawal & Srikant, 1994; Pei, Han & Lakshmanan, 2001; Witten & Frank, 2005) is a popular and efficient algorithm for deriving such association rules from large data sets.[3]

In this context, the decision tree algorithm would probably be the most popular technique for predictive modeling. The following example explains some of the basics of the decision tree algorithms. Table 1 shows a set of NGO data that could be used to predict credit risk. In this example, fictionalized information was generated on loan seekers that included debit level, income level, what type of employment they had and whether they were a good or bad credit risk.

In the example illustrated in Figure 1, the decision tree algorithm might determine that the most significant attribute for predicting credit risk is debt level. The first split in the decision tree is, therefore, made on debt level. One of the two new nodes (debt = low) is a leaf node, containing two cases with bad credits and three cases with good credit. In this example, a high debt level is a perfect predictor of a bad credit risk. The other

Figure 1. A partial decision tree derived that might be created from the Table 1.

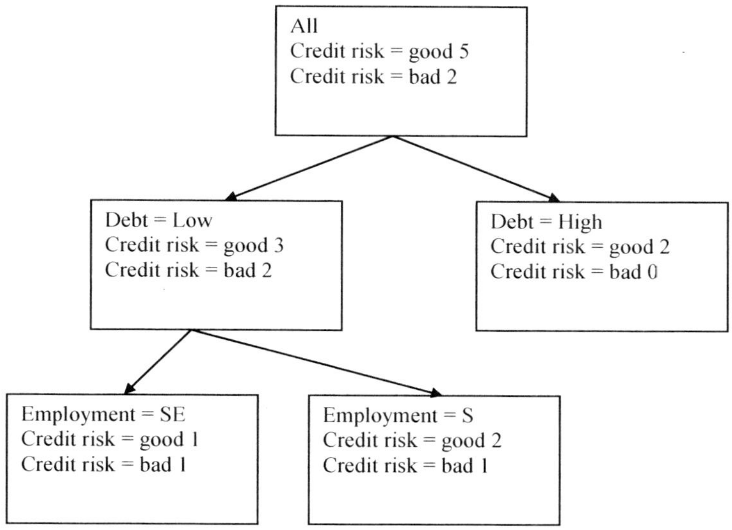

Table 2. A combination of case table and nested table

Customer ID	Age group	Martial Status	Wealth group	Product Purchases	
				Product	Quantity
1	c	M	B	Washing machine	1
				TV	1
				Shampoo	2
2	e	S	C	Diet Coke	12
				TV	1
				Jelly	3
				Cake	2
3	b	M	A	Coke	3
				Cake	1
				Jelly	1

Age group: a- below 15, b- 15-20, c- 21-26, d- 27-32, e- 33-38, f- 39-44, g- 45-50, h- 51-56, i- 57-62, j- above 62; Martial status: M- married, S- separated, D- divorced, U- unmarried; Wealth group: A- have less than \$50,000, B- between \$50,000-250,000, C- between \$251,000-450,000, D- above \$451,000; the divisions are fictitious and their actual divisions depend on decision of the analyst and implementer.

node (debt = high) is still mixed, having two good credits and zero bad credit case.

Departmental stores may use data mining to understand customer's behavior, sale trend, market behavior, and predict market strategy. This can be done using the following table. Table 2 includes two forms of tables—case table and nested table. A case table contains the case information related to the non-nested part of the data, and a nested table contains information related to the nested part of the data. In the following table, there are two input tables to the mining model. One table contains information about customer demographics. It is a case table. The other table contains information

Table 3. A table showing information for predictive marketing

Question number	Question (Data mining algorithm)
1	Identifying those customers that are most likely depart based on customer demographical information (Decision tree without nested table)
2	Grouping heterogeneous customers into subgroups based on customer profile to generate a mailing list for marketing purposes (Clustering without nested table)
3	Finding the list of other products that the customer may be interested in, based on the products the customer has purchased (Cross-selling using decision tree with nested table)
4	Grouping customers into more or less homogeneous groups based on the customer profile and the list of banking products they have subscribed to (Clustering with nested table)

Figure 2. Data mining algorithm for a supermarket

```
Supermarket "My_Choice_DT_Nonnested" Execute:
CREATE MINING MODEL [My_Choice_DT_Nonnested]
([Cust_ID] LONG KEY,
[Income] DOUBLE CONTINUOUS,
[Other_Income] DOUBLE CONTINUOUS,
[Loan] DOUBLE CONTINUOUS,
[Age_Group] DOUBLE CONTINUOUS,
[Area_Residence] TEXT DISCRETE,
[Home_Years] DOUBLE CONTINUOUS,
[Value_House] DOUBLE CONTINUOUS,
[Home_Type] TEXT DISCRETE,
[Insured] TEXT DISCRETE,
[Type_of_Insurance] TEXT DISCRETE,
[Education_Level] TEXT DISCRETE,
[Others] depends
[Leave_Yes_No] TEXT DISCRETE PREDICT)
USING Any_Decision_Trees
```

about customer purchases. It is a nested table. In database technology, a nested table is similar to a transaction table.

In the example, age group division may be made more broad sacrificing accuracy of the result, though smaller age groups segregation results in complicated algorithms. This applies to other parameters too.

To illustrate another example of data mining, hidden patterns inside data have been considered. It is a fact that, data mining finds hidden patterns inside datasets, and these patterns can be used to solve many business problems. The following table presents a few business questions that are difficult to answer without data mining, and at the same time answers to these questions are essential for making decisions on predictive marketing (Ville, 2001; Ville, 2006; Weiss & Indurkhya, 1997).

Fields for Table 3 could be Cust_ID, Income, Other_Income, Loan, Age_Group, Area_Residence, Home_Years, Value_House, Home_Type, Insured, Type_of_Insurance, Education_Level, Leave_Yes_No, and others.

The algorithm could use the CREATE statement for data mining application shown in Figure 2.

Association rule mining is another fundamental technique in data mining. In some real-life applications, for example, market basket analysis in super market chain stores, data sets can be too large for manual analysis, and potentially valuable relations among attributes may not be evident at a glance. An association rule-mining algorithm can find frequent patterns (sets of database attributes) in a given data set and generate association rules among database attributes. For example, some items can be frequently sold together, for example, milk and cereal, or bread and butter. Such items can be displayed together to improve the convenience of shopping. Association rule mining is generally be applicable to those applications in which the data set is large and it is useful to find frequent patterns and their associations, for example, market basket analysis, medical research, and intrusion detection.

Similarly, algorithms may be devised for various other social activities like, readymade garments databank (bridging the gap between developed and developing countries), NGO networks engaged in social development works, skill and capacity development databank (migration of skilled workers), jobs databank (for youths and jobless), online blood bank (during emergencies and disasters), and microcredit databank for the overall benefit of the society. Now, a case study on knowledge centers will be discussed in the next subsection.

Case Study: Knowledge Center

In recent years, information centers (designating them as knowledge centers) have been established in many countries with great enthusiasms. From developed to developing countries, they have been highly appreciated by not only the development partners, but also by all members of the communities. They evolved as telecenters, community information centers, cyber centers, village information centers, kiosks, and other familiar names as accepted by the communities. Depending on their connectivity to the Internet, many remain connected, or many remain off-line, providing various ICT supports to the community. Furthermore, depending on the availability of the Internet they use VSAT (SCPC, MCPC, TDMA, FTDMA), radio (microwave; mostly 2.4GHz, 5.8GHz free frequency), broadband (DSL, ADSL, SDSL, fiber, ISDN), dial-up (mainly PSTN) and other varieties of LAN/WAN formation. Very recently, using emerging cellular phones, GPRS and Edge have been extensively used to connect to the Internet.

However, as it has been observed, most of the knowledge centers have been established through donor funding or subsidies from national governments. While many of them have succeeded to come up with expected outcome, but at the same time, many failed to produce any visible output or outcome. Nor, any quantified evaluation been conducted by any donor agency to justify their existence, or any measurable indicators been developed to measure their performances. Throughout the years, a few research studies have been conducted (IDRC, World Bank, EU), but availability of those reports at the end-user level has remained meager. Furthermore, migration of a telecenter into the daily life of a common citizen remains detached due to many visible and nonvisible parameters. They need extensive research in an organized fashion. Most recent approach by the Telecenter.org, hopefully could accommodate a separate unit in this aspect, regardless of building new telecenters without their feasibility study.

Coming to the point of establishing knowledge centers, many of them emerged as stand alone units in a remote rural set up. These were desired, but without a loopback or feedback on their performance the operation and maintenance of those centers become stringent, day-by-day. A preferred approach in this direction is to establish them under generic clusters unified as blocks of centers in a region. Though it is extremely difficult to patent this sort of intricate networks, but in the longer run there is no alternate to built knowledge centers piggybacking on each other and operating within a cluster. This is needed to manage them properly, and at the same time, it is easier to manipulate them through proper data mining and applying data mining algorithms.

Clustering

Clustering is a mechanism for data analysis, which solves classification problems. Its object is to distribute cases (people, objects, events, dealings, etc.) into groups, so that the degree of association to be strong among members of the same cluster and weak among members of different clusters. This way each cluster describes, in terms of data collected, the class to which its members belong. Clustering is also a discovery tool. It may reveal associations and structure in data, which apparently remain nonevidence, but become sensible and useful once found. The results of cluster analysis may append to the definition of a formal classification scheme, such as a nomenclature for related animals, insects or plants; or suggest statistical models with which to describe populations; or indicate rules for assigning new cases to classes for identification and diagnostic purposes; or provide measures of definition, size, and change in what previously were only broad concepts; or find exemplars to represent classes (Mirkin, 2005). Whatever an establishment is built in now, the chances are that sooner or later it will run into a classification problem. Cluster analysis might provide the methodology to help solve it properly.[4] Data clustering methods have been proven to be a successful data mining technique in the analysis of discrete data.

The data-clustering algorithm can be used to build a datagram (Alon et al., 1999). Each node, N, may be represented by a vector, $v_N = (x_1, x_2, x_3, \ldots, x_n)$, whose components, x_1 to x_n, corresponded to expression levels in each sample cluster. The vectors can then be normalized, so that the sum over their components equaled zero, that is, $\sum_m x_m = 0$, and the magnitude equaled one, $|v_N| = 1$. The clusters may be split into two groups by first defining two cluster centroids, Kj, where j= 1, 2. However, a probability of belonging to each cluster can be determined for each node:

$$B_j(v_N) = \frac{e^{(-\alpha|v_N - K_j|2)}}{\sum_i e^{(\alpha|v_N - K_j|2)}}$$

where the cluster centroids producing knowledge outcome can be determined by the equation:

$$K_j = \frac{\sum_N v_N B_j(v_N)}{\sum_N B_j(v_N)}$$

which was solved by iterations. For $\alpha = 0$ there is only one cluster $K_1 = K_2$. α can be increased in small steps until two distinct, converged centroids be formed. Then each node may be assigned to a cluster with the larger $B_j(v_N)$. The process may be repeated to split each one of the new clusters. The algorithm may run against the cluster samples, where each cluster sample, N, may be represented by the vector, v_N.

Formation of Wi-Fi Clusters

Wi-Fi clusters are becoming more popular among the development agencies while establishing knowledge centers, as majority of places where there are demands of knowledge center are lack of robust Internet infrastructure and these demand expansion of Wi-Fi bases along the geographically dispersed peripheries. In this context, it is ideal to build Wi-Fi bases to form a symmetric matrix with homogeneous dispersion, as such a mid-range Canopy/ Smartbridge/ Bridgeaccess (using IEEE 811.a, 811.b) can cover an air distance of around 15-25Km, or an average Airpro Gold (using IEEE 811.g, 802.16) may cover an air distance of around 20-35Km. However, the cost of the radio varies with the nature of the terrain and data throughput of the knowledge center (e.g., short range, mid-range, long range, very long range; data throughput of 64Kbps to 10Mbps).

The cluster affinity search technique (CAST) and enhanced cluster affinity search technique (E_CAST) algorithms take as input an *n*-by-*n* similarity matrix M where ($M(i, j) \in [1, 0]$) and an affinity threshold A_T is defined. A_T is used to determine node membership to a cluster. For sake of calculation a few definitions are being introduced next:

- **Definition 1:** The affinity of a node x to a cluster K is defined as: $a(x) = \sum_{N \in K} M(x,N)$
- **Definition 2:** The connectivity threshold, X, of a cluster K is: $X = A_T |K|$ where $|K|$ is the cardinality of K.
- **Definition 3:** A high connectivity node is a node that will be included in a cluster. Its affinity satisfies the following: $a(i) \geq X$ where a(i) is the affinity of i.
- **Definition 4:** A low connectivity node is a node that will be removed from a cluster. Its affinity satisfies the following: $a(i) < X$ where a(i) is the affinity of i.

Each cluster is formed by alternating between adding and removing nodes from the current cluster until such time that changes no longer occur or a maximum of iterations has been executed.

- **Node addition:** Add nodes with high connectivity to the nodes in the open cluster.
- **Node removal:** Remove any nodes in the open cluster with low connectivity to the other nodes in the cluster.
- **Cluster cleaning:** Make sure all nodes are in clusters with highest affinity

However, CAST algorithm relies on the affinity threshold, A_T, which is an input variable defined by the user before initiating the clustering process. It could create a problem because of the size and quantity of the clusters produced by the algorithm may directly affect this parameter (Ben-Or, 1983; Ben-Dor, Shamir & Yakhini, 1999; Diplaris, Tsoumakas, Mitkas & Vlahavas, 2005). For this reason, to carry out a mathematical analysis on this, thorough knowledge of the data set is required before the clustering can be performed. Some may enhance the algorithm to calculate this threshold. Also, the threshold can be calculated dynamically based only on the objects in that have yet to be assigned a cluster, $U' = U \setminus (C0 \cup C1 \cup \ldots \cup Cn)$, before each cluster is created, which provide a means of fine-tuning during clusters formation. The threshold parameter, D_T, is then calculated based on the similarity values of the nodes left to be clustered.

This dynamic threshold is computed as follows:

$$D_T = \left(\frac{\sum_{i,j \in U \text{ and } M(i,j) \geq 0.5} M(i,j) - 0.5}{\left| \{u : u \in U' \text{and } a(u) \geq 0.5 \right|} \right)$$

After deducing the mathematical modeling of dynamic threshold, a pseudo-code for calculating threshold, cluster formation and remodeling step has been illustrated in Figure 3. The dynamic

Figure 3. Pseudo-codes for dynamic threshold calculation

```
Dynamic Threshold:
/// D_T is an input parameter

CAST:
D_T = fixed value (ideally, 1)

// executed before each new K_open is created

E_CAST:
a = 0;
count = 0;
for all u ∈ U such that a(u) ≥ 0.5 {
      a+ = a(u)-0.5
      count++
}
D_T = (a/count) + 0.5

Cluster Formation Algorithm Pseudo-Code:
while ( U <> ∅){
      E_CAST: Calculate Threshold, D_T
      for all u ∈ U set a(u) = 0
      create empty cluster K_open
      Pick an element u ∈ U such that M(u,x)=max{M(w,x)|w and x ∈ U}
      K_open = K_open ∪ u
      U = U \ u
      For all x ∈ U set a(x) = a(x) + M(x,u)
      while (changes in K_open occur) or (iterations < max iterations){
            //Addition Step
            while max{a(w)|w ∈U} ≥ χ {
                        Pick an element u U such that a(u)=max{a(w)|w ∈ U}
                        K_open ← K_open ∪ {u}
                        U ← U \ {u}
                        // Update affinity of all nodes
                        For all x ∈U ∪ K_open set a(x) = a(x) + M(x,u)
                        }
                        //Removal Step
                        while min{a(w)|w ∈ K_open} < X{
                                    Pick an element u ∈ K_open such that a(u)=min{a(w)|w ∈ K_open}
                                    K_open ← K_open \ {u}
                                    U ←U ∪{u}
```

continued on following page

Figure 3. continued

```
                    // Update affinity of all nodes before returning a final value
                    For all x ∈ U ∪ K_open set a(x) = a(x) - M(x,u)
            }
        }
    }

Remodeling Step:
while (changes in any K_i occur) or (iterations < max_iterations){
    // cleaning step may not converge
    for each c ∈ K_i and K_i s K and K_j ∈ K {
        Compute a normalized affinity of c to each cluster K_j such
        that a_j(c)= (Σ_{k∈Kj} M(c,k))/(|K_j|)
    }
    if max{ a_j (c) } > a_i , for all K_j ∈K and i ≠ j {
        K_i= K_i \ c
        K_j= K_j ∪ c
    }
}
```

threshold assignment has been shown here to obviate the need for the "cleaning" step as proposed in the original algorithm (Alon et al., 1999). The cleaning step is used to move any vector from its current cluster to one that it may have a higher affinity for and has a time complexity on the order of $O(n^2)$ (Bellaachia, Portnoy, Chen & Elkahloun, 2002).

Implementing Models for Knowledge Centers

Similar to designing of knowledge centers, implementation of them in a wholesome form demands extensive study on their patterns, funding conditions, localizations, implementing agencies, and foremost ultimate objectives of the implementers. Without running into complicated materials in this aspect, this subsection will look into three forms of implementing models/patterns in terms of framing a viable algorithm or mathematical formulation. They are randomized model, homogeneous model, and additive model.

Randomized Models

Most available and popular forms of implementation model so far, but at the same time most vulnerable to perish at their earlier stage, as majority of them have been implemented without any study about the sustainability parameters before implementation. Over 50% of them are sure to die. Rigorous observations reveal that, without knowing the baseline, there are at most 50% of any telecenter have a chance to survive. However, a mathematical model may be derived from the probability theory.

The modern definition of discrete probability distribution starts with a set called the sample space, which relates to the set of all *possible outcomes* in classical sense, denoted by $\delta = \{x1,x2,\ldots\}$. It is then assumed that for each element $x \in \delta$, an intrinsic "probability" value f(x) is designated, which satisfies the following properties:

1. $f(x) \in |0,1|$ for all $x \in \delta$
2. $\sum_{x \in \delta} f(x) = 1$

That is, the probability function $f(x)$ lies between zero and one for every value of x in the sample space δ, and the sum of $f(x)$ over all values x in the sample space δ is exactly equal to 1. An event may then be defined as any subset E of the sample space δ. The probability of the event E is:

$$P(E) = \sum_{x \in E} f(x)$$

so that, the probability of the entire sample space is *unity*, and the probability of the null event is *zero*. Furthermore, the function $f(x)$ mapping a point in the sample space to the "probability" value is called a probability mass function (pmf).

However, the modern definition does not try to answer how probability mass functions are

Figure 4. Algorithm of Ben-Or's consensus protocol (Adapted from Aspncs, 2002).

```
Input: Boolean value from input register
Output: Boolean value stored in output register
Data: Boolean preference, integer round
begin
    preference ← input
    round ← 1
    while true do
        send (1, round, preference) to all nodes
        wait to receive n − t (1, round, *) messages
        if received more than n/2 (1, round, v) messages then
                send (2, round, v, ratify) to all nodes
        else
                send (2, round, ?) to all nodes
        end
        wait to receive n − t (2, round, *) messages
        if received a (2, round, v, ratify) message then
                preference ← v
                if received more than t (2, round, v, ratify) messages then
                        output ← v
                end
        else
                preference ← CoinFlip(); CoinFlip returns 0 if messageis not received ;
                                            else returns 1 if message is received
        end
        round ← round + 1
    end
end
```

obtained; instead it builds a theory that assumes their existence. Observed communities are generally more stable than randomly constructed communities with the same number of species. This greater stability of observed communities is partially due to the low values of both the mean and variance of their alpha distributions. It has also been observed that, randomization of consumer resource utilization rates almost always increased the mean but not the variance of the calculated consumer similarities. Therefore, in comparison to randomly constructed communities, the lower similarities and greater stability of the observed communities suggest that competitive processes are important in shaping real communities (Lawlor, 1980).

Moreover, randomized models provide lower probabilities for some transitions, which means instead of looking at a single worst-case execution, one must consider a probability distribution over bad executions. If the termination requirement is weakened to require termination only with probability 1, the nonterminating executions continue to exist, but they may collectively occur only with probability 0. In this case, there are two ways that randomness can be brought into an acceptable model. One is to assume that the model itself is randomized; instead of allowing arbitrary valid

Figure 5. Algorithm of Bracha and Rachman's voting protocol (Adapted from Aspnes, 2002; Bracha & Rachman, 1992)

```
Input:          none
Output:         Boolean output
Local data:     Boolean preference n; integer round r; utility variables
                c, total, and stable
Shared data:    single-writer register r[n] for each node n, each of
                which holds a pair of integers (flips, stable), initially
                (0,0)
begin
    repeat
        for i ← 1 to n/log n do
1               c ← CoinFlip()
2               r[n] ← (r[n].flips + 1, r[n].stable + c)
        end
3               Read all registers r[n]
                total ← ∑n r[n].flips
        until total > n²
4       Read all registers r[n]
        total ← ∑n r[n].flips
        stable ← ∑n r[n].stable
5       if total/stable ≥ 1/2 then
                return 1
        else
                return 0
        end
end
```

operations to occur in each state, particular operations only occur with some probability. Though, randomized scheduling allows for very simple algorithms, but it depends on assumptions about the behavior of the environment that may not be justified in practice. Thus it has not been as popular as the second approach (homogeneous), in which randomness is located in the processes themselves.

Figures 4 and 5 incorporate algorithm for a consensus protocol (determine whether a message can reach all nodes) and a voting protocol (determine whether a node can be sustainable or not) that use randomized pattern.

Homogeneous Models

As, voting protocol has been depicted in Figure 3, it has been observed that unweighted and weighted voting are two of the simplest methods for combining not only randomized but also homogeneous models. In voting, each model outputs a class value (or ranking, or probability distribution) and the class with the most votes (or the highest average ranking, or average probability) is the one proposed by the community. Note that this type of voting is in fact called plurality voting, in contrast to the frequently used term majority voting, as the latter implies that at least 50% (the majority) of the votes should belong to the winning class (Diplaris, Tsoumakas, Mitkas & Vlahavas, 2005), and in comparison to the randomized model, there is an additional probability of survival opportunity. Therefore, homogeneous models have better probability to be sustainable.

In homogeneous pattern, stacking can be introduced that combines multiple classifiers by learning a meta-level (or level-1) model, which predicts the correct class based on the decisions of the base-level (or level-0) classifiers. This model is induced on a set of meta-level data that are typically produced by applying a procedure similar to k-fold cross-validation on the available data.

Let *T* be the level-0 data set. *T* is randomly split into *k* disjoint parts $T_1 \ldots T_k$ of equal size. For each fold *i*=1..i of the process, the base-level classifiers are trained on the set $T \setminus T_i$ and then applied to the test set T_i. The output of those classifiers for a test instances along with the true class of that instance form a meta-instance. A metaclassifier is then trained on the metainstances and the base-level classifiers are trained on all training data *T*. When a new instance appears for classification, the output of all base-level classifiers is first calculated and then propagated to the metalevel classifier, which outputs the final result. Thus, always there are opportunities of providing a higher value as the output.

These illustrations support that, homogeneous patterns are expensive to establish and at the same time to maintain, but in the longer run always competitive in terms of providing better services and stronger existence than the other two models, as discussed in this chapter.

Additive Models

In terms of mathematical formulation, additive models represent a generalization of multiple regressions (a special case of general linear models). In linear regression, a linear least-squares fit is compound for a set of predictor or X variables, to predict a dependent Y variable. Thus, to predict a dependent variable Y the well known linear regression equation with m predictors, can be stated as:

$$Y = b_0 + b_1 * X_1 + \ldots + b_m * X_m,$$

where Y stands for the (predicted values of the) dependent variable, X_1 through X_m represent the m values for the predictor variables, and b_0, and b_1 through b_m are the regression coefficients estimated by multiple regression. A generalization of the multiple regression model would be to maintain the additive nature of the model, but to replace the simple terms of the linear equation $b_i * X_i$ with $f_i(X_i)$ where f_i is nonparametric function of the predictor X_i. In additive models, to achieve the best prediction of the dependent variable values[5] (Hastie & Tibshirani, 1990; Schimek, 2000) an unspecific (non-parametric) function is estimated for each predictor, instead of a single coefficient for each variable (additive term).

In terms of continuous probability distributions, if the sample space is comprised of real numbers (A), then a function called the cumulative distribution function (cdf) where F is assumed to exist, which gives $P(X \leq x) = F(x)$ for a random variable X. That is, $F(x)$ returns the probability that X will be less than or equal to x.

However, the cdf is supposed to satisfy the following properties:

1. F is a monotonically non-decreasing, right-continuous function
2. $\lim_{x \to -\infty} F(x) = 0$
3. $\lim_{x \to -\infty} F(x) = 1$

If F is differentiable, then the random variable X is said to have a pdf or simply density $f(x) = dF(x)/dx$.

For a set $E \subseteq A$, the probability of the random variable X being in E is defined as:

$$P(X \in E) \int_{x \in E} dF(x)$$

In case the probability density function exists, then it can be written as:

$$P(X \in E) \int_{x \in E} f(x)\, dx$$

Whereas the *pdf* exists only for continuous random variables, the *cdf* exists for all random variables (including discrete random variables) that take values on A. These concepts can be generalized for multidimensional cases on A^n and other continuous sample spaces.

This theory reveals that, additive models are easier to establish, simpler to calculate and provide multiplier effect if chosen with better probability values. Additive models can always piggyback on existing successful ones without enough implementation costs, costlier maintenance and unknown experimentations. There is a proverb of "learning from experiences" applies in this type of establishment patterns.

Technical Issues and Recommendations

In terms of evaluating the performance indicator through data mining the process can be very complicated and time consuming. The execution of knowledge discovery using SQL (KDS) on a real world large database of 1.6 million records, 6 independent variables for a total of 4,334 different values required a total of about 5 hours on a dual Pentium Pro computer with 128Mb of RAM and a 40GB HD (Giuffrida, Cooper & Chu, 1998). The data, someone may handle or intend to handle in data mining is usually of such orders of magnitude that a human being can not in fact comprehend. In such circumstances, even an algorithm with the simplest complexity can be too expensive in terms of computation. In these contexts, algorithms with linear or log-linear complexity are needed to adopt for performing the data mining tasks.

Furthermore, according to the information theory, there is a certain limit to which a particular large body of data can be condensed without incurring loss of information. This limit is the entropy or information content of the data. Even in practice, if this theoretical limit of compression could be reached, the resulting size of the data would still be far too large for a human being to examine. Hence, the effective mining of large data sets must permit and live with loss of information and the impact may remain dependent largely on the data mining performance.

The most commonly used approach to this issue is to set a frequency threshold (benchmarking) and mine only those, which have a frequency of occurrence above this threshold (additive modeling). Such an approach arises out of the belief that if one must sacrifice some understanding of the domain, it would be best to sacrifice understanding of the least frequently occurring aspects (loosing the unsuccessful ones, and piggybacking on success cases).

This is the very practical approach that data mining was initially tagged with and is increasingly referred to as statistical data mining. However, recently there has been interest in mining some of the less frequent aspects of a data set, but certain things, such as the threat of a terrorist attack or the existence of a rare breed of poisonous mushroom, seem worthy of attention even if they occur only once and are buried inside of a large body of data. But, to mine such infrequent patterns places a large burden on the performance of traditional statistical data mining techniques. To address this issue, a number of data mining algorithms that are not statistical in nature are required. Furthermore, this then brings the question of where and how far to permit the necessary data loss to perform effective data mining if one wants to mine the infrequent rules from a data set (Drewry et al., 2002).

Monitoring the Impact of Knowledge Centers

In terms of developing a mathematical formula for monitoring the impact of knowledge centers, two forms of impact have been discussed here.

Abrupt temporary impact. In a time series, the abrupt temporary impact pattern implies as initial abrupt increase or decrease due to the intervention which then slowly decays, without permanently changing the mean of the series. This type of intervention can be summarized by the expressions:

Prior to intervention: $\text{Impact}_t = 0$
At time of intervention: $\text{Impact}_t = \lambda$
After intervention: $\text{Impact}_t = \theta * \text{Impact}_{t-1}$

This impact pattern is again defined by the two parameters λ (lambda) and θ (theta). As long as the θ parameter is greater than 0 and less than 1 (the bounds of system stability), the initial abrupt impact will gradually decay. If θ is near 0 (zero) than the decay will be very quick, and the impact will have entirely disappeared after only a few observations. If θ is close to 1 then the decay will be slow, and the intervention will affect the series over many observations. Furthermore, when evaluating a fitted model, it is again important that both parameters are statistically significant; otherwise one could reach paradoxical conclusions. For example, suppose the λ parameter is not statistically significant from 0 (zero) but the θ parameter is; this would mean that an intervention did not cause an initial abrupt change, which then showed significant decay.

Abrupt permanent impact: In a time series, a permanent abrupt impact pattern simply implies that the overall mean of the times series shifted after the intervention, and the overall shift is denoted by λ (lambda).[6]

Measuring Ripple Effect Impact of Knowledge Centers

In practice, when analyzing actual data, it is usually not that crucially important to identify exactly the frequencies for particular underlying sine or cosine functions. Rather, because the periodogram values are subject to substantial random fluctuation, one is faced with the problem of very many "chaotic" periodogram spikes. In that case, one would like to find the frequencies with the greatest *spectral densities*, that is, the frequency regions, consisting of many adjacent frequencies, which contribute most to the overall periodic behavior of the series. This can be accomplished by smoothing the periodogram values via a weighted moving average transformation.[7]

In time series, the Hamming window is a weighted moving average transformation used to smooth the periodogram values. In the Hamming (named after R. W. Hamming) window or Tukey-Hamming window (Blackman & Tukey, 1958), for each frequency, the weights for the weighted moving average of the periodogram values are computed as:

$$w_j = 0.54 + 0.46*\text{cosine}(\pi*j/p) \quad (\text{for } j=0 \text{ to } p)$$
$$w_{-j} = w_j \quad (\text{for } j \neq 0)$$

where $p = (m-1)/2$ and it is supposed that the moving average window is of width m (which must be an odd number).

This weight function will assign the greatest weight to the observation being smoothed in the center of the window, and increasingly smaller weights to values that are further away from the center.[8] In this way, ripple effect impact of knowledge centers can be calculated.

Implementing an Ideal Homogeneous Pattern

Considering all the above justifications, a homogenous pattern is suggested for implementation. However, in case of augmented product moment matrix, for a set of p variables, a $(p+1)$ X $(p+1)$ square matrix evolves. The first p rows and columns contain the matrix of moments about zero, while the last row and column contain the sample means for the p variables. Ideally, the development matrix can be shown in the following form:

$$D_M = \begin{vmatrix} M & \chi \\ \overline{\chi}' & 1 \end{vmatrix}$$

where M is a matrix with element. Thus, value of the matrix is:

$$M_{jk} = \frac{1}{N} \sum_{i=1}^{N} X_{ij} X_{jk}$$

and $\overline{\chi}$ is a vector with the means of the variables.

Another indicator about the relationship among the member of a network can be derived, if the edges of the network (Figure 4) can be set in a symmetrical matrix, like:

Figure 6. Person a is virtually linked to persons b, c, d and e (Adapted from Rahman, 2004)

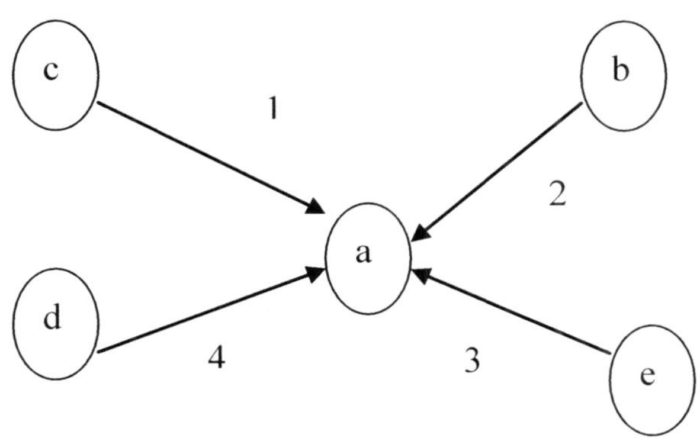

a	-	1	1	1	1
b	1	-	0	0	0
c	1	0	-	0	0
d	1	0	0	-	0
e	1	0	0	0	-

while in Figure 6, *a* knows *b*, *c*, *d* and *e*. But, the relationship between *b*, *c*, *d* and *e* may not be known (Rahman, 2004). These relationships will establish the ICT matrix and with point-to-point relationships among the network entities, the ideal relationship value will be 1 (unity). The ICT development matrix evolves from this unity relationship. Even if the network entity may follow point-to-multipoint, or multipoint-to-point paths, for a development matrix it must be upgraded to provide unity relationship value (either a zero communication, or unity communication; in other words either yes communication or no communication) (Rahman, 2006).

Implementing a Generic Model

In general, a time series consists of four different components; (a) a seasonal component (denoted as S_t, where t stands for the particular point in time), (b) a trend component (T_t), (c) a cyclical component (C_t), and d) a random, error, or irregular component (I_t). The difference between a cyclical and a seasonal component is that the latter occurs at regular intervals, while cyclical factors usually have a longer duration that varies from cycle to cycle. The trend and cyclical components are customarily combined into a *trend-cycle component* (TC_t). The specific functional relationship between these components can assume different forms.

However, two straightforward possibilities are that they combine in an *additive* or a *multiplicative* fashion:

Additive Model: $X_t = TC_t + S_t + I_t$
Multiplicative Model: $X_t = T_t*C_t*S_t*I_t$

where, X_t represents the observed value of the time series at time t.

Given some *a priori* knowledge about the cyclical factors affecting the series (e.g., business cycles), the estimates for the different components can be used to compute forecasts for future observations. However, the *exponential smoothing* method, which can also incorporate seasonality and trend components can be used as the preferred technique for forecasting purposes (Hale, Threet & Shenoi, 1994; Han, Kamber & Chiang, 1997).

FUTURE ISSUES AND CHALLENGES

Data mining algorithms in future should consider incorporation of larger databases, high dimensionality, overfitting, assessing of statistical significance, dynamic database, adaptation of knowledge theory, treatment of missing and noisy data, complex relationships between fields, understandability of tattered patterns, user interaction and prior knowledge, and integration, and versatility with other systems (Wang, 2003).

While measuring performance impact of social development activities, future research should formulate a homogeneous pattern of implementation, provided varying nature of environment, economy, culture and other parameters exist at the peripheries. Specifically, in terms of knowledge centers, there should be a symmetric matrix to follow as a guideline, over which each node, subnode, or any discrete existence of knowledge center could be established. This will reduce the design cost, operating expenditure, monitoring complexity and assist in measuring the performance quantitatively.

Given the three patterns of implementation model, yet numerous debates are running

Figure 7. Vertical pattern of data transformation (Adopted from Fayyad, Piatetsky-Shapiro & Smyth, 1996)

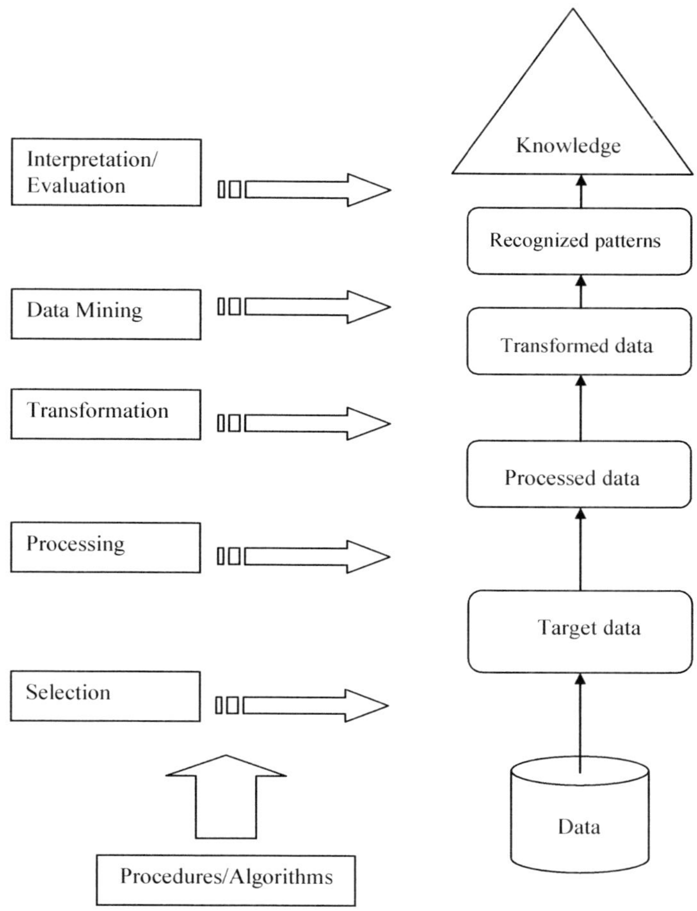

across the globe about their advantages and disadvantages. A systematic approach, in terms of establishing a mathematical formula and its consequential algorithm will ease debacles of enormous nature and lead to deduce a verified threshold as output. Furthermore, quantification of knowledge development from the immensely discrete activities of qualitative nature will remain as challenge to the future researchers.

Finally, utilizing data mining algorithms for measuring performance impact demand huge storage of data of varying nature; many of them have not been archived during the last decade of implementation phases (collection and archival of existing data) and by far most of them need to be transformed into recognized data sets, so that they can be used by verified data readers (transformation to any recognized database structure).

Now, before concluding, two patterns of data transformation are portrayed here (Figures 7 and 8), those seem relevant to the main theme of the book. If a community would like to synthesize data and transform them into knowledge, two types of transformation pattern are visible. The

Figure 8. Pyramid pattern of data transformation (A proposed pattern)

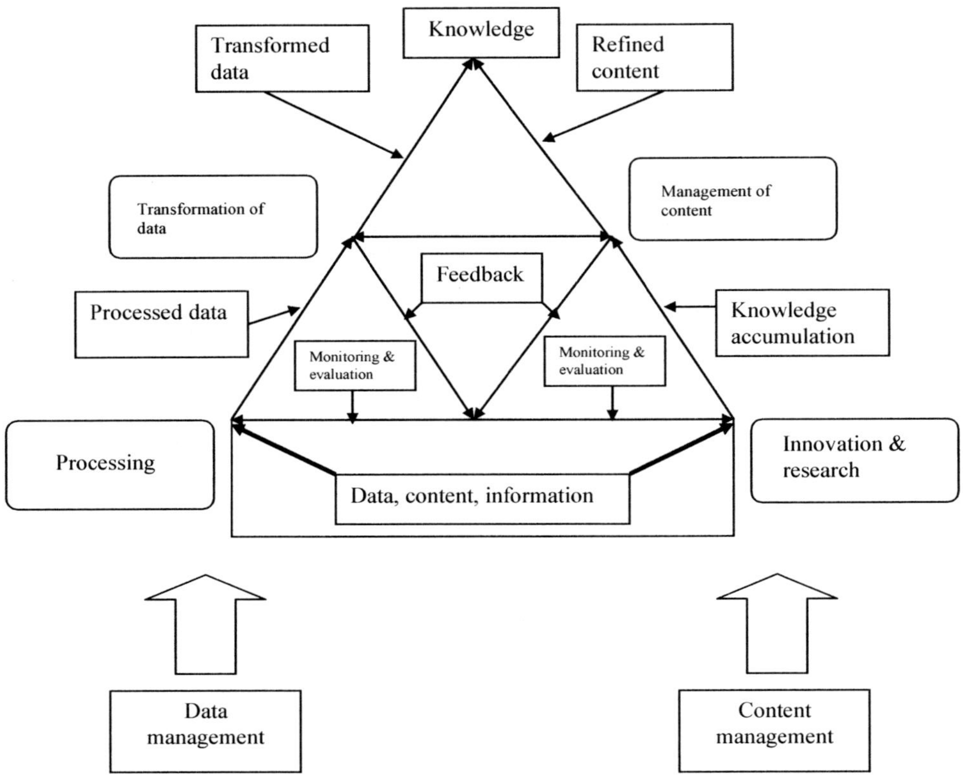

vertical one is more or less thorough and involves several stages of action during the transformation process though deserves rigorous study and closed observation. There is another form of transformation that is pyramid pattern, with processing, innovation, and knowledge at its three edge, and data remains at the core. The author, proposes a modified pyramid pattern of data transformation, although seems to be difficult in implementing, but the transformation process would take shorter period to settle down and be sustainable in the longer run. Researchers may derive separate algorithms for this transformation process, so that an acceptable measuring indicator may evolve in future.

CONCLUSION

It is well recognized, that the real-world knowledge-measurement applications obviously vary in terms of underlying data, complexity, the amount of human involvement required, and their degree of possible automation of parts of the discovery process. In most applications, however, an indispensable part of the measurement process is that the analyst explores the data and sifts through the raw data to become familiar with it and to get a feel for what the data may cover. Furthermore, very often an explicit specification of what one actually is looking for only arises during an interactive process of data exploration, analysis, and segmentation (Stumme, Wille & Wille, 1998). Therefore, proper data mining techniques with timely feedback analysis on the executed results deserves immediate attention for accurate result.

It is a difficult task to eliminate theories of probability, redundancies of efforts and abundances of varying data in determining reasonable mathematical formulae to measure the impact of social development processes. Complexity accumulates further, when it comes to projects or programmes that are related to newly evolved ICTs. Many developing and transitional economies are entangled with severe social problems within the vicious poverty cycle; thereby evolution of ICT emulated performance indicators are extremely difficult to resonate. They are diverse, deem to diverge and tend to become vulnerable in the longer run, without a verified mathematical model.

Moreover, data mining algorithms should incorporate design, development, implementation and operational factors, in addition to developing mathematical models on cost-benefit analysis. Foremost, utilizing data mining, success cases should come out at the forefront with rigorous analysis, so that they could be easily replicated elsewhere, with minimum adjustments.

REFERENCES

Abbass, H. A., Sarker, R. A., & Newton, C. S. (Eds.) (2002). *Data mining: A heuristic approach*. Hershey, PA: IGI Global.

Adamo, Jean-Marc (2001). *Data mining for association rules and sequential patterns: Sequential and parallel algorithms*. Springer Verlag

Agrawal, R. & Srikant, R. (1994). Fast algorithms for mining association rules in large databases. In *Proceedings of the 20th International Conference on Very Large Data Bases* (pp. 487-499), Santiago, Chile.

Agrawal, R., Imielinski, T. & Swami, A. (1993). Mining association rules between sets of items in large databases. In *Proceedings of the 1993 ACM SIGMOD Special Interest Group on Management of Data* (pp. 207-216), Washington, DC.

Alon et al. (1999). Broad patterns of gene expression revealed by clustering analysis of tumor and normal colon tissues probed by oligonucleotide arrays. In *Proceedings of the National Academy of Sciences.*

Aspncs, J. (2002). Randomized protocols for asynchronous consensus, Ref.

Bellaachia, A., Portnoy, D., Chen, Y. & Elkahloun, A. G. (2002) E-CAST: A data mining algorithm for gene expression data. In *Proceedings of the BIOKDD02: Workshop on Data Mining in Bioinformatics (with SIGKDD02 Conference)*, Edmonton, Alberta, Canada.

Ben-Dor, A., Shamir, R. & Yakhini, Z. (1999). Clustering gene expression patterns. *Journal of Computational Biology, 6*(3/4), 281–297.

Ben-Or, M. (1983). Another advantage of free choice: Completely asynchronous agreement protocols (extended abstract). In *Proceedings of the Second Annual ACM SIGACT-SIGOPS Symposium on Principles of Distributed Computing* (pp. 27–30), Montreal, Quebec, Canada.

Berthold, M. & Hand, D. J. (1999). *Intelligent data analysis: An introduction*. Springer Verlag.

Blackman, R. B. and Turkey, J. W. (1958). *The measurements of power spectra*. New York: Dover Publications, Inc.

Boulicaut, Jean-Francois, Esposito, F., Giannotti, F. & Pedreschi, D. (Eds.) (2004). Knowledge discovery in databases. In *Proceedings of the PKDD 2004: 8th European Conference on Principles and Practice of Knowledge Discovery in Databases*, Pisa, Italy.

Bramer, M. A. (Ed.) (1999). *Knowledge discovery and data mining: Theory and practice*. IEE Books.

Bracha, G. & Rachman, O. (1992). Randomized consensus in expected O(n2log n) operations. In S. Toueg, P. G. Spirakis & L. M. Kirousis (Eds.), *Lecture notes in computer science* (Vol. 579, pp. 143–150) Delphi, Greece: Springer. Retrieved April 12, 2008, from http://www.cs.yale.edu/homes/aspnes/randomized-consensus-survey.pdf

Chakrabarti, S. (2002). *Mining the Web: Discovering knowledge from hypertext data*. Morgan Kaufmann.

Cox, E. (2004). *Fuzzy modeling and genetic algorithms for data mining and exploration*. Morgan Kaufmann.

Curotto, C. L. & Ebecken, N. F. F. (2005). *Implementing data mining algorithms in Microsoft® SQL Server™*. WIT Press.

de Ville, Barry. (2001). Microsoft data mining, Integrated business intelligence for e-commerce and knowledge management.

de Ville, Barry (2006). *Decision trees for business intelligence and data mining: Using SAS enterprise miner*. SAS Press.

Diplaris, S., Tsoumakas, G., Mitkas, P. A., & Vlahavas, I. (2005). Protein classification with multiple algorithms. In *Proceedings of 10th Panhellenic Conference in Informatics*. Volos, Greece: Springer-Verlag.

Drewry et al. (2002). *Current state of data mining*. Department of Computer Science, University of Virginia.

Fayyad, U., G. Piatetsky-Shapiro, & P. Smyth. (1996). From data mining to knowledge discovery in databases (a survey). *AI Magazine, 17*(3), 37-54.

Freitas, A. A. (2002). *Data mining and knowledge discovery with evolutionary algorithms*. Springer-Verlag.

Giuffrida, G., Cooper, L. G., & Chu, W. W. (1998). A scalable bottom-up data mining algorithm for relational databases. In *Proceedings of the Tenth International Conference on Scientific and Statistical Database Management* (pp. 206-209)

Hale, J., Threet, J., & Shenoi, S. (1994). A practical formalism for imprecise inference control. *Ifip Trans. A-Computer Science And Technology, 60*, 139-156.

Han, J., Kamber, M. & Chiang, J. (1997). Metarule-guided mining of multi-dimensional association rules using data cubes. In *Proceedings of international conference on knowledge discovering and data mining (KDD'97)*, pp. 207-210.

Hand, D. J., Mannila, H., & Smyth, P. (2001). *Principles of data mining (Adaptive computation and machine learning).* The MIT Press.

Hastie, T.J., & Tibshirani, R.J. (1990). *Generalized additive models*. New York: Chapman and Hall.

Kantardzic, M. (2002). *Data mining: Concepts, models, methods, and algorithms*. Wiley-IEEE Press.

Kargupta, H. & Chan, P. (Eds.) (2001). *Advances in distributed and parallel knowledge discovery*. MIT/AAAI Press.

Kloesgen, W. & Zytkow, J. (Eds.) (2002). *Handbook of data mining and knowledge discovery*. Oxford University Press.

Larose, D. T. (2004). *Discovering knowledge in data: An introduction to data mining*. Wiley-Interscience.

Lawlor, L. R. (1980). Structure and stability in natural and randomly constructed competitive communities. *The American Naturalist, 116*(3), 394-408.

Liu, H. & Motoda, H. (1998a). *Feature selection for knowledge discovery and data mining*. Kluwer.

Liu, H. & Motoda, H. (1998b). *Feature extraction, construction and selection: A data mining perpective*. Kluwer

Maimon, O. & Last, M. (2000). *Knowledge discovery and data mining - The Info-Fuzzy Network (IFN) Methodology*. Kluwer Publishers, Massive Computing.

Mattison, R. M. (1997). *Data qarehousing and data mining for telecommunications*. Artech House.

Miller, H. & Han, J. (Eds.) (2001). Geographic data mining and knowledge discovery. *Research monographs in geographic information systems*. Taylor and Francis.

Mirkin, B. (2005). *Clustering for data mining: A data recovery approach*. CRC Press.

Myatt, G. J. (2006). *Making sense of data: A practical guide to exploratory data analysis and data mining*. John Wiley.

Nanopoulos, A., Katsaros, D., & Manolopoulos, Y. (2003). A data mining algorithm for generated Web prefetching. *IEEE Transactions on Knowledge and Data Engineering, 15*(5), 1155-1169.

Pal, S. K. & Mitra, P. (2004). *Pattern recognition algorithms for data mining*. Chapman & Hall/CRC.

Pawlak, Z. (1982). Rough sets. *Journal of Computer and Information Science, 11*(5), 341-356, 1982.

Pei, J., Han, J., & Lakshmanan, L. V. S. (2001). *Mining frequent itemsets with convertible constraints*. Paper presented at the Proceedings of the 17th International Conference on Data Engineering (pp. 433–332), Heidelberg, Germany.

Perner, P. & Petrou, M. (Eds.). *Machine learning and data mining in pattern recognition*. Springer Verlag.

Pyle, D. (1999). *Data preparation for data mining*. Morgan Kaufmann.

Rahman, H. (2004). Information dynamics in developing countries. In *Proceedings of the 5th International Conference on IT in Regional Areas*, Caloundra, Queensland, Australia.

Rahman, H. (2006). Role of ICTs in socio-economic development and poverty reduction. In H. Rahman (Ed.), *Information and communication technologies for economic and regional developments*

Stumme, G., Wille, R. & Wille, U. (1998). *Conceptual knowledge discovery in databases using formal concept analysis methods*. Berlin-Heidelberg, Germany: Springer, Verlag.

Thuraisingham, B. (1999). *Data mining: Technologies, techniques, tools, and trends*. CRC Press.

Utimaco (2005). *Data encryption: The foundation of enterprise security*. Foxboro, MA: Utimaco Safeware, Inc.

Wang, J. (Ed.) (2003). *Data mining opportunities and challenges*. IRM Press.

Weiss, S. M. & Indurkhya, N. (1997). *Predictive data mining: A practical guide*. Morgan Kaufmann.

Witten, I. & Frank, E. (2005). *Data mining: Practical machine learning tools and techniques* (2nd ed.). Morgan Kaufmann.

Yoon, J. P. & Kerschberg, L. (1993). A framework for knowledge discovery and evolution in databases. *IEEE Trans. On Knowledge And Data Engineering, 5*(6), 973-979.

Zhou, C., Li, Z., Meng, Y. & Meng, Q. (2004). A data mining algorithm based on rough set theory.

In *Proceedings of International Conference on Information Acquisition 2004* (pp. 413-416).

ENDNOTES

[1] http://www.microsoft.com/technet/prodtechnol/sql/2000/maintain/dmperf.mspx accessed on March 24, 2007

[2] http://www.statsoft.com/textbook/glosa.html#Algorithm accessed on March 23, 2007

[3] http://www.statsoft.com/textbook/glosa.html#Data_mining accessed on May 29, 2007

[4] http://www.bandmservices.com/Clustering/Clustering.htm accessed on March 25, 2007

[5] http://www.statsoft.com/textbook/glosa.html accessed on March 23, 2007

[6] http://www.statsoft.com/textbook/glosa.html accessed on March 23, 2007

[7] http://www.statsoft.com/textbook/sttimser.html accessed on March 28, 2007

[8] http://www.statsoft.com/textbook/glosh.html#Heuristic accessed on March 28, 2007

Section III
Applications of Data Mining

Chapter IX
Prospects and Scopes of Data Mining Applications in Society Development Activities

Hakikur Rahman
Sustainable Development Networking Foundation (SDNF), Bangladesh

ABSTRACT

Society development activities are continuous processes that are intended to uplift the livelihood of communities and thereby empower the members of communities. Along the way of socialization, these sorts of activities have become intrinsic phenomenon of a society, though, day-by-day their developments are intricately adopting innovative scientific techniques. Innovations and technologies, especially, the information and communication technologies have graced the development actors with dynamically improved tools and techniques to design, develop and implement diversified performances globally. Rapidly developed new ICTs gave the development initiators tremendous boost to take many indigenous that are geographically dispersed, but could easily be monitored. However, many of the development projects lack of proper management, thorough analysis, appropriate need assessment, and seemingly could not sustain. In most cases, development partners blame each others, among them are the initiators, designers, implementers, or the donors. Subsequently, in many countries, most innovative success cases could not see the light of sustainability, due to improper reporting, monitoring, and feedback. In consequences, projects fail. This chapter tries to establish methodologies for establishing successful development initiatives, synergizing a few success cases. Furthermore, utilizing recently available means, as such data mining, projects and activities around each corner of the globe can be easily recorded, adequately analyzed, monitored, and reported for their successful replication in other countries with necessary favorable condition exists. This chapter also highlighted a few areas of development aspects and hints application of data mining tools, through which decision-making would be easier. Along this perspective, this chapter has put forward a few potential areas of society development initiatives, where data mining applications can be engaged. The focus area varies from basic education, health care, general commodities, tourism, ecosystem management to a few advanced uses, including database tomography. Finally, the chapter provides some future challenges and recommendations in terms of using data mining applications for empowering knowledge society.

INTRODUCTION

Data mining is an interdisciplinary field of study and it is driven by various multidimensional applications. At one hand it involves techniques for machine learning, pattern recognition, statistics, algorithm, database, linguistic and visualization; and at the other hand, one applies its applications to understand human behavior, such as that of the end user of an enterprise (Ebecken, Brebbia & Weigend, 2000; Han & Kamber, 2000; ICDM, 2003). It also assists entrepreneurs to understand the nature of transactions involved, including those needed to evaluate any risk factor or detect fraud.

Apart from the intricate technology context, the applications of data mining methods deserve specials attention while to be applied in the development context. Lack of data has been found to inhibit the ability of organizations to fully assist clients, and lack of knowledge made the government vulnerable to the influence of outsiders who did have access to data from countries overseas. Furthermore, disparity in data collection need for a coordinated data archiving and data sharing, and it is extremely crucial for promoting, launching and sustaining development projects especially in developing countries (Berry & Linoff, 2000; Codata, 2002; COL, 2003).

At the same time, the technique of data mining enables governments and private organizations to carry out mass surveillance and personalized profiling, in most cases without any controls or right of access to examine this data. However, while utilizing data mining applications in terms of development contexts, the main focus should be on sustainable use of resources and the associated systems under specific context (producing ecological, limnological, climatic, social and economic benefits) of developing countries. Research activities should also focus on sustainable management of vulnerable resources and apply integrated management techniques, with a view to support optimization and sustainable use of existing resources.

In addition, the scientific issues and aspects of archiving scientific and technology data include the discipline specific needs and practices of scientific communities as well as interdisciplinary values and methods. Data archiving is primarily a program of practices and procedures that support the collection, long-term preservation, and low cost access to, and dissemination of scientific and technology data. The tasks of the data archiving include: digitizing data, gathering digitized data into archive collections, describing the collected data to support long term preservation, decreasing the risks of losing data, and providing easy ways to make the data accessible. Data archiving and the associated data centers need to be part of the day-to-day practice of science. This is particularly important now that much new data is collected and generated digitally, and regularly (Codata, 2002; Dunham, 2003; Quéau, 2001).

So far, data mining has existed in the form of discrete technologies. Recently, its integration into many other formats of information and communication technologies (ICTs) has become attractive as various organizations possessing huge databases began to realize the potential of information hidden there (Fayyad, Piatetsky-Shapiro, Smyth & Uthurusamy, 1996; Hernández, Göhring & Hopmann, 2004). The Internet can be a tremendous tool for the collection and exchange of information, best practices and vast quantities of data. But it is also becoming increasingly congested and its popular use raises issues about authentication and evaluation of information and data. The growing number and volume of data sources, together with the high-speed connectivity of the Internet and the increasing number and complexity of data sources, are making interoperability and data integration an important research and industry focus. Incompatibilities between data formats, software systems, methodologies and analytical models are barriers to easy flow and creation of data, information and knowledge (Carty, 2002). All these demand, not only technology revolution, but also tremendous uplift of human capacity as a whole. Therefore, the challenge of human development taking into account the social and economic background while protecting the environment confronts decision makers like government, local communities and development organizations. How can new technology for information and communication be applied to fulfill this task (Hernández, Göhring & Hopmann, 2004; Han & Kamber, 2006)?

This chapter focuses on areas and scopes of data mining application in general and a few decision support techniques to achieve sustainable outcomes for the society. This chapter does not go in detail about theoretical issues of data mining, nor would like to provide in-depth analysis on data mining techniques, rather it gives a brief overview of data mining in its introduction and background that are necessary to justify data mining aspects and features, and focuses deeply into data mining's contemporary and prospective application areas around the society development processes. The author mainly depends on popular data mining books and accessible literatures on WWW, as it has been observed that real world applications are mostly available in numerous Websites of many projects around the globe, including their success cases. Perhaps, there need a collective publication of success stories in this aspect, so that readers can gather knowledge from similar single sources of information. However, as far as this chapter's title concerns, there has been extensive literature review along this context to enrich the subject matter and theme of this chapter.

Furthermore, the chapter looks into authenticated global approaches and shows the capabilities of data mining as an effective instrument on the basis of its application in real projects in the developing countries. The applications could be on development of algorithms, computer security, open and distance learning, online analytical processing, scientific modeling, simple data warehousing, or interactive collocations. However, this chapter emphasizes on effective scopes and prospects of data mining application to improve social instruments, as such community development, environmental improvement and life-long learning systems; improvement of small and medium entrepreneurships, balancing of ecological patterns, biodiversity equilibrium, spatial database management, disaster management, and some more advanced uses. This chapter also put forwards a few success cases across the globe for its readers, and researchers in this field. Along the arguments, the author has tried to adopt and derive a knowledge hierarchical pyramid that may be evolved from data/content and finally this chapter recommends a few future research issues with their challenges before concluding.

BACKGROUND

Data mining and data warehousing techniques are becoming indispensable parts of almost all corporate intelligence programs (Berry & Linoff, 1997; Intransa, 2005). Data mining has been loosely defined as the process of extracting information from large amounts of data and it is becoming a pervasive technology in activities as diverse as using historical data to predict the success of a awareness raising campaign, or a promotional operation looking for patterns of sequences to act as a monitoring tool, or analyzing genome chains.

Data mining or data discovery is the process of autonomously extracting useful information or knowledge from large data stores or data sets. It can be performed on a variety of data stores, including the World Wide Web, relational databases, transactional databases, internal legacy systems, pdf documents, and data warehouses. Furthermore, data mining is the ability to query very large databases in order to satisfy a hypothesis (top-down data mining); or to interrogate a database in order to generate new hypotheses based on rigorous statistical correlations (bottom-up data mining) (Hand, Mannila & Smyth, 2000; Rud, 2001; Tan, Steinbach & Kumar, 2005; Thearling, 1995).

Sometimes, data mining could be derogatory, as it involves sorting of accurate information through a huge volume of data, and the extracting decision rules may favor one or disfavor another, without considering any cause-and-effect relationship. It seems as a technique of betting or letting a few monkeys jump on a keyboard, and perhaps at a point of time in future a sonnet may evolve. But, using modern day's techniques and tools, data mining is no more so gloomy, and it is becoming an important component of knowledge science through accumulation of accurate data and by taking intelligent decisions out of them.

Furthermore, in many countries, governments' restriction on persistent data mining and protection of privacy in e-transactions without encroachment of the principle of free access throws important challenge to the policy makers. This deepens further while ensuring the legitimate users' rights to access information, as well as legal rights for privacy. Moreover, policies need to be adopted to ensure protection of sensitive information and law enforcement on the networks. In addition to these, complicacies are there to incorporate commercial, ethical and social version of data mining in terms of providing technology solutions during cryptography or digital signature. Specifically, the business enterprise data requirements grow at 50 – 100 percent a year and creating a constant storage infrastructure management challenge (Intransa, 2005). Therefore, formulation of a generalized code of conduct across the cross-sectoral approach to provide fairness, equality, justice and morality in handling collection, repackaging, modification and sale of public/private data produces a new version of challenge. Finally, a collaborative partnership among private, commercial, government and civil society organizations always remain a far cry.[1]

However, despite all these, utilization of data mining tools and techniques for preservation of raw data to make them useful, and utilization of acquired data in making knowledgeable decision support systems have not been overshadowed. The promising side is that, not only researchers and academicians, but also business entrepreneurs are becoming interested in data mining applications. As newer storage techniques are brought into the environment, it is typically added within inflexible single server "silos." Moreover, to ensure that data is available to all users usually means moving data from server to server to make it readily accessible. Though this is time-consuming and results in multiple pools of data that need to be managed, neither of which improves return of investment (ROI). In this context, the best practice to improve the ROI of computing and storage resources is to efficiently consolidate and share these resources. An IP-SAN (Internet protocol- storage area network) provides a highly cost-effective storage solution that leverages an enterprise's existing IT expertise and resources. In addition to these, a shared file system eliminates the inflexible "single server data silos" and makes data readily available to those who need it (Intransa, 2005).

In recent days, the Internet has become an increasingly important tool for the collection and exchange of information, best practices and vast quantities of data. But at the same time, it is also becoming increasingly congested and its popular use raises issues about authentication and evaluation of information and data. The growing number and volume of data sources, together with the high-speed connectivity (narrow speed in case of developing countries) of the Internet and the increasing number and complexity of data sources, are making interoperability and data integration an important research and industry focus. Incompatibilities among data formats, software systems, methodologies and analytical models create barriers to the easy flow and creation of data, information and knowledge (Carty, 2002).

Around these perspectives, this chapter has tried to focus on the best utilization of data mining applications for the benefit of the society and in these contexts; the following section deals with the methodologies of data mining applications and their uses to empower the knowledge societies.

MAIN THRUST

Methodologies

Tools of ICTs, as such radio, television, telephones, computers, and the Internet can provide access to knowledge in sectors like agriculture, microenterprise, education, and human rights by offering a new realm of choices that enable the common people to improve their quality of life. Unfortunately, however, till now not everyone could enjoy equal access to these technologies; and the resulting digital divide is found not only between industrialized and developing countries, but also within developing countries. Moreover, as the divide grows wider, it aggravates the existing divisions of power and inequities in access to resources even between men and women, the literate and nonliterate, and urban and rural populations (CIDA, 2005; Witten & Frank, 1999; Witten & Frank, 2005).

To improve the situation, diversified strategies have been adopted in many countries. For the purposes of this strategy, knowledge for development has been integrated into development programs so that the beneficiaries can access, utilize, and disseminate information and knowledge. This is done with a view to promote socioeconomic development through appropriate ICTs, coupled with the development of required associated skills. Moreover, ICTs offer new ways of providing access to information and knowledge, and thereby create significant opportunities for learning; networking, social organization and participation; and improving transparency and accountability. For example, grass roots work by nongovernmental organizations and civil society organizations has greatly benefited from media such as the Internet (CIDA, 2005). However, for sake of focus of this chapter, a few methods of ICT for development have been described in this section. The author argues that the following methodologies might be adopted at national level to improve access, utilization and dissemination of information for promoting knowledge society.

Greater Role of Public Authorities in Access to Information

In many cases, industries and entrepreneurs (telcos, private entrepreneurs, and others) are providing infrastructure support in addition to the government initiatives for access to information resources, as well as contents. In these cases, there are need to define the concept of public domain and universal access by promoting common public welfare in a global context, and at the same time encourage private initiatives by protecting information rights and economic interests. In each case, balance of information right leading to intellectual property rights, ethical integrity, cultural diversity, localized discrimination and return of investment (ROI) have to be taken in consideration during program development.

Broader and More Efficient Provision of Public Contents

Though much of the global knowledge is not related to intellectual property rights, but efforts should be given to avoid under-provision of this knowledge. Clear understanding should be formulated between national and global public goods. If necessary, appropriate ramifications should be adopted on the concept of public domain about classical and anonymous works and information produced with public funds. They should be categorized as copy-left information and should be made available at free of cost as open content with provisions of open source. Necessary policies need to be adopted comprising economic, political, ethical, social and educational boundaries, including their sustainability (operational, economical; and/or technical, nontechnical).

Facilitating Improved Access to Networks and Services

In many countries, still the most important economic obstacles in accessing information are telecommunication tariffs, Internet access fees, licensing fees, taxes, duties, and many other factors. These issues need to be resolved as soon as possible. In this aspect, there should be a balance between public administration costs (compromise, incentives, or even subsidy for a short period) and regulations (not being regulatory, but being the facilitator) between commercial interests and national interest by keeping civil and moral obligations intact and promoting equitable access. A competent financial mechanism should be put into place to ensure universal access to information by providing cross subsidies, preferential taxation, and other type of incentives. Telecommunication regulatory and tariff policies should have soft corner with mechanisms providing Internet access to general communities. Perhaps, some tangible and intangible products should be recognized and they would be excluded from tariff enclosure.

Furthermore, preferential rates should be introduced for educational, research and cultural organizations. All public service institutions should have their own public access center from where general public should be able to access relevant information (forms, rates, procedures, rules and others) at free of costs. These can be compensated during the submission of the forms, rather than during purchase of the forms. Public institutions with regional and local level branches may open similar outlets in phases with the same service provision to the common public. In addition to these, public institutions can form common networks of infrastructure comprising existing civil society networks and private/commercial networks by forming consortia, community outlets, freenet centers, public access centers, and so forth. Finally, role of media should be put into the picture; a vibrant media is an essential ingredient for a democratic, responsive and transparent civil society.[2]

Technology Standards and Privacy Issues

These two elements are extremely essential in establishing content repositories. So far data mining concerns, competing private firms have little interest in preserving the open standards that are really essential to a fully functioning interactive network, as well as formation of open content that would eventually become public goods. Markets may encourage innovation, but they do not necessarily insure the public interest. In this case, governments could decide to encourage and support the developments of public domain content (data, information and software) and freewares (LINUX, Apache, etc.). This goal is becoming absolutely vital, considering the importance of equipping the schools of the world with basic computer facilities (OLPC, one laptop per child). Similarly, privacy issues are also of strategic importance. The protection of privacy has become one of the most important human rights issues of this new millennium. Many search engines or agencies are accumulating/ monitoring/ mining their subscribers' information with/without the consent of the subscriber. This could be valuable in the sense of knowledge gathering and eventually be used in intelligent decision making processes, however, to do this the level of anonymity and privacy protection need to be clearly defined. Furthermore, any ethical or political issues are also need to be covered before taking similar decisions.

An on-going pact, named Ukusa (binding the United States, Great Britain, Canada, Australia, New Zealand) uses the ECHELON network that is supervised by the US National Security Agency in order to monitor and process more than 3 billion phone calls, faxes, e-mails per day throughout the world. Similarly, on a mere click on a hypertext link, the most casual consultation of a site on the World Wide Web generates cookies that feed uncontrollable databases. The technique of data mining (exploitation of data) enables governments, other agencies and private organizations to carry out mass surveillance and personalized profiling, and in most cases without any controls or right of access to examine this data. They vary from medical care to transport systems, financial transfers to commercial and banking transactions, and thereby enormous quantities of information are accumulated every day. Moreover, commercial interests want to exploit powerful data-mining resources for marketing research or for information reselling to data brokers and to the "individual reference service" industry. They give less preference to questions like, should personal information copyright belong to the persons concerned or to the data miners processing electronic transactions or what level of anonymity and privacy protection is desirable? These are essentially philosophical and political issues. (Chakrabarti, 2002; MITRE, 2001)

Development of Consolidated National E-Strategies and E-Policies

Developing countries are increasingly in quest of designing and implementing national strategies to manage the development of appropriate ICT regulatory, legislative and policy frameworks (UNCTAD, 2004). In this context, appropriate decision making bodies have to be formulated at national level with concrete plan of actions and agendas.

UNCTAD is becoming a partner of the Global ePolicy Resource Network (ePol-NET) by providing its expertise in the design of e-strategies, e-commerce, legal and regulatory issues, e-measurement, e-finance and e-government to enhance efficiency and effectiveness of the governance system. ePol-NET functions as a virtual network, and partners of this network include the government of Ireland, which is providing the secretariat for the partnership, as well as the governments of Canada, France, Italy, Japan and the United Kingdom; ECA; ITU; UNDP; OECD and the Commonwealth Telecommunications Organization. (UNCTAD, 2004)

Congenial Atmosphere for E-Business and E-Finance

Flourishing of small and medium enterprises (SMEs) is another precondition to empower a community. They stay between micro- and macroeconomics and act as the boosting power for a nation by raising the economic capacity of the community. ICT, as usual, stands there to support, develop and protect their interests. However, in many cases, lack of adequate information at the disposal of financial service providers on SMEs and their payment performance goes against SME financing. By adequate information, it is meant that, national policy makers should have sufficient information about the formation of SMEs, their operations, their outcomes and their effective support for the development of economically sustained society.

A partnership has been designed to explore the opportunities arising from innovative Internet-based electronic finance methods and their data mining capacities and find ways of improving the small and medium entrepreneurs (SMEs) access to trade-related finance and especially, e-finance. Leading partners are from international and local financial service providers, enterprise associations, Governments and other public entities, international organizations including the World Bank, WTO and ITC, as well NGOs such as the World Trade Point Federation. (UNCTAD, 2004)

E-Measurement and ICT Indicators

Finally, it is essential to measure the achievement of ICT initiatives in each country by taking care of the local environment. Till now, not much progress has been made in developing necessary measuring tools for quantifying the progress in ICTD initiatives. Neither are there any acceptable indicators to quantify the impact of ICT4D implementations at the grass roots. Therefore, e-measurement is essential for assessing the state of advancement, especially in developing countries in the use and impact of ICTs.

The WSIS Plan of Action (Geneva and Tunis phases of the Summit) calls for the development of indicators to monitor progress in the use of ICTs for development. The principal stakeholders have agreed to identify a set of core indicators that could be collected by all countries and harmonized at the international level so as to facilitate the measurement of the achievement of international development goals, including those contained in the Millennium Declaration; to assist developing countries in building capacity to monitor ICT developments at the national level; and to develop a global database on ICT indicators. Partnership activities in this aspect include the WSIS member states, OECD, ITU, UNESCO and the UN ICT task force, as well as the UN regional commissions and other relevant regional bodies working on e-measurement issues. (UNCTAD, 2004).

One may argue that inclusion of these methodologies is redundant here, but, the author feels that without necessary preconditions of ICT flourishing at the grass roots, a nation/ a society/ an enterprise can not proceed further to reap out its ultimate benefits. After appropriate preconditions for enlightened ICT prevails, application of necessary tools, as such data mining cannot be applied in coherence at various stages to retrieve/store information and content for making intelligent decision. Furthermore, data mining is a newly evolved process in the ICT sector and need to be keenly nurtured with proper monitoring utilities. To support these arguments, a few uses of ICT have been put forward in the next subsection that may be improved further by using data mining.

However, before proceeding to the next subsection on data mining applications, the author would like to draw attention regarding transformation of data into knowledge. These include entrepreneurs' data mining (see Table 1). In addition to this data backup strategy, Table 2 and 3 are being derived from various researches that are implying as how stored data can be converted to become knowledge. Finally, a knowledge hierarchy diagram has been introduced in Figure 1, which shows how data/content ultimately become wisdom.

Table 1. Data backup strategy and low cost operation in SMEs

Small and medium sized businesses' data protection pain points	How a Software Can Relieve the Downtime Pain
Limited IT resources for data backup and recovery	Storage server can provide, continuous, automatic backup and incredibly easy recovery
Al critical data on one server	Storage server and download servers can be restored in a flash. They are capable to work as very low-cost servers and thus making backup servers affordable to companies of all sizes
Regulatory pressures and complex processes	Storage server employees can find individual backup files quickly – without tying up expensive IT resources
Cash flow disruptions are very damaging	Storage server can provide failover capabilities to a backup server, so that the business can keep running

Table 2. Physical outcomes of data towards creation of knowledge and their management issues

Physical outcomes (bottom-up preference)	Management issues
Knowledge portals	Regular update, security, sustained support
Innovative techniques on faster data search, mass storage and accurate data analysis	QoS with economic value and proper monitoring
Optimum resource scheduling with dynamic adaptability	Interfaces, Open source techniques
Extended connectivity with possibility to connect home users	Interconnectivity, interoperability

Table 3. Applications of data towards creation of knowledge portals and their management issues

Outcome ⇒ Knowledge portals	
Applications	**Management issues**
Knowledge management and visualization systems	QoS, workflow techniques
Search engines	Service interaction models
Data mining	Collaborative service composition
Fail safe recovery	Visualization
Authorization	Grid-aware simulation
Encryption	Access Interfaces and technologies

In this way, it has been observed that data/content can enlighten human skill to transform them into a knowledge society. According to Ackoff (1989), the content of the human mind can be classified into five categories:

1. **Data:** Facts or figures
2. **Information:** Useful data; answers to "who," "what," "where," and "when"
3. **Knowledge:** Application of information; answers "how"

Figure 1. Knowledge hierarchy (Adopted from Ackoff, 1989)

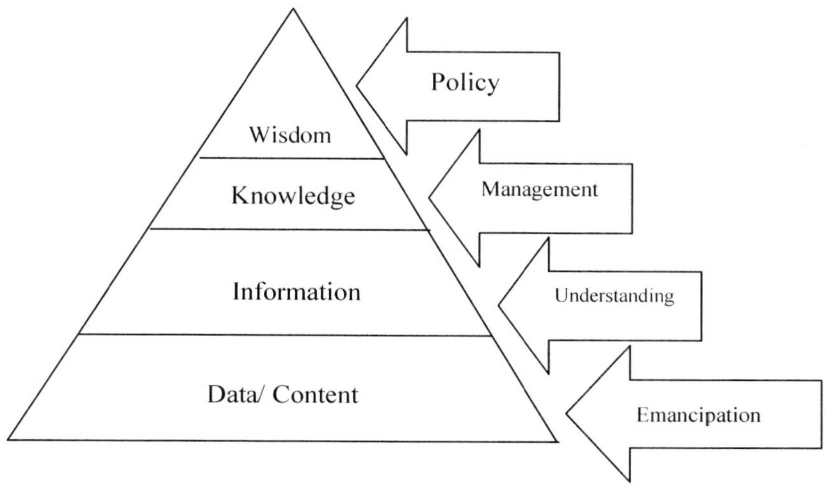

4. **Understanding:** Appreciation of "why"
5. **Wisdom:** Evaluated understanding (Markus, 2005).

Data Mining Applications in Social Development Implications

Use of data mining techniques ranges from its diversified applications in learning systems, knowledge discoveries, web intelligence, entrepreneurship management, data visualization, pattern recognition, statistical analysis, production control and machine learning to collaborative filtering, bioinformatics, ecosystem management, spatial data mining and knowledge economy (Berry & Linoff, 1999; Berry & Linoff, 2002; Berry & Linoff, 2004; Berson & Smith, 1997; Berson, Smith, & Thearling, 1999; Bozdogan, 2004; Braha, 2001; Bramer, 1999; Cerrito, 2006; Cios, 2000; Cios, Pedrycz, & Swiniarski, 1998; Delmater & Hancock, 2001; Fayyad, Grinstein, & Wierse, 2001; Ville, 2001). However, for sake of illustrating data mining applications for social development purposes, its use in financial market study, earth science, ecosystem management, health networks, tourism industries, general commodities, small and medium sized enterprises, e-learning, decision support systems, knowledge centers and a few advanced uses are described in this portion. This sub-section discusses a few prospective areas, where data mining researchers may continue their comprehensive studies to develop justified and proper applications that would assist in establishing sound social development systems. Along this context, several case studies have been introduced to assist the readers and researchers to have a glimpse on those scopes.

Education

Transforming computing knowledge into education is necessary for empowering the knowledge society. Community members should develop computing competencies so that they can help to advance the social progress of the nation by raising their knowledge level. Transformation is needed at the grass roots, computing professionals and higher educations institutes (HEIs). Many of the disciplines at those outlets are increasingly dependent on computing to manage data and information necessary to support decision-making. Furthermore, there are advances in computing services that comes from R&D in computing-related science, technology, and engineering. Simultaneously, advances are notable in other disciplines with the desire to solve grand challenges (using simulations, supercomputers, data mining, or virtual environments), and to enhance the quality of life for individuals and social groups. Therefore, transforming education, in computing-related disciplines, with technical innovation is needed to improve the future of a country. In addition, by extending knowledge of key computing concepts across the range of core curriculum areas inherent in undergraduate education will prepare a nation to leap into the knowledge society.[3,4]

ICT is already an imperative factor at all levels of education, though secondary education is in a decisive stage in many developing countries. But, it is a proven fact that learning and studying at this stage has potential impact on the new members in the community-of-knowledge society. Towards implementing educational policies by promoting sustainable ICT infrastructures for secondary schools, the first step is the penetration of Internet-connected computers in the schools. The second step is even more crucial; the continuous evolution of didactic methods so that young learners will actually learn to learn within the WWW-based infrastructures. Thirdly, application of data mining will turn these institutions into learning organizations and eventually they will become catalysts in the innovative processes of education system. (Kommers, Kinelev & Kotsik, 2003)

Distance Education

Simultaneously, info-miners in the distance education community can use one or more of infomining tools to offer high quality open and distance learning information retrieval and search service. ICT based infomining services could include producing digital libraries of open content, as such e-books, journals, reports and databases on DVD and similar high-density information storage media. These could be in online or off-line formats and should be made PC/laptop (perhaps, for one laptop per child-OLPC) accessible and store considerably more information per unit than a CD ROM. This form of learning is commonly known as e-learning, and relates directly to ICT for development and formation of knowledge economy (COL, 2003).

Knowledge Center

It is considered that the interactive relationship between knowledge, attitude and behavior is the basic parameters of social bonding in a community and needs to be specifically investigated while considering the issues of sustainable development focused towards improving the quality of life of villagers. The group of villages sharing the same geography shares almost similar problems; as such problems related to agriculture (corps, harvesting, food security, supply chain), environment (natural disaster, calamity, draught, rain), health (nutrition, disease, medicine, treatment, physician, hospital), education (school, college, university, student, teacher, support center, tutorial center, learning center) and public services (government and nongovernment agencies along with other development partners). Most of them depend on traditional forms of support and existing resource infrastructure (land, skill, knowledge, capital, technology, etc.). Applying a baseline field survey pertaining to such conditions followed by a thorough analysis of the relations specific to the conditions pertaining to achieve sustainable development can generate policy implications and develop and implement policies towards improving the quality of life of villagers. Simultaneously implemented multiple data collection techniques (documentary—historical data and observation, and survey—questionnaire and interviews) can be used for the knowledge and information accumulation through utilizing village centers/knowledge centers/kiosks/information centers.

Decision Support System

To handle huge amounts of data for making fast and easy decision it is not recommended to work with the same structures as for online transaction processing (OLTP), that are traditionally supported by the operational databases. In OLTP, tasks are structured for isolated transactions, and transactional database are designed to reflect the operational semantics of known applications, and, in particular, to minimize concurrency conflicts. On the other hand, the summarization and consolidation of data in data warehouses is targeted for decision support. High workloads arise with mostly ad hoc, complex queries to a huge number of records and a big amount of operations. In this form of query request, where query performance and response times are more important than transaction throughput, is known as online analytical processing (OLAP). In this context, data mining can perform automated search for hidden patterns in typically large and multidimensional databases. The results of data mining techniques are abstract behavior models that can be used to explain and predict consequences, for example, to support risk management and mitigation of natural hazards. So far, data mining and geographical information system (GIS) have existed as two separate technologies. Recently, their integration has become attractive as various organizations possessing huge databases began to realize the potential of information hidden there (UNDP, 2001).

Once a big data warehouse is being filled, obtaining the concealed causal relationships will help to answer many important questions. For example, in some places in Nicaragua the concept of backyards was introduced in order to provide people with fresh vegetables. This concept failed in many places, or in many places it succeeded. Then the question arises, as, are there any hidden reasons for this paradoxical result? Mining the data in terms of socioeconomic aspects may provide an explanation. (UNDP, 2001)

By applying mathematical modeling or data mining through a careful data collection a system can take care of making intelligent decision. However, the data characteristics, their physical dimensions, their chemical, biological, social, or financial implications must be taken into consideration.

Data compilation and acquisition means measurements, experiments, and communication. These require modern technical equipment, a clear understanding of nature, appropriate financial sources and, last but not least, for an excellent understanding of human being, a psychological and social sensitivity, experience and devotion. In addition to simple data collection, the basis of mathematical modeling has both its different targets and its various constraints, that is, it implies a number of optimization problems. This thorough process of optimization underlying the data collection is known as experimental design. Having made this design and found the data, then, these data have to be well interpreted. Finally, the system should dynamically incorporate a data analysis into the whole entity of mathematical modeling and problem solution. Thereby, this dynamics will be termed as learning. (Gökmen, 2004)

General Commodity

Currently, there is no comprehensive and systematic consultative framework that enables the sharing of information and the use of complementary expertise among representatives of all key actors involved in the review of the general commodity situation and the operation of commodity markets. Most importantly these phenomenon are totally absent in many developing countries. Moreover, the efforts of interested stakeholders should thus be put together and directed towards a pragmatic approach aimed at bringing both focus and priority to break the cycle of poverty in which many commodity producers and commodity-dependent countries are apparently locked-in.

Such a consultative process need to address the commodity informatics in a concerted manner by proposing specific action with respect to the following issues: facilitating collaboration among stakeholders and accomplishing greater coherence in the integration of commodity issues in development portfolios; collecting and sharing best practices and lessons learned; maximizing the mobilization of resource flows; commodity sector vulnerability and risks; mechanisms to facilitate the participation of developing country farmers in international markets; distribution of value-added services in the commodity value chain; promotion of economically, socially and environmentally sustainable approaches in production and trade of individual commodities of interest; promoting business networks within developing countries and between developing and developed country enterprises; and established commodity information and knowledge management networks (UNCTAD, 2004). Utilizing data mining techniques, much of these issues can be resolved through simple data algorithms and cost effective solutions can easily be devised.

WinIDAMS (1.2a) issued in April 2006, features interactive data import/export; wide range of data analysis techniques such as: table building, regression analysis, one-way analysis of variance, discriminant analysis, cluster analysis, principal components factor analysis and analysis of correspondences, partial order scoring, rank ordering of alternatives, segmentation and iterative typology,[5] *and can be used in establishing a marketing and monitoring system for general commodities.*

Small and Medium Sized Entrepreneurs

Small and medium sized entrepreneurs (SMEs) often face a conundrum in archiving data/information. Though tape backup systems are inexpensive and fairly reliable, but they offer poor recovery point objective (RPO)[6] and recovery time objective (RTO)[7] for critical applications, and they are usually ineffective for remote data backup. Hardware mirroring technology (which use remote copy technology to provide synchronous mirroring between two sites) offers excellent RPO and RTO, but they are highly expensive for a small or midsize business to buy and manage. Moreover, they are less than ideal for backing up remote locations, which often have low-bandwidth connections.

New solutions based on asynchronous software-based replication can achieve the acceptable RTO and RPO objectives for small or midsize business' critical applications without adding cost and complexity of the synchronous replication approach. Furthermore, in software-based replication, only the bytes that are actually changed by each write (not the entire block of information or the whole file) are replicated. Therefore, in comparison to synchronous replication solutions, asynchronous approach offers lower load on the production servers, faster updates, and the ability to send replication updates across low-bandwidth Internet networks (Intransa, 2005).

E-Commerce

Utilizing information and data grid major elements of e-commerce applications can easily be developed. These involve, electronic data interchange (EDI) and allowing more flexibility, negotiated exchange and encouraging entrepreneurial activity, e-marketplace is beginning to be enormous repository of data and information. Use of WWW metadata technologies, as RDF (resource description framework) and XML (extensible markup language) in technical, scientific and engineering applications are becoming paramount and they are being used in e-commerce applications too nowadays.

It is a fact that, the knowledge-store of an entrepreneurship is commonly available in gray literature. However, recent researches integrated within the scientific process associated with metadata encourage researchers assisting in development of wealth creation processes, including technology transfer, and effectively transmutes across directly into e-Commerce. Furthermore, beyond metadata-based cataloguing of knowledge assets, utilizing recent data mining techniques, it is possible to do text—or multimedia—data mining to extract further refined knowledge (Jeferry, 2000).

Healthcare Sector

The advantages of the ICTs in the complex health-care sector are already well known and well stated. It is, nevertheless, paradoxical that although the medical community has embraced with satisfaction most of the technological findings allowing the improvement of patient care, but health-care informatics has not been advanced that much.

An information model for knowledge management (KM) based upon the use of key performance indicators (KPIs) in health-care systems has been developed. Founded on the use of balanced scorecard (BSC) framework (Kaplan/Norton) and quality assurance techniques in health care (Donabedian), this research is carrying out a patient journey centered approach that drives information flow at all levels of the day-to-day process of delivering effective and managed care, toward information assessment and knowledge discovery. Furthermore, in order to persuade health-care decision-makers to assess the added value of KM tools, its methodologies include new performance measurement and performance management techniques at all levels of a health-care system. Thus the KPIs are forming a complete set of metrics enabling the performance management of a regional health-care system. In addition, the performance framework is being technically applied by the use of state-of-the-art KM tools, such as data warehouses and business intelligence information systems. Hence, in technological sense, the infrastructure is becoming an important KM tool that enables knowledge sharing amongst various health-care stakeholders and between different health-care groups. (Berler, Pavlopoulos & Koutsouris, 2005)

Moreover, health research including health systems research nowadays has become an essential component in achieving the health related MDG's, especially the targets set out in reducing maternal and child mortality. In this context, to improve universal access to sustainable quality health care implies a coherent approach to the health system including research into health information systems and the organization and management of functional and cost-effective health services that are socially equitable and financially sustainable. Data mining application can enrich the health-care informatics and improve the necessary support services by increasing the availability of reliable data at minimum effort.

Tourism

Tourism is, without a reservation, the single largest economic sector in many countries and treated as a significant element of economic activity in almost every country. For many developing countries, tourism is of strategic importance and became a major source of foreign exchange earnings. However, this economic importance is not always effectively reflected in public policy. Good tourism policy must incorporate a dynamic approach to tourism development rather than a passive reaction to externally created prospects.

Thus, given tourism's central role in the economy of many countries (specially, Caribbean and South East Asian countries), tourism authorities must study various aspects of the industry from both the demand and supply sides. The information collected from surveys, desk and online research could be used to customize policies and strategies aimed at redressing any supply problems or enhancing the tourism product and providing a more competitive and attractive destination to the visitor. Research must, however, be based on timely, reliable and accurate information, which is essential for effective policy formulation and decision-making. Thus tourism enterprises had to depend on accurate tourism statistics, which help them to undertake research, prepare business plans and assist in the design of promotion and marketing strategies and campaigning. Similarly, tourism policy-makers require critical information to develop proper forward planning of the sector. This involves effective anticipation and necessary management change in order to maximize the economic and financial benefits of tourism. In this aspect, the important concern is that critical information can only be obtained through high-quality research which facilitates proper monitoring of tourism-related policies, evaluation of the effectiveness of specific tourism initiatives, benchmarking the performance of a particular destination, comparative analyses, and observing trends of visitors (Andrew, 2005).

As tourism is an information-intensive service, the UNCTAD e-tourism initiative has been designed to provide developing countries the technical means to promote, market and sell their tourism services online. Partners in e-tourism are the UNCTAD member States, the World Tourism Organization, UNESCO, national tourism authorities and many universities. Other potential partners include regional associations of developing countries, transport operators and IT companies. (UNCTAD, 2004)

Anti-Drug Network

Accumulating information about drug manufacturers, their traffic route, usage pattern, supply chain, and at the same time integrating related databases of users, resellers and vendors can establish a useful drug network (rather, antidrug network) in detecting illegal drug trafficking. Data mining makes it possible.

Anti Drug Network (ADNET): A middle-aged man in a light blue Mustang is on the way to enter the United States from Mexico through one of the numerous customs checkpoints along the southwest border. He is confident no one will suspect that he is transporting more than 10 pounds of heroin in secret compartments inside his vehicle; he has done it before and he plans to do it again and again.

But, a customs system operator at a site near El Paso, Texas, uses the Anti-Drug Network (ADNET) system to access data on the driver and his car via his license plate. It's just routine check and takes a few moments.

The agent quickly learns through a system that accesses a large data warehouse of information on crossings, seizures, and motor vehicles that the driver makes this trip on a regular basis, at a regular time, but this trip is different. She decides it is worth her time and trouble to continue the inspection. Ten minutes later, she finds more than a dozen small packages of white powder inside the vehicle. The drugs were seized and the driver was arrested.

Situations like this occur almost daily across the many ports of entry along the Mexican/U.S. border and other entry points into the United States. It may happen to many other countries and continents. Sophisticated data-sharing systems developed by the ADNET community (i.e., Department of Defense, U.S. Coast Guard, Department of Justice, Department of State, Department of Treasury, Federal Communications Commission, and the intelligence community) give U.S. drug and law enforcement officials a cache of information needed to track the flow of illegal narcotics and other dangerous substances into the country. (MITRE, 2001)

Network Business Intelligence

Network business intelligence leverages the traffic management system to mine information about exactly what is flowing on the network, generating reports that combine application, user, and server information and technical views. Network operators can then determine the network's application and protocol mix, as how applications are performing, what impact they are having on other traffic streams, and what the network requirements are for all traffic (Allot, 2005).

Deep packet inspection (DPI) identifies application types when port number alone is not enough by looking further inside the packet header. This is particularly helpful for applications using dynamic port numbers, such as voice over IP (VoIP), hyper text transfer protocol (HTTP), Citrix-based remote-access applications, and the Microsoft NetMeeting conferencing application. HTTP consistently uses port 80; but at the same time many web applications and traffic types use HTTP. So merely a port number is not adequate for identifying specific HTTP applications. Added with information about user, application, protocol, and machine behavior on the network, one can configure the traffic management system to automatically classify and shape all traffic in a way that optimizes the network usage to maximize the return on investment (ROI). (Allot, 2005).

Environment

Information technologies are important not only for their growing use in decision-making processes and knowledge management systems, but also their use diversifications. Their use has yielded significant improvements in the efficiency of energy and materials use and contributed to economic expansion without the increases in environmental impacts leading to efficiency improvements. Advances in information technology are likely to continue to provide opportunities for the development of improved environment and ambient livelihood. Data mining assists to accumulate historical data and change pattern of local environment parameters to make intelligent decision for protecting the degradation of environment, especially if they are human made (Allenby, Compton & Richards, 2007).

Amongst the variety of datasets that are involved in the knowledge society, spatial (or map) information takes a major place in terms of content. These spatial information sets are essential to make sound decisions at the local, regional, and central level planning, implementation of action plans, infrastructure development, disaster management support, and even in business development. Natural resources management, flood mitigation, environmental restoration, land use assessments and disaster recovery are just a few examples of areas in which decision-makers are benefiting from mining spatial information.

With the accessibility of satellite-based remote sensing data and the association of spatial datasets around geographical information systems (GIS), united with the global positioning system (GPS), the processes of semantic spatial information systems are now a reality. The advent of GIS technology has transformed spatial data handling capabilities and in many cases made it essential for reexmining the roles of government with respect to the supply and availability of geographic information. Using GIS technology, users are now able to process maps, both individually and along with tabular data and mix them together to provide a new perception, the spatial visualization of information. (Rao, 2002)

Ecosystem Management

Data mining application with databases on sustainable use of natural resources and the associated ecosystems (ecological, hydrological, limnological, climatic, social, and economic conditions), especially in developing countries are critical to ecosystem management, and can be very effective for optimization of resources to improve the ecosystem.

Research activities on sustainable management of three most vulnerable ecosystems "humid and semi-humid ecosystems," "coastal zones" and "arid and semi-arid ecosystems," can take care of integrated water management on river basin scale (recommended by the European Union Water Initiative as well as Article 10 of the Convention on Biological Diversity [CBD]) and support the implementation of the provisions related to research and sustainable use included in the CBD workprograms on Dry and Subhumid lands, Inland water and Marine and Coastal Biodiversity. The ecosystem approach is a strategy for the management of land, water and livelihood resources that promote conservation and sustainable use in an equitable way and places people and their natural resource use practices at the core of decision making. (European Commission, 2005)

Management dynamics of humid and semi-humid ecosystems under optimized treatment may lead to sustainable use of renewable natural resources; identification of policy options and/or management strategies for harnessing judicious use of such resources focused on integrated approach and analysis of natural and agro-resource use at local and/or regional levels (sustainable water management, forest ecosystem reclamation, biodiversity management). However, this sort of longer term skill development and decision-making processes require intensive implementation of data mining schemes.

Mineral Resources

Mineral resources are important for all the nation's citizens and these are essential for individuals, companies, and communities that depend on minerals production for income and broader economic development. Like food, air, and water, minerals are fundamental ingredients of human life. However, science and information on mineral resources underpin private and public decisions that determine whether, under what conditions, and at what costs minerals become available to producers and consumers.

Use of data mining through adaptive learning and accumulating information along the years on their availability, deposit, nature and process of extraction, and analyzing their pertinent values proper resource management can be performed. A good news is that, using recent technologies advances in minerals science and improvements in minerals information contribute to greater availability of minerals, extraction at lower cost and with less environmental damage; help society respond to the depletion of known mineral deposits and contribute to the substitution of relatively abundant minerals for increasingly scarce ones; and help develop alternative sources of supply for minerals subject to unexpected supply disruptions (BESR, 2004).

Earth System

Understanding the Earth system is essential to augment human health, safety, and welfare, alleviating human suffering including poverty by protecting the global environment, reducing disaster losses, and achieving sustainable development.[8] In this aspect, appropriate data mining applications may be implemented to achieve comprehensive, coordinated and sustained observations of the Earth system, in order to improve monitoring of the state of the earth, boost understanding of earth processes, and improve prediction on the behavior of the earth system. These would eventually accumulate to produce a long-term quality data bank to make intelligent decision and provide benefit to the society by reducing loss of life and property from natural and human-induced disasters; understanding environmental factors affecting human health and well-being; improving management of energy resources; understanding, assessing, predicting, mitigating, and adapting to climate change; improving water resources management through better understanding of the water cycle; improving weather information, forecasting and warning; improving the management and protection of terrestrial, coastal, marine ecosystems; supporting sustainable agriculture; combating desertification; and understanding, monitoring and conserving biodiversity (UNDP, 2001).

A Few Advanced Uses

Apart from the fields that have been described so far, the field of data mining is exploding in many aspects. New techniques in data mining are accelerating research across almost all the scientific disciplines. Data mining techniques are being used to facilitate research and experimentation in the development of advanced materials. Researchers are even using quantum mechanical methods to mine crystal structure and property databases to calculate and predict the properties of new ternary and quaternary materials. In this way, each new experiment expands the database and provides new insights into the laws of physics and chemistry (Carty, 2002; Halpern, 2003; Kargupta, Joshi, Sivakumar & Yesh, 2004). Moreover, new lines of research are addressing the links between natural systems and socioeconomic systems, as well as sustainable consumption and production patterns (Science Blog, 2002; Nwabueze, 2003).

Lixin Zhang, President of the World High Technology Society stated that science and technology contributed towards economic growth without sacrificing a great loss of natural resources. Through appropriate data mining, information technology could dig out a lot of information encouraging further penetration of new ICT into developing countries and countries with economies in transition aimed to combine traditional knowledge with knowledge provided by the scientific community. (WSSD, 2002)

There are advanced uses of data mining in the field of development, integration and validation of GRID technologies and their applications in research, industry and for addressing societal challenges. Research in this aspect ranges from GRID technology building blocks to grid-related middleware and large-scale applications. Research also includes test beds for computational GRIDs that are the basic layer for harnessing processing power by distributing massive computational tasks to numerous resources (compute cycles and data storage) over matching communication links. Consequently, information and knowledge GRIDs allow access to dispersed information, and knowledge discovery and extraction from spread knowledge resources. In this context, they make use of cognitive techniques and tools such as data mining, machine learning, content semantics, ontology engineering, information visualization, and intelligent agents. (Alpaydin, 2004; CORDIS, 2006).

Internet database that contains human drug metabolism data and in turn, be made available to users across the globe via a nonprofit basis. Depending on the chemical structures for both the parent drug or xenobiotic and the various metabolic biotransformation products, the Human Drug Metabolism Database (hDMdb) will be extremely useful to both the medicinal and toxicological chemistry. During the production of open databases with chemical structures connected to biological properties demands new tools of data mining (IUPAC, 2005).

Similarly, database tomography is a textual database analysis system consisting of two major components: (a) algorithms for extracting multiword phrase frequencies and phrase proximities (physical closeness of the multiword technical phrases) from any types of large textual database, to augment, and (b) interpretative capabilities of the expert human analyst; mostly dependent on the application of appropriate data mining techniques (Kostoff, Tshiteya, Pfeil & Humenik, 2002).

Other Potential Usage Areas

Managing Arid and Semi-Arid Ecosystems

Research on arid and semiarid ecosystem dynamics under varying degrees of human activity pressure lead to more sustainable use of renewable natural resources in natural, rural and peri-urban areas. Data mining applications can be applied to identify opportunities for enhanced economic and sustainable production by analyzing natural and agro-rescurce use systems at local, regional level and/or international levels through an integrated approach. Here, the resource management practices with historical data from indigenous people plays critical role in planning and implementing sustainable management strategies of renewable natural resources. Appropriate tools, including information systems, decision support tools, criteria or indicators of sustainability and rehabilitation, past and present examples of participatory approaches, based on data mining and existing datasets could be a cost-effective support for dry land ecosystem management and policies.

Furthermore, it is complemented that, the longer-term outcome of data mining applications would lead to technological, management and policy research, in the following focal areas:

1. **Improved agriculture and agroforestry systems:** By taking into account traditional knowledge, the database system should also look at opportunities beyond farm boundaries in order to diversify income generation and sustain rural and periurban livelihoods. In addition, land improvement or rehabilitation strategies should look into the broader socioeconomic, inclusive demographic, institutional and political dimensions of desertification or ecosystem degradation.
2. **Sustainable, integrated water resource management (IWRM) at river-basin:** The data mining system should address such dimensions as; increasing use of efficiency, particularly in irrigated agriculture; agroforestry, increasing recycling and reuse for tree growing, (periurban plantations, etc.), including innovative multipurpose utilization requiring integrated management attentive to quantity and quality aspects; control of sediment load, erosion, flash floods, control of private use, pollution and water logging; water supply/resource management at basin level in order to meet competing demands including up-stream and down-stream effects in relation to periurban areas and groundwater management in terms of quantity, quality and change in water table.
3. **Research in forest ecosystem restoration and reclamation techniques:** Research in terms of data mining should include afforestation, vegetation rehabilitation techniques especially using native species of economic value (to mitigate or to halt soil, water and land cover degradation caused by unsustainable forestry and farming practices or unsuitable urban settlements). Research on restoration or enrichment of degraded lands and secondary forests with a concern for conservation of biodiversity, can tap new market opportunities or mitigate the negative environmental impacts of market systems through proper data analysis. The ecosystem approach, where huge concentration of data sets are visible, should seek to develop tools to make appropriate balance between conservation and the use of biologi-

cal diversity while taking full account of the cultural, social roles (gender concerns) and the function in biodiversity conservation and land rehabilitation (European Commission, 2005).

4. **Biodiverse, biosafe, and value added crops:** Research to increase the sustainable use and productivity of annual and perennial under-utilized tropical and subtropical crops and species is important for the livelihoods of local populations. These crops have potential for wider use and could significantly contribute to food security, agricultural diversification and income generation. Innovative tools and data mining techniques for the characterization, development and use of crops with enhanced tolerance to abiotic stress and in particular:
    - o Tolerance to drought, salinity, heat, cold
    - o Enhanced nutrient uptake
    - o Enhanced tolerance to heavy metals and acid soil; will enrich the related information management systems.
5. **Aquatic farming systems:** "Farming down aquatic food webs" with particular attention to economic viability, social acceptability and an enabling institutional environment are being regarded as key dimensions to sustainable aquatic farming systems. This combination is expected to improve the conditions for food-insecure households through new knowledge products, processes and policy relevant dialogue. In this aspect, particular emphasis should be given to enhanced participatory approaches with strong possibilities of generating an impact in society and promoting social empowerment through knowledge (European Commission, 2005). Accumulation of intrinsic, but apparently invisible information and data at local levels are necessary preconditions to develop data mining solutions in this dimension.

FUTURE ISSUES AND CHALLENGES

The issue of scientific data collection and management has traditionally been addressed on an informal basis, initially by individual scientists, who generally felt that they had to collect their own research data.[9] Later, groups of scientists in the same or related disciplines began to collaborate on the development of larger databases for general-purpose use. This approach had the added scientific advantage of encouraging research testing hypotheses on the same body of data by using different data mining tools (Earth Institute News, 2005; Grossman, Kamath, Kegelmeyer, Kumar, & Namburu, 2006; Kargupta, Joshi, Sivakumar, & Yesh, 2004).

The reality of data explosion in multidimensional databases is an astounding and at the same time, an extensively misunderstood phenomenon. To implement proper data mining utility, it is essential to understand what data explosion is, what causes it, and how it can be avoided, because the consequences of ignoring data explosion can be very costly, and in most cases, result in project failure (Potgieter, 2003). Moreover, the exponential growth of business information generated every day means even more and more data has to be backed up. Regardless of the circumstances customers expect services to resume instantly after any disruption. In addition, the increasing need to access data almost around the clock has dramatically shrunk the time permitted to backup data. Foremost, today's data protection challenge poses substantial risks to companies of all sizes, but they pose the greatest risk to small and midsize businesses (NSI Software, 2004; Ville, 2007).

However, one of the problems of data explosion is that it results in a massive database and the size of the database in one product can literally be hundreds and even thousands of times greater than the same database in another product. Taking this as an opportunity, rather than admitting about the problems of data explosion, the vendor with the massive database argues that his database is handling large data sets, while the vendor will imply that the vendor of the smaller database (a database without data explosion) cannot address large enterprise datasets. This is a wrong concept. Though the correct analysis should be to compare sizes with equal volumes of base data, but due to the size of the databases are so profoundly different, prospective clients find it hard to believe that such dramatic differences are possible with similar datasets (Dorian, 1999; Potgieter, 2003; Tan, 2006).

This creates confusion at the client's end through misinterpretation by the vendor. Proper data mining tools can eradicate this problem of data explosion, though many other factors are involved in this. Future research work may be carried out in this facet for a longer period of time. Research work should be continued in case of massive databases to restrict the data explosion. This will reduce introduction of expensive hardware in the procurement process, reduce the loading and calculation times, reduce establishment and operational costs, curtail any hidden cost that may arise within the undesired processes, provide intelligent enterprise solutions at reduced cost and effort, and foremost will be able to save projects from being total failures (Potgieter, 2003).

Future research should also focus advanced techniques of stream data mining and its applications that includes data compression, data visualization, intelligent logging systems, sensor network systems, integrated sensor devices, secure storage and firewalls for cracking and SPAM mails. At the same time, research work should progress in the field of information visualizing tools and advanced applications for spatial data, spatio-temporal data, high dimensional data and graph-structured data.

Research work should also continue to reduce challenges in case of data interpretation, data integrity, data compartmentalization, and data archival by keeping transparency of data chain. There must be a clear relationship among the data generator, data collector, data processor, data archiver and data disseminator (CORDIS, 2006; Wang & Fu, 2005). Proper data mining method will be able to take care of these issues.

In case of health sector data mining, research should focus on diseases having considerable impact on the economic development perspectives of the affected communities taking into account their socioeconomic status. Research should provide new knowledge on biology, epidemiology and technologies relevant for sustainable surveillance and control of diseases on a regional scale; on innovation and improvement of existing interventions and help to implement appropriate strategies and policies for prevention, control, and treatment.

Regarding data mining of learning content, social service delivery, e-governance and e-government challenges will remain, as these processes are very dynamic in nature and mostly dependent on various factors that are inter-dependent on each other. Similarly, in case of mining agricultural data, one should learn at first about the agricultural issues, including the perspective of food security to establish data mining algorithm, and perhaps the same condition is applied to almost all field of data mining.[10]

Most of the time, cost of professional data management is not appreciated. Furthermore, diversity of technology makes data archiving practice and deployment more complex, and as more and more people expect access to scientific and technical data, science and data management become more interactive, and more complex. To face this, innovations in data archiving and management practice and technology is essential to reduce any inevitable expenses. In addition, to make them sustainable, data archival and data centers need to provide public services that add value to the data collections.

SPAM and neuromarketing form another complicated context in the area of data mining. With extended use of the Internet, the potential for subtle and not so subtle control for gross invasion of privacy, is coming at the front. However, the best-known weapon against such control, that is, encryption, creates as many problems as it solves, due to its paradoxical allowance of certain factors that further elude the law. It is a newly developed situation in the domain of Internet and as days are passing, it is becoming critical, and perhaps sometimes harmful, when networks are threatened by SPAM or undesired emails/contents.

This impact is usually associated with the demand of so called, universal access. But universal access/service alone does not suffice (Servaes, 2004). In order to develop the proper rights and responsibilities in the conditions and complexities of a knowledge society, demand and provision of information need to be compromised. The situation aggravates further with the cognitive abilities that are necessary to navigate in such a complex information space. All these problems are compounded in the underprivileged parts of the world, but they need to deal squarely with this challenge. Mining parameters pertaining to web browsing, balancing the demand-supply chain, behavior pattern of the end user, justifying the nature of content and search optimization can ease the problem in a very small sense. However, there are many other complex issues involved in this context and need further research.

CONCLUSION

Due to the dynamic nature of digital content development and delivery methods, selection of proper data mining methods remains critical, especially when the contents are related to; scientific, technical, medical, agricultural, social, music, or online computer and video games. Recently emerged network convergence and rapid diffusion of high-speed broadband has shifted attention towards broadband content and applications that promise new business opportunities, growth and employment. Moreover, the potential for digital content growth is very high and growth is only just the beginning. But, technologies to assure the diffusion of content and content products are not increasing in that pace, as still the processes remain R&D-intensive to establish a common platform for faster networks, new standards, software intensive products, virtual reality applications, data-base management and others. In addition, mobile content and applications in the field of mobile telecommunication service and content industry are also generating huge contents for innovative data mining.

In these revised situations, relationships between content originators and final users are changing. Further to these changes, certain intermediaries are being created and attitudes to content ownership and acquisition are also changing. To suffice more, complete disintermediation and direct contact between content creators and content users has not yet been developed to a significant extent.

A favorable scenario is that with the advent of ICTs, and particularly the Internet, people who can afford the hardware and connectivity can gain access to a wealth of content. Much of them are free, or obtained at nominal cost. But, particularly end users in the developing countries cannot even afford to pay the nominal fees. So, there is an appeal for open and free content provision. Those who advocate the creation of these open and free educational (or knowledge) resources believe in the principal that education (or knowledge development) is indeed a basic human right. But, a fundamental tension prevails in fostering the development of such resources as; how will they ultimately be paid for? At a time when knowledge is increasingly becoming commoditized, and universities are running more like businesses; there is a significant counter-flow of arguments that are limiting the institutional capacity to produce open educational resources (OER) (Unwin, 2006). These features are in a way disrupting R&D in this track of data mining.

It is true that, advances in technology have changed data collection methods and popularized large-scale data sets, especially in higher education. But, to turn the abundant raw data into valid knowledge, researchers need to realize that traditional statistical techniques have weaknesses when used to study large volumes of data. In this context, more effective and balanced analytical tools are necessary (Larose, 2004; Smart, 2005; Wang, 2003). The concept of the "learning organization," including "lifelong learning" for staff, is now recognized as a key element in corporate strategies. This will reinforce consistency, common identity, shared corporate culture, common actions, clear responsibilities, coordination and dissemination of good practice (Markus, 2005). In the field of natural science, the efficient handling, updating and maintenance of the spatial data infrastructure need highly qualified, properly trained staff.

Similarly, in the field of agriculture development of innovative and efficient data mining for environment-friendly, post-harvest, storage, processing and marketing methods for products derived from such crops, with the objective of increasing market accessibility and product-added value by promoting the development of niche export markets remains a challenge. At the same time, development and dissemination of sustainable improved production and management practices taking into account traditional knowledge and innovative methods for the conservation and use of genetic resources in food and agriculture looks for pragmatic data mining techniques. Moreover, policy, regulatory and institutional issues related to coexistence of multisource crops in the agricultural and food chain, including trade and food security issues throw further challenges in terms of improved and accurate data mining techniques.

In health sector, challenges for research include the need to develop cross-sectorial policies to ensure sustainable measures to fight diseases with a specific focus on poverty reduction through health improvement. Fundamental issues such as gender, ethics and equity must be taken into account. However, critical research and analysis call for data mining solutions to preserve related and relational data for a longer period of time.

Furthermore, to collect the data and to offer it up-to-date to a vast number of organizations, public investors, stakeholders and common citizens require technical infrastructures and organized processes. Though the use of ICT is increasingly acknowledged as a strategy to assist in this aspect, but without an adequate study on the applicability and the natural and social context of ICT, it may become useless or even counterproductive. As such, installing a telecenter in an urban zone with little or telecommunication infrastructure may sound like an improvement on the quality of life, but if the increasing cost for electricity and networking cannot be afforded in the long run, the incentive will fail (UNDP, 2001). Therefore, not only the data mining solutions are facing challenges, but also, the total context of content accumulation system is not beyond the challenge barrier.

Finally, through all these success cases, it is observed that partnership among implementing agencies in scientific and technical matters should address key societal issues through interdisciplinary research approaches by combining the natural and social sciences. In this context, the overall objective of the research goal should be to develop equitable and strong scientific partnerships among developing countries in order to contribute to their sustainable development by means of human capital development, mobility and institution building (European Commission, 2005). Therefore, applying data mining tools to empower knowledge society, or rather strengthen human capacity demands intricate methodologies and extensive researches through in-depth studies.

REFERENCES

Ackoff, R. L. (1989). From data to wisdom. *Journal of Applied Systems Analysis, 16*, 3-9.

Allenby, B. R., Compton, W. D., & Richards, D. J. (2007). *Information systems and the environment overview and perspectives.* Retrieved April 13, 2008, from http://books.nap.edu/openbook.php?record_id=6322&page=1

Allot (2005). The traffic management handbook. MN: Allot Communications Ltd.

Alpaydin, E. (2004). *Introduction to machine learning (adaptive computation and machine learning).* The MIT Press.

Andrew, M. (2005). *The role of research in sustainable tourism policy-making.* Paper presented at the First Regional Sustainable Tourism Policy and Intersectoral Planning Workshop Grand Barbados Hotel, Barbados, West Indies.

Berler, A., Pavlopoulos, S., & Koutsouris, D. (2005). Using key performance indicators as knowledge-management tools at a regional health-care authority level. *IEEE Trans Inf Technol Biomed, 9*(2), 184-192.

Berry, M. J. A. & Linoff, G. S. (1997). *Data mining techniques for marketing, sales and customer support.* John Wiley & Sons.

Berry, M. J. A. & Linoff, G. S. (1999). *Mastering data mining: The art and science of customer relationship management.* John Wiley & Sons.

Berry, M. J. A. & Linoff, G. S. (2000). *Mastering data mining.* John Wiley & Sons.

Berry, M. J. A. & Linoff, G. S. (2002). *Mining the Web: Transforming customer data.* John Wiley & Sons.

Berry, M. J. A. & Linoff, G. S. (2004). *Data mining techniques: For marketing, sales, and customer relationship management.* Wiley Computer Publishing.

Berson, A. & Smith, S. J. (1997). *Data warehousing, data mining, and OLAP.* McGraw Hill.

Berson A., Smith, S. J., & Thearling, K. (1999). *Building data mining applications for CRM.* McGraw Hill.

BESR (2004). Board on Earth Sciences and Resources (BESR), *Future challenges for the U.S. Geological survey's mineral resources program (2004).* Washington, D. C.: The National Academies Press.

Bozdogan, H. (Ed.) (2004). *Statistical data mining and knowledge discovery.* CRC Press.

Braha, D. (Ed.) (2001). *Data mining for design and manufacturing: Methods and applications.* Kluwer Publishers.

Bramer, M. A. (Ed.) (1999). *Knowledge discovery and data mining: Theory and practice.* IEE Books.

Carty, A. J. (2002). Scientific and technical data: Extending the frontiers of research. In *Proceedings of the Opening Address at CODATA 2002: Frontiers of Scientific and Technical Data,* Montréal, Canada.

Cerrito, P. (2006). *Introduction to data mining using SAS enterprise miner.* SAS Press.

Chakrabarti, S. (2002). *Mining the Web: Discovering knowledge from hypertext data* (1st ed.). Morgan Kaufmann.

CIDA (2005). *CIDA's strategy on knowledge for development through information and communication technologies (ICT).* Canadian International Development Agency. Retrieved April 13, 2008, from http://www.acdi-cida.gc.ca/ict

Cios, K. J. (Ed.) (2000). *Medical data mining and knowledge discovery.* Physica-Verlag (Springer).

Cios, K., Pedrycz, W., & Swiniarski, R. (1998). *Data mining methods for knowledge discovery.*

Codata (2002). Committee on data for science and technology (CODATA). In *Proceedings of the*

Workshop Synthesis on Archiving Scientific and Technical Data, Pretoria, South Africa. Retrieved April 13, 2008, from http://www.tgdc-codata.org.cn/english/Html/SA-CT.html

COL (2003). *Find information faster: COL's "Info-mining" tools*. Retrieved April 13, 2008, from http://www.col.org/colweb/site/pid/2927

CORDIS (2006). *GRID technologies and applications through CORDIS*. Community Research & Development Information Service. Retrieved April 13, 2008, from http://www.environmonument.com/projects.htm

Delmater, R. & Hancock, M. (2001). *Data mining explained: A manager's guide to customer-centric business intelligence*. Digital Press.

Dorian, F. (1999). *Data preparation for data mining*. Morgan Kaufmann.

Dunham, M. (2003). *Data mining introductory and advanced topics*. Prentice Hall.

Earth Institute News (2005). Scientific community must develop cross-disciplinary standards and practices in academia. Retrieved April 13, 2008, from http://www.earthinstitute.columbia.edu/news/2005/story05-01-05c.html

Ebecken, N. F. F., Brebbia, C. A., & Weigend, A. (2000). *Data mining II* (1st ed.). Computational Mechanics, Inc.

European Commission (2005). *Specific programme for research technological development and demonstration: Integrating and strengthening the European research area, 2005 Work Programme (SP1-10)*.

Fayyad, U., Piatetsky-Shapiro, G., Smyth, P., & Uthurusamy, R. (Eds) (1996). *Advances in knowledge discovery and data mining*. AAAI/MIT Press.

Fayyad, U., Grinstein, G. & Wierse, A. (2001). *Information visualization in data mining and knowledge discovery*. Morgan Kaufmann.

Gökmen, A. et al. (2004). *Balaban Valley Project: Improving the quality of life in rural area in Turkey, 7*(Dec 2004). Retrieved April 13, 2008, http://www.geocities.com/doriendetombe/detombevol7menmbalabanabstract.html

Grossman, R. L., Kamath, C., Kegelmeyer, P., Kumar, V., & Namburu, R. (Eds.) (2006). *Data mining for scientific and engineering applications (Massive computing)* (1st ed.). Springer.

Halpern, J. Y. (2003). *Reasoning about uncertainty*. MIT Press.

Han, J. & Kamber, M. (2000). *Data mining: Concepts and techniques* (1st ed.). Morgan Kaufmann.

Han, J. & Kamber, M. (2006). *Data mining: Concepts and techniques* (2nd ed.). Morgan Kaufmann.

Hand, D. J., Mannila, H., & Smyth, P. (2000). *Principles of data mining*. MIT Press.

Hernández, V., Göhring, W., & Hopmann, C. (2004). Sustainable decision support for environmental problems in developing countries: Applying multi-criteria spatial analysis on the Nicaragua Development Gateway niDG. *Research on computing science* (Vol. 11, pp.136-150). México: Instituto Politécnico Nacional.

ICDM (2003). ICDM 2003 tutorial. In *Proceedings of the Third IEEE International Conference on Data Mining, Sponsored by the IEEE Computer Society*, Melbourne, Florida. Retrieved April 13, 2008, from http://www.cs.sfu.ca/~ester/ICDM2003/Lazarevic.abstract.htm

Intransa (2005). *Managing storage growth with an affordable and flexible IP SAN: A highly cost-effective storage solution that leverages existing IT resources*. CA: Intransa, Inc.

IUPAC (2005). *Chemistry and human health council report: 2003-2005*. International Union of Pure and Applied Chemistry, IUPAC Division VII. Retrieved April 13, 2008, from http://www.iupac.org/news/archives/2005/43rd_council/Item_09_Div_VII.pdf

Jeffery, K. G. (2000). *The grid for e-science: E-commerce benefits, information technology department*. CLRC, ITD.

Kargupta, H., Joshi, A., Sivakumar, K., & Yesh, Y. (Eds) (2004). *Data mining: Next generation challenges and future directions*. AAAI Press.

Kommers, P., Kinelev, V., & Kotsik, B. (2003). ICT in secondary education for the knowledge society. In T. Varis, T. Utsumi & W. R. Klemm (Eds), *Global peace through the global university system*. The Finnish National Commission for UNESCO, University of Tampere, Hämeenlinna, Finland.

Kostoff, R. N., Tshiteya, R., Pfeil, K. M., & Humenik, J. A. (2002). *Power source text mining using bibliometrics and database tomography*.

Larose, D. T. (2004). *Discovering knowledge in data: An introduction to data mining*. Wiley-Interscience.

Markus, B. (2005). *Building spatial knowledge infrastructure*. Paper presented at the ISPRS Workshop on Service and Application of Spatial Data Infrastructure, XXXVI, Hangzhou, China. Retrieved April 13, 2008, from http://www.commission4.isprs.org/workshop_hangzhou/papers/65-70%20Bela%20markus-A103.pdf

MITRE (2001). *Stopping traffic: Anti drug network (ADNET)*. MITRE Digest Archives. Retrieved April 13, 2008, from http://www.mitre.org/news/digest/archives/2001/adnet.html

NSI Software (2004). *Six tips small and midsize businesses can use to protect their critical data*. NJ: NSI Software.

Potgieter, J. (2003). *OLAP data scalability: Ignore OLAP data explosion at great cost*. NSW Australia: SPF Pty Ltd.

Quéau, P. (2001). *The information society and the global good*. Retrieved April 13, 2008, from http://goanna.cs.rmit.edu.au/~aym/rinseap/bali/QueauTalk.html

Rao, M. (2002). *Systems design of a national spatial data*. Bangalore: Indian Space Research Organisation Headquarters.

Rud, O. P. (2001). *Data mining cookbook: Modeling data for marketing, risk, and CRM*. Wiley.

Science Blog (2002). Partnerships, finance, sustainable production and consumption patterns. Press Release: United Nations. Retrieved April 12, 2008, from http://www.scienceblog.com/community/older/archives/L/2002/A/un020319.html

Servaes, J. E. J. (2004). Knowledge is power (revisited): Internet and democracy. In P. Lee (Ed.), *Proceedings of the International Conference on Internet Communication in Intelligent Societies* (pp. 1 – 16). Chinese University of Hong Kong, Hong Kong. Retrieved April 13, 2008, http://www.com.cuhk.edu.hk/conference/2004/)

Smart, J. C. (Ed.) (2005). *Higher education: Handbook of theory and research* (Vol. 20). Virginia Tech: Springer.

Tan, Pang-Ning, Steinbach, M., & Kumar, V. (2005). *Introduction to data mining*. Pearson Addison Wesley.

Tan, Pang-Ning (2006). *Introduction to data mining*. Addison Wesley Publication.

Thearling, K. (1995). *From data mining to database marketing*. DIG White Paper 95/02. Retrieved April 13, 2008, from http://www.cs.uvm.edu/~xwu/icdm/cfp-03.shtml

Nwabueze, K. (November 30, 2003). A case study: Role of technology venture capitalist market in developing countries, data mining, integration, and analysis. *Timbuktu Chronicles*. Retrieved April 13, 2008, http://timbuktuchronicles.blogspot.com/2003_11_01_archive.html

UNCTAD (2004). UNCTAD XI multi-stakeholder partnerships, information and communication technologies for development (ICTfD). In *Proceedings of the United Nations Conference on Trade and Development.* Retrieved April 13, 2008, http://www.unctad.org/en/docs//td-l380add1_en.pdf

UNDP (2001). *United Nations Development Program: Making new technologies work for human development.* Oxford: Oxford University Press.

Unwin, T. (2006). *Facing the challenges, dgCommunities: Open educational resources.* Retrieved April 13, 2008, from http://topics.developmentgateway.org/openeducaion

Ville, Barry de (2001). *Microsoft data mining: Integrated business intelligence for e-commerce and knowledge management.*

Ville, Barry de (2007). *Microsoft data mining: Integrated business intelligence for e-commerce and knowledge management.* Digital Press.

Wang, J. (2003). *Data mining: Opportunities and challenges.* IRM Press.

Wang, L. & Fu, X. (2005). *Data mining with computational intelligence (advanced information and knowledge processing)* (1st ed.). Springer.

Witten, I. & Frank, E. (1999). *Data mining, Practical machine learning tools and techniques with Java implementations.* Morgan Kaufman.

Witten, I. & Frank, E. (2005). *Data mining, practical machine learning tools and techniques* (2nd ed.). Morgan Kaufman.

WSSD (2002). *Press release for fifth partnership plenary, world summit on sustainable development.* Johannesburg, South Africa.

ENDNOTES

1 http://webworld.unesco.org/infoethics2000/themes.html accessed on June 14, 2006

2 http://webworld.unesco.org/infoethics2000/themes.html accessed on June 14, 2006

3 Retrieved March 06, 2007 from Retrieved April 20, 2006 from http://www.cathalac.org/index.php?option=com_content&task=view&id=173&Itemid=256

4 Retrieved May 06, 2007 from http://curric.dlib.vt.edu/DLcurric/proposalsummary04.pdf

5 http://portal.unesco.org/ci/en/ev.php-URL_ID=2070&URL_DO=DO_TOPIC&URL_SECTION=201.html accessed on June 15, 2006

6 RPO: It is the point in time to which data must be restored in order to resume processing transactions. RPO is the basis on which a data projection strategy is developed.

7 RTO: It is a disaster recovery concept in information technology. The RTO is determined based on the acceptable down time in case of a disruption of operations.

8 Retrieved March 06, 2007 from http://curric.dlib.vt.edu/DLcurric/proposalsummary04.pdf

9 Retrieved April 21, 2006 from http://www.codata.org/archives/2002/ArchivingWG-PretoriaRpt.pdf

10 *The Club of Amsterdam Journal,* August 2005, Issue 51, available at http://www.clubofamsterdam.com/press.asp?contentid=494&catid=85

Chapter X
Business Data Warehouse:
The Case of Wal-Mart

Indranil Bose
The University of Hong Kong, Hong Kong

Lam Albert Kar Chun
The University of Hong Kong, Hong Kong

Leung Vivien Wai Yue
The University of Hong Kong, Hong Kong

Li Hoi Wan Ines
The University of Hong Kong, Hong Kong

Wong Oi Ling Helen
The University of Hong Kong, Hong Kong

ABSTRACT

The retailing giant Wal-Mart owes its success to the efficient use of information technology in its operations. One of the noteworthy advances made by Wal-Mart is the development of the data warehouse which gives the company a strategic advantage over its competitors. In this chapter, the planning and implementation of the Wal-Mart data warehouse is described and its integration with the operational systems is discussed. The chapter also highlights some of the problems encountered in the developmental process of the data warehouse. The implications of the recent advances in technologies such as RFID, which is likely to play an important role in the Wal-Mart data warehouse in future, is also detailed in this chapter.

INTRODUCTION

Data warehousing has become an important technology to integrate data sources in recent decades which enables knowledge workers (executives, managers, and analysts) to make better and faster decisions (SCN Education, 2001). From a technological perspective, Wal-Mart, as a pioneer in adopting data warehousing technology, has always adopted new technology quickly and successfully. A study of the applications and issues of data warehousing in the retailing industry based on Wal-Mart is launched. By investigating the Wal-Mart data warehouse from various perspectives, we review some of the critical areas which are crucial to the implementation of a data warehouse. In this chapter, the development, implementation, and evaluation of the Wal-Mart data warehouse is described, together with an assessment of the factors responsible for deployment of a successful data warehouse.

Data Warehousing

Data warehouse is a subject-oriented, integrated, time-variant, non-updatable collection of data used in support of management decision-making (Agosta, 2000). According to Anahory and Murray (1997), "a data warehouse is the data (meta/fact/dimension/aggregation) and the process managers (load/warehouse/query) that make information available, enabling people to make informed decisions." Before the use of data warehouse, companies used to store data in separate databases, each of which were meant for different functions. These databases extracted useful information, but no analyses were carried out with the data. Since company databases held large volumes of data, the output of queries often listed out a lot of data, making manual data analyses hard to carry out. To resolve this problem, the technique of data warehousing was invented. The concept of data warehousing is simple. Data from several existing systems is extracted at periodic intervals, translated into the format required by the data warehouse, and loaded into the data warehouse. Data in the warehouse may be of three forms — detailed information (fact tables), summarized information, and metadata (i.e., description of the data). Data is constantly transformed from one form to another in the data warehouse. Dedicated decision support system is connected with the data warehouse, and it can retrieve required data for analysis. Summarized data are presented to managers, helping them to make strategic decisions. For example, graphs showing sales volumes of different products over a particular period can be generated by the decision support system. Based on those graphs, managers may ask several questions. To answer these questions, it may be necessary to query the data warehouse and obtain supporting detailed information. Based on the summarized and detailed information, the managers can take a decision on altering the production volume of different products to meet expected demands. The major processes that control the data flow and the types of data in the data warehouse are depicted in Figure 1. For a more detailed description of the architecture and functionalities of a data warehouse, the interested reader may refer to Inmon and Inmon (2002) and Kimball and Ross (2002).

BACKGROUND

Wal-Mart is one of the most effective users of technology (Kalakota & Robinson, 2003). Wal-Mart was always among the front-runners in employing information technology (IT) to manage its supply chain processes (Prashanth, 2004). Wal-Mart started using IT to facilitate cross docking in the 1970s. The company later installed bar codes for inventory tracking, and satellite communication system (SCS) for coordinating the activities of its supply chain. Wal-Mart also set-up electronic data interchange (EDI) and a computer terminal

Figure 1. Process diagram of a data warehouse (adapted from Anahory and Murray [1997])

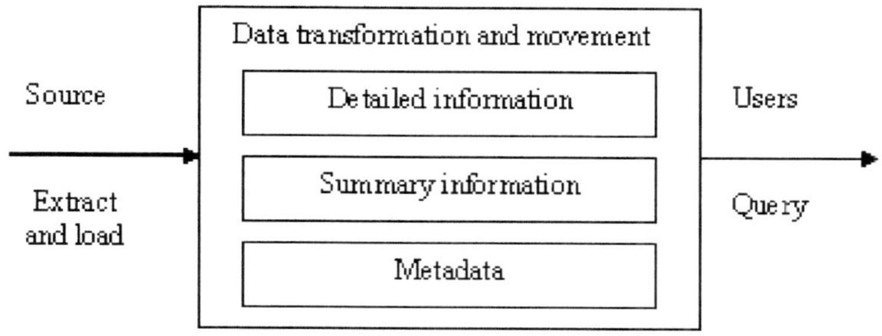

network (CTN), which enabled it to place orders electronically to its suppliers and allowed the company to plan the dispatch of goods to the stores appropriately. Advanced conveyor system was installed in 1978. The point of sale (POS) scanning system made its appearance in 1983, when Wal-Mart's key suppliers placed bar-codes on every item, and universal product code (UPC) scanners were installed in Wal-Mart stores. Later on, the electronic purchase order management system was introduced when associates were equipped with handheld terminals to scan the shelf labels. As a result of the adoption of these technologies, inventory management became much more efficient for Wal-Mart. In the early 1990s, Wal-Mart information was kept in many different databases. As its competitors, such as Kmart, started building integrated databases, which could keep sales information down to the article level, Wal-Mart's IT department felt that a data warehouse was needed to maintain its competitive edge in the retailing industry.

Since the idea of data warehouse was still new to the IT staff, Wal-Mart needed a technology partner. Regarding data warehouse selection, there are three important criteria: compatibility, maintenance, and linear growth. In the early 1990s, Teradata Corporation, now a division of NCR, was the only choice for Wal-Mart, as Teradata was the only merchant database that fulfilled these three important criteria. Data warehouse compatibility ensured that the data warehouse worked with the front-end application, and that data could be transferred from the old systems. The first task for Teradata Corporation was to build a prototype of the data warehouse system. Based on this prototype system, a business case study related to the communication between the IT department and the merchandising organizations was constructed. The case study and the prototype system were used in conjunction to convince Wal-Mart executives to invest in the technology of data warehouse.

Once approved, the IT department began the task of building the data warehouse. First, information-based analyses were carried out on all of the historical merchandising data. Since the IT department did not understand what needed to be done at first, time was wasted. About a month later, there was a shakedown. The IT department focused on the point-of-sales (POS) data. Four teams were formed: a database team, an application team, a GUI team, and a Teradata team. The Teradata team provided training and overlooked everything. The remaining teams held different responsibilities: the database team designed, created, and maintained the data warehouse, the application team was responsible for loading, maintaining, and extracting the data, and the GUI team concentrated on building the interface for the data warehouse. While working on different parts of the data warehouse, the teams supported the operations of each other.

Hardware was a limitation in the data warehouse implementation at Wal-Mart. Since all data needed to fit in a 600 GB machine, data modeling had to be carried out. To save up storage space, a technique called "compressing on zero" was used (Westerman, 2001). This technique was created by the prototype teams. The technique assumed that the default value in the data warehouse was zero, and when this was the case, there was no need to store this data or allocate physical space on the disk drive for the value of zero. This was quite important since it required equal space to store zero or any large value. This resulted in great disk space savings in the initial stages of the database design. Data modeling was an important step in Wal-Mart data warehouse implementation. Not only did it save up storage but was responsible for efficient maintenance of the data warehouse in the future. Hence, it is stated by Westerman (2001), "If you logically design the database first, the physical implementation will be much easier to maintain in the longer term."

After the first implementation, Wal-Mart data warehouse consisted of the POS structure (Figure 2). The structure was formed with a large fact-base table (POS) surrounded by a number of support tables.

The initial schema was a star schema with the central fact table (POS) being linked to the other six support tables. However, the star schema was soon modified to a snowflake schema where the large fact-table (POS) was surrounded by several smaller support tables (like store, article, date, etc.) which in turn were also surrounded by yet smaller support tables (like region, district, supplier, week, etc.). An important element of the POS table was the activity sequence number which acted as a foreign key to the selling activity table. The selling activity table led to performance problems after two years, and Wal-Mart decided to merge this table with the POS table. The next major change that took place several years later was the addition of the selling time attribute to the POS table. The detailed description of the summary and fact tables can be obtained from Westerman (2001).

MAIN THRUST

Approximately one year after implementation of the data warehouse in Wal-Mart, a return on

Figure 2. Star schema for Wal-Mart data warehouse (Source: Westerman, 2001)

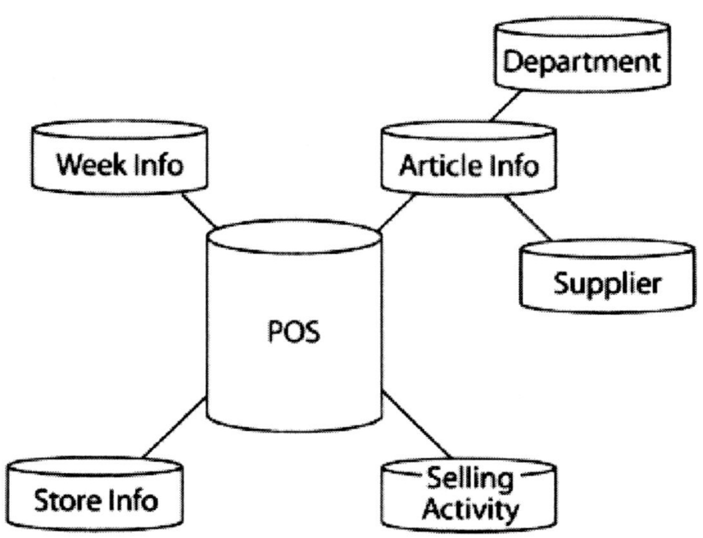

investment (ROI) analysis was conducted. In Wal-Mart, the executives viewed investment in the advanced data warehousing technology as a strategic advantage over their competitors, and this resulted in a favorable ROI analysis. However, the implementation of the data warehouse was marked by several problems.

Problems in Using the Buyer Decision Support Systems (BDSS)

The first graphical user interface (GUI) application based on the Wal-Mart data warehouse was called the BDSS. This was a Windows-based application created to allow buyers to run queries based on stores, articles, and specific weeks. The queries were run and results were generated in a spreadsheet format. It allowed users to conduct store profitability analysis for a specific article by running queries. A major problem associated with the BDSS was that the queries run using this would not always execute properly. The success rate of query execution was quite low at the beginning (i.e., 60%). BDSS was rewritten several times and was in a process of continual improvement. Initially, the system could only access POS data, but in a short period of time, access was also provided to data related to warehouse shipments, purchase orders, and store receipts. BDSS proved to be a phenomenal success for Wal-Mart, and it gave the buyers tremendous power in their negotiations with the suppliers, since they could check the inventory in the stores very easily and order accordingly.

Problems in Tracking Users with Query Statistics

Query Statistics was a useful application for Wal-Mart which defined critical factors in the query execution process and built a system to track the queries. Tracking under this query statistics application revealed some problems with the warehouse. All users were using the same user-ID and password to log on and run queries, and there was no way to track who was actually running the specified query. Wal-Mart did manage to fix the problem by launching different user-IDs with the same password "walmart". But this in turn led to security problems as Wal-Mart's buyers, merchandisers, logistics, and forecasting associates, as well as 3,500 of Wal-Mart's vendor partners, were able to access the same data in the data warehouse. However, this problem was later solved in the second year of operation of the data warehouse by requiring all users to change their passwords.

Performance Problems of Queries

Users had to stay connected to Wal-Mart's bouncing network and database, throughout its entire 4,000-plus store chain and this was cost-ineffective and time-consuming when running queries. The users reported a high failure rate when the users stayed connected to the network for the duration of the query run time. The solution to this problem was deferred queries, which were added to enable a more stable environment for users. The deferred queries application ran the query and saved the results in the database in an off-line mode. The users were allowed to see the status of the query and could retrieve the results after completion of the query. With the introduction of the deferred queries, the performance problems were solved with satisfactory performance, and user confidence was restored as well. However, the users were given the choice to defer the queries. If they did not face any network-related problems they could still run the queries online, while remaining connected to Wal-Mart's database.

Problems in Supporting Wal-Mart's Suppliers

Wal-Mart's suppliers often remained dissatisfied because they did not have access to the Wal-Mart data warehouse. Wal-Mart barred its suppliers from viewing its data warehouse since they did not want suppliers to look into the Wal-Mart inventory warehouse. The executives feared, if given access to the inventory warehouse, suppliers would lower the price of goods as much as they could, and this in turn would force Wal-Mart to purchase at a low price, resulting in overstocked inventory. Later on, Wal-Mart realized that since the goals of the supplier and the buyer are the same (i.e., to sell more merchandise), it is not beneficial to keep this information away from the suppliers. In fact, the information should be shared so that the suppliers could come prepared. In order to sustain its bargaining power over its suppliers and yet satisfy them, Wal-Mart built Retail Link, a decision support system that served as a bridge between Wal-Mart and its suppliers. It was essentially the same data warehouse application like the BDSS but without the competitors' product cost information. With this the suppliers were able to view almost everything in the data warehouse, could perform the same analyses, and exchange ideas for improving their business. Previously, the suppliers used to feel quite disheartened when the buyers surprised them with their up-to-date analyses using the BDSS. The suppliers often complained that they could not see what the buyers were seeing. With the availability of the Retail Link, the suppliers also began to feel that Wal-Mart cared about their partners, and this improved the relationship between the suppliers and the buyers.

Once the initial problems were overcome, emphasis was placed on integration of the data warehouse with several of the existing operational applications.

Integration of the Data Warehouse with Operational Applications

When it comes to integration, the main driving force for Wal-Mart was the ability to get the information into the hands of decision makers. Therefore, many of the applications were integrated into the data warehouse (Whiting, 2004). As a result, the systems were able to feed data into the data warehouse seamlessly. There were also technical reasons for driving integration. It was easier to get data out of the integrated data warehouse, thus making it a transportation vehicle for data into the different computers throughout the company. This was especially important because this allowed each store to pull new information from the data warehouse through their replenishment system. It was also very effective since the warehouse was designed to run in parallel, thus allowing hundreds of stores to pull data at the same time. The following is a brief description of Wal-Mart's applications and how they were integrated into the enterprise data warehouse.

Replenishment System

The process of automatic replenishment was critically important for Wal-Mart since it was able to deliver the biggest ROI after the implementation of the data warehouse. Since the replenishment application was already established, the system was quite mature for integration. The replenishment system was responsible for online transaction processing (OLTP) and online analytical processing (OLAP). It reviewed articles for orders. The system then determined whether an order was needed and suggested an order record for the article. Next these order records were loaded into the data warehouse and transmitted from the home office to the store. The store manager then reviewed the suggested orders, changed prices, counted inventory, and so on. Before the order was placed, the store managers also reviewed the flow of goods by inquiring about article sales trends, order trends, article profiles, corporate information, and so on. These were examples of OLAP activities. This meant that the order was not automatically placed for any item. Only after the store manager had a chance to review the order and perform some analyses using the data warehouse was it decided whether the order was

Table 1. An example store trait table

Store ID	Pharmacy	Fresh Deli	Bakery	Beach	Retirement	University	<60K Sqft	>120K Sqft	Kmart Comp	Target Comp	Real Comp	etc.
2105	N	N	N	N	Y	N	N	Y	Y	N	N	...
2106	Y	Y	Y	N	N	Y	N	Y	N	Y	N	...

going to be placed or not. The order could either be placed if the order could be filled from one of the Wal-Mart warehouses, or the order could be directed to the supplier via electronic data interchange (EDI). In either of the two cases, the order would be placed in the order systems and into the data warehouse.

Distribution via Traits

The traiting concept was developed as an essential element of the replenishment system. The main idea was to determine the distribution of an article to the stores. Traits were used to classify stores into manageable units and could include any characteristics, as long as it was somewhat permanent. Furthermore, these traits could only have two values: TRUE and FALSE. Table 1 is an example of what a store trait table might look like.

Traits could also be applied to articles in a store where a different table could be created for it. These different trait tables were used as part of the replenishment system. The most powerful aspect of this traiting concept was the use of a replenishment formula based on these traits. The formula was a Boolean formula where the outcome consisted of one of two values. If the result was true, the store would receive an article and vice versa. This concept was very important for a large centrally-managed retail company like Wal-Mart, since the right distribution of goods to the right stores affected sales and hence the image of the company. A distribution formula might look like this:

*Store distribution for Article X = (pharmacy * fresh deli * bakery * — < 60K sq. ft.).*

This formula indicated that a store which had a pharmacy, a fresh deli, a bakery, and had a size of more than 60,000 sq. ft., should receive the order. From Table 1, we can see that store 2106 satisfies all these conditions and hence should receive the article X. In this manner, each article had its own unique formula, helping Wal-Mart distribute its articles most effectively amongst its stores.

All this information was very valuable for determining the allocation of merchandise to stores. A data warehouse would provide a good estimate for a product based on another, similar product that had the same distribution. A new product would be distributed to a test market using the traiting concept, and then the entire performance tracking would be done by the data warehouse. Depending on the success or failure of the initial trial run, the traits would be adjusted based on performance tracking in the data warehouse, and this would be continued until the distribution formula was perfected. These traiting methods were replicated throughout Wal-Mart using the data warehouse, helping Wal-Mart institute a comprehensive distribution technique.

Perpetual Inventory (PI) System

The PI system was used for maintenance of inventory of all articles, not just the articles appearing in the automatic replenishment. Like the replenishment system, it was also an example of an OLAP and OLTP system. It could help managers see the entire flow of goods for all articles, including replenishment articles. This data was available in the store and at the home office. Thus, with the use of the replenishment and PI systems, managers could maintain all information related to the inventory in their store electronically. With all this information in the data warehouse, there were numerous information analyses that could be conducted. These included:

- The analysis of the sequence of events related to the movement of an article
- Determination of operational cost
- Creation of "plan-o-grams" for each store for making planning more precise. This could allow buyers and suppliers to measure the best selling locations without physically going to the store.

The PI system using the enterprise data warehouse could also provide benefits to the customer service department. Managers could help customers locate certain products with certain characteristics. The system could allocate the product in the store, or identify if there were any in storage, or if the product was in transit and when it could arrive or even if the product was available in any nearby stores. This could be feasible due to the data provided by the PI system and the information generated by the data warehouse.

FUTURE TRENDS

Today, Wal-Mart continues to employ the most advanced IT in all its supply chain functions. One current technology adoption in Wal-Mart is very tightly linked with Wal-Mart's data warehouse, that is, the implementation of Radio Frequency Identification (RFID). In its efforts to implement new technologies to reduce costs and enhance the efficiency of supply chain, in July 2003, Wal-Mart asked all its suppliers to place RFID tags on the goods, packed in pallets and crates shipped to Wal-Mart (Prashanth, 2004). Wal-Mart announced that its top 100 suppliers must be equipped with RFID tags on their pallets and crates by January, 2005. The deadline is now 2006 and the list now includes all suppliers, not just the top 100 (Hardfield, 2004). Even though it is expensive and impractical (Greenburg, 2004), the suppliers have no choice but to adopt this technology.

The RFID technology consists of RFID tags and readers. In logistical planning and operation of supply chain processes, RFID tags, each consisting of a microchip and an antenna, would be attached on the products. Throughout the distribution centers, RFID readers would be placed at different dock doors. As a product passed a reader at a particular location, a signal would be triggered and the computer system would update the location status of the associated product. According to Peak Technologies (http://www.peaktech.com), Wal-Mart is applying SAMSys MP9320 UHF portal readers with Moore Wallance RFID labels using Alien Class 1 passive tags. Each tag would store an Electronic Product Code (EPC) which was a bar code successor that would be used to track products as they entered Wal-Mart's distribution centers and shipped to

Figure 3. RFID label for Wal-Mart (Source: E-Technology Institution (ETI) of the University of Hong Kong [HKU])

individual stores (Williams, 2004). Figure 3 is an example of the label. The data stored in the RFID chip and a bar code are printed on the label, so we know what is stored in the chip and also the bar code could be scanned when it became impossible to read the RFID tag. According to Sullivan (2004, 2005), RFID is already installed in 104 Wal-Mart stores, 36 Sam's Clubs, and three distribution centers, and Wal-Mart plans to have RFID in 600 stores and 12 distribution centers by the end of 2005.

The implementation of RFID at Wal-Mart is highly related to Wal-Mart's data warehouse, as the volume of data available will increase sufficiently. The industry has been surprised by estimates of greater than 7 terabytes of item-level data per day at Wal-Mart stores (Alvarez, 2004). The large amount of data can severely reduce the long-term success of a company's RFID initiative. Hence, there is an increasing need to integrate the available RFID data with the Wal-Mart data warehouse. Fortunately, Wal-Mart's data warehouse team is aware of the situation and they are standing by to enhance the data warehouse if required.

CONCLUSION

In this chapter we have outlined the historical development of a business data warehouse by the retailing giant Wal-Mart. As a leader of adopting cutting edge IT, Wal-Mart demonstrated great strategic vision by investing in the design, development and implementation of a business data warehouse. Since this was an extremely challenging project, it encountered numerous problems from the beginning. These problems arose due to the inexperience of the development team, instability of networks, and also inability of Wal-Mart management to forecast possible uses and limitations of systems. However, Wal-Mart was able to address all these problems successfully and was able to create a data warehouse system that gave them phenomenal strategic advantage compared to their competitors. They created the BDSS and the Retail Link which allowed easy exchange of information between the buyers and the suppliers and was able to involve both parties to improve sales of items. Another key achievement of the Wal-Mart data warehouse was the Replenishment system and the Perpetual Inventory system, which acted as efficient decision support systems and helped store managers throughout the world to reduce inventory, order items appropriately, and also to perform ad-hoc queries about the status of orders. Using novel concepts such as traiting, Wal-Mart was able to develop a successful strategy for efficient distribution of products to stores. As can be expected, Wal-Mart is also a first mover in the adoption of the RFID technology which is likely to change the retailing industry in the next few years. The use of this technology will lead to the generation of enormous amounts of data for tracking of items in the Wal-Mart system. It remains to be seen how Wal-Mart effectively integrates the RFID technology with its state-of-the-art business data warehouse to its own advantage.

REFERENCES

Agosta, L. (2000). *The essential guide to data warehousing*. Upper Saddle River, NJ: Prentice Hall.

Alvarez, G. (2004). What's missing from RFID tests. *Information Week*. Retrieved November 20, 2004, from http://www.informationweek.com/story/showArticle.jhtml?articleID=52500193

Anahory, S., & Murray, D. (1997). *Data warehousing in the real world: A practical guide for building decision support systems*. Harlow, UK: Addison-Wesley.

Greenburg, E. F. (2004). Who turns on the RFID faucet, and does it matter? *Packaging Digest*, 22.

Retrieved January 24, 2005, from http://www.packagingdigest.com/articles/200408/22.php

Hardfield, R. (2004). The RFID power play. *Supply Chain Resource Consortium*. Retrieved October 23, 2004, from http://scrc.ncsu.edu/public/APICS/APICSjan04.html

Inmon, W. H., & Inmon, W. H. (2002). *Building the data warehouse* (3rd ed.). New York: John Wiley & Sons.

Kalakota, R., & Robinson, M. (2003). *From e-business to services: Why and why now?* Addison-Wesley. Retrieved January 24, 2005, from http://www.awprofessional.com/articles/article.asp?p=99978&seqNum=5

Kimball, R., & Ross, M. (2002). *The data warehouse toolkit: The complete guide to dimensional modeling* (2nd ed.). New York: John Wiley & Sons.

Prashanth, K. (2004). *Wal-Mart's supply chain management practices (B): Using IT/Internet to manage the supply chain*. Hyderabad, India: ICFAI Center for Management Research.

SCN Education B. V. (2001). *Data warehousing — The ultimate guide to building corporate business intelligence* (1st ed.). Vieweg & Sohn Verlagsgesellschaft mBH.

Sullivan, L. (2004). Wal-Mart's way. *Information Week*. Retrieved March 31, 2005, from http://www.informationweek.com/story/showArticle.jhtml?articleID=47902662&pgno=3

Sullivan, L. (2005). Wal-Mart assesses new uses for RFID. *Information Week*. Retrieved March 31, 2005, from http://www.informationweek.com/showArticle.jhtml?articleID=159906172

Westerman, P. (2001). *Data warehousing: Using the Wal-Mart model*. San Francisco: Academic Press.

Whiting, R. (2004). Vertical thinking. *Information Week*. Retrieved March 31, 2005, from http://www.informationweek.com/showArticle.jhtml?articleID=18201987

Williams, D. (2004). The strategic implications of Wal-Mart's RFID mandate. *Directions Magazine*. Retrieved October 23, 2004, from http://www.directionsmag.com/article.php?article_id=629

This work was previously published in Database Modeling for Industrial Data Management: Emerging Technologies and Applications, edited by Z.M. Ma, pp. 244-257, copyright 2006 by Information Science Publishing (an imprint of IGI Global).

Chapter XI
Medical Applications of Nanotechnology in the Research Literature[1]

Ronald N. Kostoff
Office of Naval Research, USA

Raymond G. Koytcheff
Office of Naval Research, USA

Clifford G.Y. Lau
Institute for Defense Analyses, USA

ABSTRACT

The medical applications literature associated with nanoscience and nanotechnology research was examined. About 65,000 nanotechnology records for 2005 were retrieved from the Science Citation Index/ Social Science Citation Index (SCI/SSCI) using a comprehensive 300+ term query. The medical applications were identified through a fuzzy clustering process. Metrics associated with research literatures for specific medical applications/ applications groups were generated.

INTRODUCTION

During 2003–2005, a comprehensive text mining study was performed to overview the technical structure and infrastructure of the global nanotechnology research literature, as well as the seminal nanotechnology literature (Kostoff, Stump, Johnson, Murday, Lau & Tolles, 2005a; Kostoff, Murday, Lau & Tolles, 2005b; Kostoff, Stump, Johnson, Murday, Lau & Tolles 2006a; Kostoff, Murday, Lau & Tolles, 2006b). Based on the global interest generated by these reports, it was decided to update and expand the study using more recent data, a much more comprehensive query, and more sophisticated analytical tools. A detailed report from the updated study is contained in Kostoff, Koytcheff, and Lau (2007).

In the updated study, text mining was used to extract technical intelligence from the open source global nanotechnology and nanoscience research literature (Science Citation Index/Social Science Citation Index (SCI/SSCI) databases (SCI, 2006)). Identified were: (1) the nanotechnology/ nanoscience research literature infrastructure (prolific authors, key journals/ institutions/ countries, most cited authors/ journals/ documents); (2) the technical structure (pervasive technical thrusts and their interrelationships); (3) nanotechnology instruments and their relationships; (4) potential nonmedical nanotechnology applications, and (5) potential health applications.

Most importantly, in the updated study, all of the technical structural analyses of the total nanotechnology database show medical and nonmedical applications being a key driver in nanoscience and nanotechnology research. The objectives of this paper are to examine the nanotechnology medical applications literature in depth, and especially show medical applications relationships to each other and to the underlying science disciplines.

In order to place the nanotechnology medical applications analyses and findings in their proper context, the overall nanotechnology study will first be summarized.

SUMMARY OF OVERALL NANOTECHNOLOGY STUDY

Bibliometrics

Global nanotechnology research article production has exhibited exponential growth for more than a decade. The most rapid growth over that time period has come from East Asian nations, notably China and South Korea. While the U.S. remains the leader in aggregate nanotechnology research article production, in some selected nanotechnology subareas China has achieved parity or taken the lead in research article production.

The main institutional copublishing groups are East Asian: one each from China, Japan, and South Korea. However, publication connectivity among institutions is much weaker than common interest or citation connectivity. Correlation of institutions by the journals they cite reveals four nationality-based (or locality-based) clusters: Chinese, Japanese, American, and European. Institutions from the same nationality group cite the same focused journals (primarily, but not exclusively, domestic). Correlation of institutions by documents they cite reveals that *only the Chinese institutions constitute a strongly-connected network.*

The dominant country copublishing network is a complex web of mainly European nations roughly following geographic lines: Nordic, Central Europe, Eastern Europe, and a Western Europe/Latin American group of romance language nations. There is also a UK component country network, but it is not linked to the interconnected continental members of the European Union. Correlation of countries by common thematic interest shows two major poles: U.S. and China. The U.S. pole is strongly connected thematically to a densely connected network of English-speaking North American representatives, Western/ Central European nations, and most of the East Asian allies. China is relatively isolated except for India, and the Eastern European and Latin American representatives are outside the main network as well.

There is a clear distinction between the publication practices of the three most prolific Western nations and the three most prolific East Asian nations. The Western nations publish in journals with almost twice the weighted average impact factors (Journal Impact Factor is a metric that reflects the average citations received by papers published recently in the journal) of the East Asian nations. However, much of the difference stems from the East Asian nations publishing a nonnegligible amount in domestic low Impact Factor journals, while the Western nations publish in higher impact factor international journals.

Additionally, some of the Asian countries (e.g., China) are publishing in journals whose initial access date in the SCI/SSCI is relatively recent. If China is publishing a nonnegligible fraction of its research output in newly-accessed relatively low impact factor journals, then some of its apparently rapid growth will not be in the traditional sense of increased sponsorship or productivity, but rather due to the SCI/SSCI's decision to access existing journals' articles. From another perspective, China's research article production may have been somewhat more competitive for decades, but was artificially suppressed by many of its journals' noninclusion in the SCI/SSCI until only recently.

Of the 30 institutions publishing large numbers of nanotechnology papers, only 4 are from the U.S., whereas of the 25institutions producing highly cited nanotechnology papers, an astonishing 21 papers are from the U.S. The two journals that contain the most cited nanotechnology papers since 1991 are *Science* and *Nature*, and the two countries that lead in production of the most cited papers are the U.S. and Germany, with the U.S. having four times the number of most cited nanotechnology papers as Germany. The U.S. and Germany account for 40% of the most cited nanotechnology papers, while the high paper volume production East Asian countries of China and South Korea account for only two percent of the most cited papers.

Computational Linguistics

The retrieved records for 2005 can be divided into two main categories with similar numbers of records (first level). One category focuses on phenomena and thin films (32,983 records), and the other category focuses on materials/structures (31,742 records). The "phenomena" component of the first category is roughly four times the size of the "films" component, whereas the "nanotubes" component of the materials/structures category is about one ninth the size of the nonnanotubes component.

Furthermore, maps were constructed to show groupings of related nonmedical Applications into broader thematic areas. An autocorrelation map of the most widely referenced nonmedical applications showed five weakly-connected subnetworks:

- Electronic devices and components
- Optical switching
- Tribology and corrosion
- Optoelectronic sensors
- Electrochemical conversion and catalysis

In addition to the maps, factor analyses were performed to show nonmedical thematic areas from a slightly different perspective. A six-factor analysis showed the following themes:

- **Factor 1:** Optoelectronics
- **Factor 2:** Tribology
- **Factor 3:** Lithography
- **Factor 4:** Control systems
- **Factor 5:** Devices
- **Factor 6:** Microsystems

Finally, for the most frequently mentioned nonmedical applications, the following observations were made:

- *TiO2, Pt, Si, gold, and polymers* tend to stand out as the most pervasive associated material types.
- *Morphology, thickness /diameter/particle size, optical properties, catalytic performance, and electrochemical properties* tend to stand out as the most pervasive associated material properties.
- *Deposition, absorption, oxidation, immobilization, catalysis, degradation, and self-assembly* tend to stand out as the most

pervasive associated nanoscale phenomena.

- *Thinfilms, nanowires, nanotubes (especially carbon), and self-assembled monolayers* tend to stand out as the most pervasive associated nanostructures.

BACKGROUND

The two main components of the present nanotechnology medical applications study are the text mining analytical procedure and the nanotechnology topical literature. The text mining background will be summarized briefly. The nanotechnology background has been described in detail (Kostoff et al., 2005b, 2006b), and will not be repeated here. The nanotechnology background is updated and expanded in Kostoff et al. (2007). Finally, the relevant technology transfer background issues will be summarized.

Text Mining

A typical text mining study of the published literature develops a query for comprehensive information retrieval, processes the retrieved database using computational linguistics and bibliometrics, and integrates the processed information. In this section, the computational linguistics and bibliometrics are overviewed.

Science and technology (S&T) computational linguistics (Hearst, 1999; Kostoff, 2003a; Losiewicz, Oard & Kostoff, 2000; Zhu & Porter, 2002) identifies pervasive technical themes in large databases from technical phrases that occur frequently. It also identifies relationships among these themes by grouping (clustering) these phrases (or their parent documents) on the basis of similarity. Computational linguistics can be used for:

- Enhancing information retrieval and increasing awareness of the global technical literature (Greengrass, 1997; Kostoff, Eberhart & Toothman,1997a; TREC, 2004)
- Potential discovery and innovation based on merging common linkages among very disparate literatures (Gordon & Dumais, 1998; Kostoff, 2003b, 2006c; Swanson, 1986; Swanson & Smalheiser, 1997)
- Uncovering unexpected asymmetries from the technical literature (Kostoff, 2003c; Goldman, Chu, Parker & Goldman, 1999). For example, Kostoff (2003c) predicted asymmetries in recorded bilateral organ (lungs, kidneys, testes, ovaries) cancer incidence rates from the asymmetric occurrence of lateral word frequencies (left, right) in Medline case study articles.
- Estimating global levels of effort in S&T subdisciplines (Kostoff, Green, Toothman & Humenik, 2000; Kostoff, Shlesinger & Tshiteya, 2004a; Viator & Pastorius, 2001)
- Helping authors potentially increase their citation statistics by improving access to their published papers, and thereby potentially helping journals to increase their impact factors (Kostoff et al., 2004a; Kostoff, Shlesinger & Malpohl, 2004b)
- Tracking myriad research impacts across time and applications areas (Kostoff, Del Rio, García, Ramírez & Humenik, 2001; Davidse & VanRaan, 1997)

Evaluative bibliometrics (Garfield, 1985; Narin, 1976; Schubert, Glanzel & Braun, 1987) uses counts of publications, patents, citations and other potentially informative items to develop science and technology performance indicators. Its validity is based on the premises that (1) counts of patents and papers provide valid indicators of R&D activity in the subject areas of those patents or papers, (2) the number of times those patents or papers are cited in subsequent patents or papers provides valid indicators of the impact or importance of the cited patents and papers, and (3) the citations from papers to papers, from patents to patents and from patents to papers provide indicators of intellectual linkages between the organizations that are producing the patents and papers, and knowledge linkage between their subject areas (Narin, Olivastro & Stevensf, 1994). Evaluative bibliometrics can be used to:

- Identify the infrastructure (authors, journals, institutions) of a technical domain
- Identify experts for innovation-enhancing technical workshops and review panels
- Develop site visitation strategies for assessment of prolific organizations globally
- Identify impacts (literature citations) of individuals, research units, organizations, and countries

Technology Transfer

In its modern form, nanotechnology has been around for about 15 years. It has the status of an emerging technology, and many papers/books have been written promoting its applications potential in many areas. One goal of this study was to document the medical applications potential. A second goal was to identify some of the science and infrastructure markers of potential nanotechnology medical applications, so that the science of nanotechnology/ nanoscience could be accelerated to advanced levels of development. This chapter is intended to facilitate the nanotechnology transition process by identifying the significant application areas.

In 1997, a special issue of the Journal of Technology Transfer (edited by the first author) addressed accelerated conversion of science to technology (Kostoff, 1997b). Its articles emphasized the importance of potential downstream users of science to become involved with the science development as early and broadly as possible, in order to direct the science toward potential user needs, and smooth the eventual transition to actual applications. The important first step in this conversion process is to identify the science relevant to specific desired applications, and identify the associated infrastructure. Once contact is made between the on-going science and the potential user, then the full technology transfer process can be initiated.

This chapter will provide such information of importance to the nanotechnology technology transfer community. It will identify the main nanotechnology health applications from today's vantage point, as well as the related science and infrastructure.

APPROACH

The following approach describes how the main nanotechnology medical applications were identified, as well as their direct and indirect relationships.

A document fuzzy clustering analysis (Karypis, 2006), where documents are divided into groups based on their text similarities and where documents can be assigned to more than one group, was performed on the ~65,000 total nanotechnology records retrieved for the overall nanotechnology study. The resulting hierarchical taxonomy was inspected visually, and the largest subnetwork that included all medical applications (hereafter called the "health subnetwork") was identified. A metalevel taxonomy of the health subnetwork (the highest two hierarchical levels) was generated, then a taxonomy of the elemental (lowest level) clusters was generated. These clusters were analyzed for infrastructure and technical content. In the remainder of this chapter, the medical applications will be referred to as "health."

RESULTS

Nanotechnology Health Types

The document clustering approach used to identify the health types was a recent algorithmic upgrade of the CLUTO software package (Karypis, 2006) called fuzzy clustering, where a record could be assigned to multiple clusters. Fuzzy clustering, compared to nonfuzzy clustering where a document is assigned to one cluster only, is important for articles that have multiple thrusts, such as health applications articles in a research database.

There were 256 elemental clusters specified for the algorithm. Of these, 22 were in the health subnetwork. Of these 22 elemental clusters, 19 related directly to health. The resultant 19 clusters are of different types. Some address specific health problems (e.g., tumor treatment, sentinal lymph node cancer), some address health treatment mechanisms (e.g., drug release, drug delivery), some address biomaterial types (e.g., cells, DNA, biofilms, virus proteins, amyloid fibrils), but most are health-related phenomena and processes (e.g., peptide sequences, binding and affinity, detection, sensing). The higher level taxonomy categories will now be discussed, followed by a discussion of the elemental clusters.

Higher Level Taxonomy Categories

Highest Level Category

Table 1 contains a summary of the infrastructure, pervasive thrusts, and related science for the 19 elemental clusters. Characteristics of the highest level category (node) in the health subnetwork are summarized in the last row on Table 1. Because about 15% of the elemental clusters in the 22 cluster health subnetwork were not strictly health-related, the results on this row should be considered a good approximation. In addition, the numbers of records listed for the highest level node (and all nodes on Table 1) include counts of records from different elemental clusters (due to the fuzzy nature of the clustering), and therefore have intrinsic multiple counts.

Country Productivity

In this highest-level category in the health subnetwork, the USA appears to have a commanding lead (a ratio of about 3 to 1) over its nearest competitor (China). However, these results must be considered in context. First, in total SCI/SSCI articles, the USA had about four times as many records as China when these data were obtained. Second, for overall nanotechnology, the USA had about 25% more records than China for 2005. Third, for nanotechnology instrumentation, China actually had 25% more records than the USA. Fourth, relative to China, the USA had a commanding lead in overall biomedical articles, as our recent text mining study on China showed (Kostoff et al., 2006). When all these facts are integrated, it appears that China is placing substantial emphasis on its nanotechnology medical research relative to its overall medical research.

A more interesting comparison is between the top Asian producers (China, Japan, South Korea) and the top European producers (Germany, England, France). The Asian group produced 1,756 papers, while their European counterparts produced 1,326 papers, a 32% difference. In aggregate, the Asian group has a population of 1.48 Billion and GDP of $14 Trillion-PPP/ $7.7 Trillion-OER, while the European group has a population of .19 Billion and a GDP of $5.78 Trillion-PPP/ $6.67 Trillion-OER. Thus, on a per capita population basis, the European group is almost an order of magnitude more productive of nanotechnology medical Applications papers, while on a PPP GDP basis, the productivity advantage shrinks to a factor of two.

However, a brief time trend analysis shows the extent of the challenge for Europe. The following short query, which represents some of the key medical terms from Table 1 and their relation to nanotechnology, was entered into the SCI/ SSCI search engine for two years: 1995 and 2006 (December 2006):

Table 1. Central health themes and infastructure

THEME/ #RECORDS/ COUNTRIES	INSTITUTIONS	JOURNALS	RELATED SCIENCE
DRUG RELEASE (235 Records) USA 58 Peoples R China 55 India 37 South Korea 31 Japan 24 Germany 23	Natl Univ Singapore 11 Zhejiang Univ 9 Korea Res Inst Chem Technol 8 Chonbuk Natl Univ 6	*Journal of Controlled Release* 30 *International Journal of Pharmaceutics* 28 *Drug Development and Industrial Pharmacy* 9 *Journal of Microencapsulation* 8 *European Journal of Pharmaceutics and Biopharmaceutics* 8	polymer, hydrogel, nanoparticles, chitosan, microsphere, molecular.weight, particle. size, water.soluble, light.scattering, ethylene glycol, cross linking, differential scanning calorimetry, scanning electron microscopy, poly lactic acid, atomic force microscopy, transmission electron microscopy, dynamic light scattering, fourier transform infrared, bovine serum albumin, poly ethylene glycol, poly lactide glycolide
DRUG DELIVERY (197 Records) USA 79 Peoples R China 36 India 31 Germany 26 Japan 23 Italy 21 France 19 South Korea 18 England 18	Natl Univ Singapore 10 Univ Michigan 8 Zhejiang Univ 7 Postgrad Inst Med Educ & Res 7	*Journal of Controlled Release* 36 *International Journal of Pharmaceutics* 23 *Journal of Drug Delivery Science and Technology* 11 *Biomaterials* 10	nanoparticles, cancer, cancer cells, cellular uptake, size distribution, tumor cells, scanning electron microscopy, poly lactide glycolide, solid lipid nanoparticles, poly ethylene glycol, blood brain barrier, transmission electron microscopy, bovine serum albumin, confocal laser scanning microscopy
TUMOR TREATMENT (208 Records) USA 107 Japan 24 Germany 22 Peoples R China 19 France 19 South Korea 18	Univ Texas 7 Univ Michigan 7 Chinese Acad Sci 7 Washington Univ 6 Ohio State Univ 5	*Journal of Controlled Release* 11 *Journal of Magnetism and Magnetic Materials* 10 *Pharmaceutical Research* 8 *Magnetic Resonance in Medicine* 6 *Biomaterials* 6	liposomes, mice, cells, nanoparticles, tumor cells, tumor growth, contrast agents, endothelial cells, flow cytometry, cell lines, magnetic resonance imaging, scanning electron microscopy, transmission electron microscopy, blood brain barrier, superparamagnetic iron oxide nanoparticles, surface plasmon resonance, tumor bearing mice, central nervous system, tumor necrosis factor, atomic force microscopy

continued on following page

Table 1. continued

SENTINEL LYMPH NODE CANCER (112 Records) USA 50 England 12 Netherlands 9 Italy 8 Germany 6 Japan 5 France 5	Massachusetts Gen Hosp 5 Harvard Univ 4 Univ Barcelona 3 MIT 3 Hosp Clin Barcelona 3 Brigham & Womens Hosp 3 Beth Israel Deaconess Med Ctr 3 Amer Biosci Inc 3	*European Journal of Nuclear Medicine And Molecular Imaging* 7 *Urology* 4 *Journal of Clinical Oncology* 4	lymphoscintigraphy, metastases, lymph node, risk factors, breast cancer, sentinel node, magnetic resonance imaging, squamous cell carcinoma, scanning electron microscopy, von willebrand factor, lymph node biopsy, low density lipoprotein, high density lipoprotein, intercellular adhesion molecule
TISSUE CELLS (269 Records) USA 92 Peoples R China 36 Japan 36 Singapore 30 Germany 26 South Korea 23 England 23	Natl Univ Singapore 24 Tsing Hua Univ 7 MIT 7 Johns Hopkins Univ 7	*Biomaterials* 45 *Tissue Engineering* 22 *Journal of Biomedical Materials Research Part A* 19 *ASBM6: Advanced Biomaterials VI* 9	cells, tissues, collagen, scaffold, bone, osteoblast, extracellular matrix, cell adhesion, cell culture, endothelial cells, cell proliferation, cell attachment, cell morphology, calcium phosphate, osteoblast cells, bone tissue, self assembly, tissue culture, phosphatase activity, cell growth, scanning electron microscopy, atomic force microscopy, transmission electron microscopy, x-ray photoelectron spectroscopy, alkaline phosphatase activity, polymerase chain reaction, mesenchymal stem cells, polylactic glycolic acid, bone marrow stromal cells
CELLS, EMPHASIZING ADHESION (605 Records) USA 254 Germany 86 Japan 82 Peoples R China 52 South Korea 46 Canada 30 England 28 France 27	Univ Calif 28 Harvard Univ 20 Johns Hopkins Univ 16 Univ Tokyo 11 Natl Univ Singapore 11 Cnrs 11 Chinese Acad Sci 11	*Biomaterials* 25 *Journal Of Biomedical Materials Research Part A* 16 *Langmuir* 14 *Biophysical Journal* 12	Cells, adhesion, apoptosis, endothelial.cells, cell lines, cell surface, cell adhesion, cancer cells, epithelial cells, cell proliferation, cell growth, cell death, extracellular matrix, stem cells, tumor cells, flow cytometry, atomic force microscopy, transmission electron microscopy, scanning electron microscopy, surface plasmon resonance, smooth muscle cells, green fluorescent protein, human umbilical vein, magnetic resonance imaging, superparamagnetic iron oxide nanoparticles

continued on following page

Table 1. continued

BIOFILMS (83 Records) USA 33 Japan 9 Germany 8 England 8 South Korea 6 Peoples R China 6 Canada 6	Montana State Univ 4 Chinese Acad Sci 3 Univ Calif 3	*Water Science and Technology* 4 *On the Convergence of Bio-Information-, et al.* 4	biofilm, muscles, bacteria, biofilm formation, infection, colon, pathogen, tissue, strain, epithelial cells, pseudomonas aeruginosa, staphylococcus epidermidis, escherichia coli, scanning electron microscopy, transmission electron microscopy, atomic force microscopy, extracellular polymeric substances, confocal laser scanning, polymerase chain reaction
VIRUS PROTEINS (205 Records) USA 228 Germany 70 Japan 66 Peoples R China 34 Italy 34 France 32 England 32	Univ Calif 30 Osaka Univ 12 Univ Texas 11 Univ Illinois 8 Linkoping Univ 8 Chinese Acad Sci 8	*Langmuir* 29 *Journal of Biological Chemistry* 27 *Biochemical and Biophysical Research Communications* 14 *Journal of Virology* 13 *Journal of Molecular Biology* 12 *Biochemistry* 12	protein, virus, capside, gene, sequence, escherichia coli, wild type, virus particles, capside assembly, capside protein, self assembly, atomic force microscopy, transmission electron microscopy, surface plasmon resonance, amino acid sequence, green fluorescent protein, tobacco mosaic virus, open reading frame, density gradient centrifugation, amino acid
PROTEIN INTERACTIONS (641 Records) USA 247 Germany 85 Japan 65 Peoples R China 55 Italy 50 England 44 France 35	Univ Calif 26 Chinese Acad Sci 19 Univ Illinois 12 Univ Texas 10 Max Planck Inst 9 Univ Washington 8 Tokyo Inst Technol 8 Osaka Univ 8 Linkoping Univ 8	*Langmuir* 37 *Analytical Chemistry* 17 *Proc NAS-USA* 16 *Biomacromolecules* 16	protein, binding, surface, membranes, unfolding, fluorescence, protein adsorption, mass spectrometry, protein surface, x-ray diffraction, atomic force microscopy, surface plasmon resonance, bovine serum albumin, scanning electron microscopy, transmission electron microscopy, differential scanning calorimetry, human serum albumin, green fluorescent protein, polyacrylamide gel electrophoresis, protein protein interactions, quartz crystal microbalance, fourier transform infrared, self assembled monolayer, poly ethylene glycol, tandem mass spectrometry
AMYLOID FIBRILS (114 Records) USA 50 England 15 Japan 14 Italy 9 Germany 8 Sweden 7	Univ Cambridge 8 Osaka Univ 6 Niddkd 5 Japan Sci & Technol Agcy 5 Fukui Univ 4 Univ Calif 4	*Biochemistry* 16 *Biophysical Journal* 8 *Journal Of Molecular Biology* 7 *Journal Of Biological Chemistry* 7	amyloid.fibrils, protein, peptide, alzheimers disease, collagen, protofibril, prion, beta sheet structure, fibril formation, self assembly, amyloid beta, neurodegenerative diseases, collagen fibrils, amyloid deposits, thioflavin fluorescence, atomic force microscopy, transmission electron microscopy, paired helical filaments

continued on following page

Table 1. continued

PEPTIDE SEQUENCES (187 Records) USA 86 Japan 28 Israel 14 Germany 14 Australia 14 Canada 12	MIT 8 Univ Calif 5	*Langmuir* 13 *Analytical Chemistry* 10 *Journal of the American Chemical Society* 9 *Journal of Biological Chemistry* 9 *Biochemistry* 7 *Biophysical Journal* 6	peptide, binding, sequences, amino acids, peptide nanotubes, neuropeptides, structure, protein, circular dichroism, antimicrobial peptides, peptide sequence, alpha helix, molecular dynamics, model peptide, surface plasmon resonance, atomic force microscopy, amino acid residues, transmission electron microscopy, amino acid sequence, matrix laser desorption, quartz crystal microbalance, tandem mass spectrometry, differential scanning calorimetry, self assembled monolayers, solid phase peptide
BINDING AND AFFINITY (415 Records) USA 211 Japan 51 Germany 47 England 45 France 34	Chinese Acad Sci 13 CNRS 12 Lund Univ 11 Univ Calif 10 Univ Penn 9 Univ Oxford 9 NCI 9 Scripps Res Inst 8	*Journal of Biological Chemistry* 44 *Biochemistry* 35 *Biochemical and Biophysical Research Communications* 19 *Journal of the American Chemical Society* 16	binding, receptors, affinity, protein, interaction, surface plasmon resonance, ligand, high affinity, binding affinity, binding sites, amino acid, active site, ligand binding, binding protein, cell surface, dissociation rate, atomic force microscopy, site directed mutagenesis, amino acid residues, human immunodeficiency virus, high affinity binding, isothermal titration calorimetry, low density lipoprotein, equilibrium dissociation constants, immobilized sensor chip, expressed escherichia coli, human serum albumin, transmission electron microscopy, quartz crystal microbalance, molecular dynamics simulations, epidermal growth factor, fluorescence resonance energy
IMMUNOSENSORS (248 Records) USA 74 Peoples R China 54 Japan 30 Germany 16 England 16 South Korea 15	Hunan Univ 10 Univ Turku 7 Kyushu Univ 7 Sw China Normal Univ 6 Sogang Univ 6	*Analytical Chemistry* 22 *Biosensors & Bioelectronics* 17 *Analytica Chimica Acta* 13 *Sensors and Actuators B-Chemical* 12 *Langmuir* 12	antibodies, antigens, assays, detection, igg, immobilization, immunoassays, binding, protein, immunosensor, gold, monoclonal antibody, immunosorbent assay, antigen antibody, assay elisa, antigen binding, gold surface, gold nanoparticles, escherichia coli, antibody binding, surface plasmon resonance, enzyme linked immunosorbent assay, atomic force microscopy, quartz crystal microbalance, self assembled monolayer, bovine serum albumin, electrochemical impedance spectroscopy, transmission electron microscopy
DETECTION, EMPHASIZING SURFACE PLASMON RESONANCE (162 Records) USA 93 Japan 47 Peoples R China 44 Germany 40 South Korea 34	Tsing Hua Univ 11 Arizona State Univ 10 Kyushu Univ 9 Chinese Acad Sci 9 Univ Calif 8 Max Planck Inst Polymer Res 7 CNR 7 Acad Sci Czech Republ 7	*Sensors and Actuators B-Chemical* 31 *Analytical Chemistry* 28 *Biosensors & Bioelectronics* 18 *Analytical Biochemistry* 10 *Analytica Chimica Acta* 9	detection, sensor, chip, biosensor, mass spectrometry, liquid chromatography, real time, sensor chip, refractive index, sensor surface, gold surface, self assembled, gold nanoparticles, metal ions, surface plasmon resonance, bovine serum albumin, laser desorption ionization

continued on following page

Table 1. continued

BIOSENSORS (92 Records) USA 38 Peoples R China 28 South Korea 9 Japan 9 Germany 9	Univ Calif 6 Chinese Acad Sci 5 Univ Twente 4 Pacific Nw Natl Lab 4 Louisiana Tech Univ 4 CSIC 4	*Biosensors & Bioelectronics* *14* *Analytical Biochemistry* *6* *Langmuir* *5* *Electroanalysis* *5* *Chemical Communications* *5* *Analytical Chemistry* *5*	enzymes, immobilization, glucose oxidase, enzyme activity, enzyme loading, glucose biosensor, immobilized enzyme, electrode surface, catalytic activity, free enzyme, glassy carbon electrode, steady state current, glucose oxidase, scanning electron microscopy, direct electron transfer, multi wall carbon nanotubes, surface plasmon resonance
DNA DETECTION (282 Records) USA 166 Peoples R China 81 Japan 67 Germany 54 France 27 England 27	Chinese Acad Sci 18 Univ Calif 17 Purdue Univ 8	*Analytical Chemistry* 21 *Nucleic Acids Research* 20 *Langmuir* 20 *Nano Letters* 16 *Journal of Nanoscience and Nanotechnology* 16 *Biosensors & Bioelectronics* 14	DNA, oligonucleotid, target DNA, DNA hybridization, gold nanoparticles, nucleic acids, single stranded DNA, surface plasmon resonance, double stranded DNA, polymerase chain reaction, atomic force microscopy, x-ray photoelectron spectroscopy, peptide nucleic acid, self assembled monolayers, quartz crystal microbalance
DNA MOLECULES (411 Records) USA 149 Japan 66 Peoples R China 64 Germany 42 France 26 England 26	Chinese Acad Sci 19 Russian Acad Sci 12 Univ Calif 10 Univ Tokyo 9 Osaka Univ 9 Delft Univ Technol 8	*Nano Letters* 20 *Langmuir* 18 *Nucleic Acids Research* 16 *Proc NAS-USA* 12	DNA molecules, DNA binding, DNA fragments, self assembly, bound DNA, DNA protein, DNA sequence, DNA complexes, DNA hybridization, target DNA, atomic force microscopy, double stranded DNA, surface plasmon resonance, single stranded DNA, transmission electron microscopy, calf thymus DNA, x-ray photoelectron spectroscopy, scanning electron microscopy
DNA, EMPHASIZING GENE DELIVERY AND TRANSFECTIION (110 Records) USA 66 Peoples R China 37 Japan 23 South Korea 15 Germany 15 England 13 France 10	Chinese Acad Sci 7 Univ Calif 5 Kyoto Univ 5 Delft Univ Technol 5	*Journal of Controlled Release* 11 *Langmuir* 9 *Bioconjugate Chemistry* 9 *Nucleic Acids Research* 7	dna, gene, transfection, chitosan, plasmid dna, gene delivery, transfection efficiency, dna complexes, dna nanoparticles, gene transfer, gene therapy, gene expression, surface charge, particle size, atomic force microscopy, transmission electron microscopy, poly ethylene glycol, gene delivery systems, polymerase chain reaction, nonviral gene delivery, green fluorescent protein, plasmid dna encoding, dynamic light scattering

continued on following page

Table 1.continued

CELLS, EMPHASIZING MEMBRANES AND BACTERIA (348 Records) USA 416 Germany 128 Japan 111 Peoples R China 97 England 66 France 56	Univ Calif 37 Harvard Univ 27 Univ Tokyo 15 Johns Hopkins Univ 14 Univ Penn 13 Natl Univ Singapore 13 Chinese Acad Sci 13	*Biomaterials* 39 *Langmuir* 25 *Biophysical Journal* 18 *Journal of Membrane Science* 17 *Journal of Biomedical Materials Research Part A* 17	cells, membranes, bacteria, vesicles, cytoplasm, cell wall, transmission electron microscopy, scanning electron microscopy, atomic force microscopy, green fluorescent protein, human immunodeficiency virus, confocal laser scanning microscopy, whole cell patch clamp, gram negative bacteria, surface plasmon resonancell wall, quantum dots, fourier transform infrared, single particle tracking, bacterial cell surface, plasma membrane, escherichia coli, bacterial cells, epithelial cells,
TOT HEALTH+ (6512) USA 2106 Peoples R China 735 Japan 696 Germany 625 England 364 France 337 South Korea 325 Italy 262 Canada 217 India 170	Univ Calif 205 Chinese Acad Sci 153 Natl Univ Singapore 94 Osaka Univ 78 Univ Texas 68 Harvard Univ 68 Univ Illinois 62 Natl Inst Adv Ind Sci & Technol 57 Russian Acad Sci 56 Tsing Hua Univ 55 Univ Tokyo 54 CNRS 54	*Langmuir* 213 *Analytical Chemistry* 127 *Biomaterials* 126 *Journal of Physical Chemistry B* 120 *Biophysical Journal* 104 *Journal of Biological Chemistry* 102 *Journal of the American* *Chemical Society* 97 *Journal of Controlled Release* 96 *Biochemistry* 88 *Proc NAS-USA* 82	cells, protein, dna, membrane, binding, drugS, fluorescence, peptides, surface, nanoparticles, detection, interaction, surface plasmon resonance, atomic force microscopy, scanning electron microscopy, transmission electron microscopy, differential scanning calorimetry, x-ray photoelectron spectroscopy, bovine serum albumin, poly ethylene glycol, single stranded dna, double stranded dna, green fluorescent protein, fourier transform infrared, quartz crystal microbalance, polymerase chain reaction, self assembled monolayer, drug delivery systems, magnetic resonance imaging, confocal laser scanning, dynamic light scattering, , enzyme linked immunosorbent assay, resonance energy transfer, cell surface, x-ray diffraction, escherichia coli, amino acid, particle size, drug release, cell line, cell adhesion, dna molecules, mass spectrometry, endothelial cells

(protein* or peptide* or DNA or drug*) AND (nano* NOT (NaNO3 or NaNO2 or nanomolar* or nanosecond* or nanogram* or nanomole*))

In 1995, there were 10 records from the Asia group in total, and 70 records from the Europe group in total. In 2006 so far (probably 80% of the 2006 records have been entered into the database), there were 1,094 records for the Asia group, and 750 records for the Europe group, as shown in Figure 1.

Given this growth differential in medically-related records in just a decade, the difference in median age of the population (Germany, 42.6 years; England, 39.3 years; France, 39.1 years; China, 32.7 years; South Korea, 35.2 years, Japan, 42.9 years), and the high growth rate in graduation of scientists and engineers in China and South Korea (over 50% combined from 1995-2002), the research paper production differential between Asia and Europe can only be expected to increase, probably substantially.

The next largest Asian producer of nanotechnology medical applications papers, India, should not be neglected in this equation. From about 1980

Figure 1. Comparison of medical nanotechnology publications between Europe and Asia

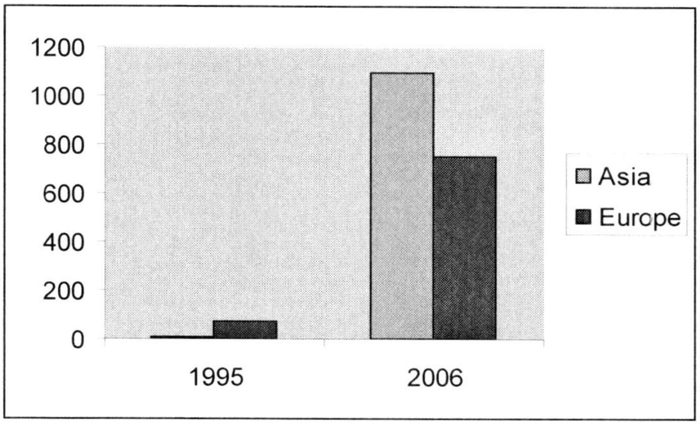

to almost 2000, India's growth rate of research publications stagnated (Kostoff, Johnson, Bowles & Dodbele, 2006d), while China's mushroomed. In the last decade, India has started to experience a resurgence of research paper productivity. Coupled with recent statements of strong support by the Indian Prime Minister for increasing research output production (Mukhergee, 2006), India could be the "dark horse" in this race. Its population (1.1 Billion) rivals China's, and its relatively low median age (24.9 years) reflects a large labor pool potentially available for increased emphasis on research production.

Institutional Productivity

The USA has substantial institutional representation in the top ten (California, Texas, Harvard, Illinois). These university publication numbers include all the state campuses. Thus, the University of California system includes University of California Berkeley (UCB), University of California Santa Barbara (UCSB), University of California San Francisco (UCSF), and so forth.

Leading Journals

While the leading journals have a strong chemistry component, a number of them cross disciplines among physics, chemistry, biology, and materials.

Associated Science

The science associated with the total health-type applications in the highest-level category can be divided into four major categories: instrumentation, materials, structures, phenomena. The key elements of each of these categories are as follows:

- **Instrumentation:** Surface plasmon resonance, atomic force microscopy, scanning electron microscopy, transmission electron microscopy, differential scanning calorimetry, x-ray photoelectron spectroscopy, fourier transform infrared spectroscopy, quartz crystal microbalance, magnetic resonance imaging, confocal laser scanning, enzyme linked immunosorbent assay, laser scanning microscopy, x-ray diffraction, mass spectrometry.
- **Materials:** Protein, DNA, peptides, drugs, bovine serum albumin, poly ethylene glycol, single stranded DNA, double stranded DNA, green fluorescent protein, lipids, human serum albumin, Escherichia Coli, antibodies, tissues, enzymes, genes, oligonucleotides, gold, nucleic acid.
- **Structures:** Cells, membranes, surfaces, nanoparticles, self-assembled monolayers, cell surfaces, endothelial cells, receptors.

- **Phenomena:** Fluorescence, interaction, polymerase chain reaction, dynamic light scattering, resonance energy transfer, particle size, drug release, cell adhesion, binding affinity, gene expression, transfection efficiency.

Second Highest Level Categories

The highest level category is divided by the fuzzy clustering algorithm into two categories, with the category centered around cells, proteins, and membranes being about seven times the size (number of records) of the category centered around DNA. The larger category's main journals (*Langmuir* 185, *Biomaterials* 120, *Journal of Physical Chemistry B* 112, *Analytical Chemistry* 108, *Journal of Biological Chemistry* 97, *Biophysical Journal* 95) focus on chemistry, physics, biology, and materials, while the smaller category's main journals (*Langmuir* 30, *Nano Letters* 29, *Nucleic Acids Research* 27, *Analytical Chemistry* 27, *Journal of the American Chemical Society* 21, *Journal of Nanoscience and Nanotechnology* 21) focus on chemistry and nanotechnology. The only journal in common at the top is *Langmuir*.

The larger category's main country performers (USA 1867, Peoples R China 620, Japan 608, Germany 561, England 323, France 301, South Korea 299) are remarkably similar to the smaller category's main country performers (USA 273, Peoples R China 123, Japan 106, Germany 78, England 46, France 43, South Korea 33). In aggregate of these main performers, Asia outproduces Europe by about 30%.

Lower Level Taxonomy Categories

Characteristics of the lower level taxonomy categories (elemental clusters) are summarized in the rows of Table 1. There are five main groupings: cancer treatment (drug release, drug delivery, tumor treatment, sentinel lymph node cancer), cells (tissue cells, cells (emphasizing adhesion), cells (emphasizing membranes and bacteria), biofilms), proteins (protein interactions, amyloid fibrils, peptide sequences, binding and affinity), sensing and detection (immunosensors, detection (emphasizing surface plasmon resonance), biosensors), and DNA (DNA detection, DNA emphasizing gene delivery and transfection). Only one group deals with a specific disease (cancer treatment), one is functional (sensing and detection), and the other three are based on fundamental biological materials at different aggregation levels (cells, proteins, DNA).

Because of the large number of elemental clusters, only the highlights or unusual features of each will be discussed, starting from the top row. Following each discussion are representative article titles from the cluster in bold italics, to illustrate the theme more concretely.

1. **Drug release:** USA, China dominant. India ranks much higher in this cluster relative to its overall health types ranking. Aggregating top producers only in Europe and Asia, Asia outproduces Europe by more than a factor of six. Even though Singapore is not listed as a leading country, the University of Singapore stands out as the institutional leader. No USA presence in leading institutions. The journals appear rather applied and focused. Materials and structures appear to be the science emphasis.
   - ***Physical characterization of controlled release of paclitaxel from the TAXUS(TM) Express(2TM) drug-eluting stent.***
   - ***Potential of guar gum microspheres for target specific drug release to colon.***
2. **Drug delivery:** USA dominant. Again, India ranks high, and University of Singapore leads. Asia outproduces Europe by 20%. Journals are again pharmaceutical oriented, and very applied. Again, no USA presence in leading institutions. Strong cancer focus in the science.

- ***Highly specific HER2-mediated cellular uptake of antibody-modified nanoparticles in tumour cells.***
- ***Developement and characterization of biodegradable nanospheres as delivery systems of anti-ischemic adenosine derivatives.***

3. **Tumor treatment:** USA has commanding lead. Asia outproduces Europe by fifty percent. American institutions dominate. Some physics journals along with pharmaceuticals. Laboratory research at cellular level, with magnetic physics emphasis, seems to dominate science.
   - ***Enhanced tumour uptake of Doxorubicin loaded poly(butyl cyanoacrylate) nanoparticles in mice bearing Dalton's lymphoma tumour.***
   - ***MRI after magnetic drug targeting in patients with advanced solid malignant tumors.***
4. **Sentinel lymph node cancer:** Again, USA and USA institutions dominant. Europe outproduces Asia by an order of magnitude. Many hospitals represented. Journals applied, and clinically oriented. Cancer detection focus in science.
   - ***SPECT-CT for topographic mapping of sentinel lymph nodes prior to gamma probe-guided biopsy in head and neck squamous cell carcinoma.***
   - ***Diagnostic performance of nanoparticle-enhanced magnetic resonance Imaging in the diagnosis of lymph node metastases in patients with endometrial and cervical cancer.***
5. **Tissue cells:** USA has commanding lead. Singapore surprisingly high. National University of Singapore again leader, by a wide margin. Asia outproduces Europe by factor of 2.5. Journals strongly biomaterials oriented. Science strongly focused on structure: cells, tissues, and bones.
   - ***Nano-fibrous scaffolds for tissue engineering.***
   - ***Self-organization of rat cardiac cells into contractile 3-D cardiac tissue.***
6. **Cells, emphasizing adhesion:** USA with commanding lead. Asia outproduces Europe by 30%. Strong USA university participation; also from National University of Singapore. Journals have strong biomaterials/biophysics orientation. Science strongly focused on cell growth, interactions, and death.
   - ***Development of a rare cell fractionation device: Application for cancer detection.***
   - ***Nanostructured designs of biomedical materials: Applications of cell sheet engineering to functional regenerative tissues and organs.***
7. **Biofilms:** USA dominant. Asia outproduces Europe by thirty percent. Montana State University not seen before. Very applied journals. Science strongly focused on films and infection.
   - ***Tooth development in a scincid lizard, Chalcides viridanus (Squamata), with particular attention to enamel formation.***
   - ***Adherence and biofilm formation of Staphylococcus epidermidis and Mycobacterium tuberculosis on various spinal implants.***
8. **Virus proteins:** USA with commanding lead. Europe outproduces Asia by 70%. Strong USA university representation, with University of California system dominant. Strong biochemistry journal emphasis. Strong virus research.
   - ***Identification of a region in the herpes simplex virus scaffolding protein required for interaction with the portal.***
   - ***Mass spectroscopic characterization of the coronavirus infectious bronchi-***

tis virus nucleoprotein and elucidation of the role of phosphorylation in RNA binding by using surface plasmon resonance.
- ***Expression of human papillomavirus type 16 L1 protein in transgenic tobacco plants.***

9. **Protein interactions:** USA with commanding lead. Europe outproduces Asia by eighty percent. USA institutions strong. Science focused on protein binding, other surface phenomena.
   - ***Analysis of protein interactions on protein arrays by a wavelength interrogation-based surface plasmon resonance biosensor.***
   - ***Biosensors: basic features and application for fatty acid-binding protein, an early plasma marker of myocardial injury.***
   - ***A central role for protein aggregation in neurodegenerative disease; Mechanistic and structural studies of human stefins.***
10. **Amyloid fibrils:** USA with commanding lead. Europe outproduces Asia by factor of three. Except for University of California system, USA universities not among most prolific. Biochemical/ biophysical journals. Science linked to Alzheimer's Disease and other neurodegenerative diseases.
    - ***Structure and function of amyloid in Alzheimer's disease.***
    - ***Surface plasmon resonance for the analysis of beta-amyloid interactions and fibril formation in 1Alzheimer's disease research.***
    - ***Structure and morphology of the Alzheimer's amyloid fibril.***
11. **Peptide sequences:** USA with commanding lead. Israel, Australia surprisingly high. Asia outproduces Europe by factor of two. MIT major institutional player, followed by University of California system. Science focused on binding, sequencing.
    - ***Novel electrochemical biosensing platform using self-assembled peptide nanotubes.***
    - ***Plasma levels of AGE peptides in type 1 diabetic patients are associated with serum creatinine and not with albumin excretion rate: Possible role of AGE peptide-associated endothelial dysfunction.***
    - ***Interactions of primary amphipathic cell penetrating peptides with model membranes: Consequences on the mechanisms of intracellular delivery of therapeutics.***
12. **Binding and affinity:** USA with overwhelming lead, solid institutional representation. Europe outproduces Asia by factor of 2.5. Biochemistry focus. Science focused on binding, reception, and affinity.
    - ***Biomacromolecule surface recognition using nanoparticles.***
    - ***Two-step mechanism of binding of apolipoprotein E to heparin.***
    - ***Formation of viscoelastic protein layers on polymeric surf aces relevant to platelet adhesion.***
13. **Immunosensors:** No infrastructure element dominant, as in previous cases. Asia outproduces Europe by factor of three. No USA institutional representation in upper tier. Strong use of immune system components in science.
    - ***Enhancement of the sensitivity of surface plasmon resonance (SPR) immunosensor for the detection of anti-GAD antibody by changing the pH for streptavidin immobilization.***
    - ***Development of functionalized terbium fluorescent nanoparticles for antibody labeling and time-resolved fluoroimmunoassay application.***

14. **Detection, emphasizing surface plasmon resonance:** USA with strong lead. Asia outproduces Europe by factor of three. Strong chemistry focus; some electronics. Science focused on sensors, use of gold.
    - ***The fabrication of protein chip based on surface plasmon resonance for detection of pathogens.***
    - ***Intracellular monitoring of superoxide dismutase expression in an Escherichia coli fed-batch cultivation using on-line disruption with at-line surface plasmon resonance detection.***
    - ***Surface plasmon resonance detection of endocrine disruptors using immunoprobes based on self-assembled monolayers.***
15. **Biosensors:** USA lead; China strong second. Asia outproduces Europe by factor of five. Research focus on enzyme-based biosensors that involve enzyme immobilization.
    - ***A novel glucose biosensor based on the nanoscaled cobalt phthalocyanine-glucose oxidase biocomposite.***
    - ***Multiwall carbon nanotube (MWCNT) based electrochemical biosensors for mediatorless detection of putrescine.***
    - ***Biosensors in drug discovery and drug analysis.***
16. **DNA detection:** USA with commanding lead. Asia outproduces Europe by forty percent. Strong USA institutional representation. Science focus is on DNA at surfaces for use in DNA biosensors.
    - ***A biosensor monitoring DNA hybridization based on polyaniline intercalated graphite oxide nanocomposite***
    - ***Detection of DNA and protein molecules using an FET-type biosensor with gold as a gate metal***
17. **DNA molecules:** USA with commanding lead. Asia outproduces Europe by thirty percent. Chinese Academy of Science institutional leader. Russian Academy of Science strong institutional presence, even though Russia not major player. Science focuses on DNA binding and DNA networks.
    - ***Atomic force microscopy study of the structural effects induced by echinomycin binding to DNA.***
    - ***Impedance sensing of DNA binding drugs using gold substrates modified with gold nanoparticles.***
18. **DNA, emphasizing gene delivery and transfection:** USA has strong lead. Asia outproduces Europe by factor of two. University of California system only USA presence in institutional leaders. Science focus on gene delivery and transfection efficiency.
    - ***Optical tracking of organically modified silica nanoparticles as DNA carriers: A nonviral, nanomedicine approach for gene delivery***
    - ***Nanoparticle based systemic gene therapy for lung cancer: Molecular mechanisms and strategies to suppress nanoparticle-mediated inflammatory response***
    - ***Calcium phosphate nanoparticles as a novel nonviral vector for efficient trasfection of DNA in cancer gene therapy***
19. **Cells, emphasizing membranes and bacteria:** Commanding USA lead. Europe outproduces Asia by 25%. Commanding USA organizational representation, with University of Californai system at forefront. Biomaterials literature emphasis. Science focuses on cell membranes and bacterial adhesion.
    - ***Microtubule-dependent matrix metalloproteinase-2/matrix metalloproteinase-9 exocytosis: Prerequisite in human melanoma cell invasion***
    - ***Long-term effects of HIV-1 protease inhibitors on insulin secretion and***

   insulin signaling in INS-1 beta cells
   - *Early stages of HIV replication: How to hijack cellular functions for a successful infection*
   - *Membrane-based on-line optical analysis system for rapid detection of bacteria and spores*

The USA is the leader in all 19 clusters. China took second place in seven clusters, Japan in six, Germany in four, and England in two. In terms of main institutions, University of California system led in five clusters, Chinese Academy of Science led in four, and surprisingly University of Singapore led in three. University of Singapore has strong presence in pharmaceuticals and biomaterials, Chinese Academy of Science has strong presence in DNA and binding, and University of California system has strong presence in cells and protein interactions.

These results require further context. The four major institutions discussed are of different size, have different funding levels, and have different manpower and other resources. For example, in 2005, there were 3,399 articles and reviews in the SCI/SSCI that contained at least one author with a National University of Singapore address, and there were a total of 6,622 authors listed on these records. The corresponding numbers for the other major institutions are: Chinese Academy of Science, 14,347 records, 19,089 authors; Russian Academy of Science, 11,216 papers, 30,137 authors; University of California system, 27,954 records, 84,667 authors.

Thus, for the National University of Singapore to be the publication leader in three thrust areas requires a considerable concentration of its modest resources relative to the other major institutions.

How do Asia and Europe differ in their research thrust areas? In the nineteen clusters listed on Table 1, the Asian leading countries in aggregate led by a factor of two or more (publications) in the following thematic areas:

- Drug release
- Tissue cells
- Peptide sequences
- Immunosensors
- Detection, emphasizing surface plasmon resonance
- Biosensors
- DNA, emphasizing gene delivery and transfection

By contrast, the European countries led in the following areas:

- Sentinel lymph node cancer
- Amyloid fibrils
- Binding and affinity

The Asian countries appear to emphasize biomaterials and detection/ sensing, while the European countries appear to emphasize treatment of specific diseases.

For the technology transfer community, these results contain some important messages. First, while there are some pervasive infrastructure results throughout the elemental clusters (e.g., USA is always most productive, China, Japan, Germany, England typically rank high), there are many individual differences. To understand the specific research infrastructure related to specific health applications, disaggregated evaluations are necessary. While the present analysis had a reasonable level of disaggregation, users interested in very specific medical applications will want to conduct much more disaggregated analyses. There are substantial differences between the overall nanotechnology health results and very specific health applications results.

Additionally, while there are some instruments that pervade the different elemental clusters, there are substantial instrumentation, material, nanostructure, and phenomenological differences among the clusters. Again, the individual cluster research can differ substantially from the overall nanotechnology health applications average. Readers who are interested in tracking the nanotechnology health-related research for technology transfer purposes are well advised to conduct specific analyses of the above type for each application. For investors, identifying which research areas pervade multiple applications would be extremely valuable, and the same recommendations are made as for technology transfer application.

Future Research Issues

The underlying science areas (such as cells, proteins, DNA, and membranes) have always been the mainstay of biomedical research. The explosive growth in nanotechnology has enabled nanotechnology research instruments to be used in medical applications. A natural question is whether the growth in medical applications is proportional to the growth in nanotechnology research? To address that question, it is necessary to study the growth in papers on nanotechnology medical applications relative to the total publications in nanotechnology. The USA continues to be the leading country in publications. However, substantial numbers of papers are from the developing Asian countries. It would be good to study the rate at which these other countries are developing and the possibility they may catch up with the US.

Additionally, the present study focused on output quantity, and was conducted at a reasonable level of disaggregation. Future studies need to address quality and resource issues as well, using much more detailed levels of disaggregation. Larger numbers of clusters should be run, and the citation impact of institutions associated with each cluster should be obtained along with numbers of publications. To understand the production efficiency better, the resources required to generate these publications should be obtained.

SUMMARY AND CONCLUSION

The study has identified the main nanotechnology medical applications as well as the related science and infrastructure. These relationships will allow the potential user communities to become involved with the medical applications-related science and performers at the earliest stages, to help guide the science conversion towards specific user needs more efficiently.

The pervasive instrumentation, materials, structures, and phenomena related to the most frequently mentioned nanotechnology health applications were identified, as follows:

- **Instrumentation:** Surface plasmon resonance, atomic force microscopy, scanning electron microscopy, transmission electron microscopy, differential scanning calorimetry, x-ray photoelectron spectroscopy, fourier transform infrared spectroscopy, quartz crystal microbalance, magnetic resonance imaging, confocal laser scanning, enzyme linked immunosorbent assay, laser scanning microscopy, x-ray diffraction, mass spectrometry.
- **Materials:** Protein, DNA, peptides, drugs, bovine serum albumin, poly ethylene glycol, single stranded DNA, double stranded DNA, green fluorescent protein, lipids, human serum albumin, Escherichia Coli, antibodies, tissues, enzymes, genes, oligonucleotides, gold, nucleic acid.
- **Structures:** Cells, membranes, surfaces, nanoparticles, self-assembled monolayers, cell surfaces, endothelial cells, receptors
- **Phenomena:** Fluorescence, interaction, polymerase chain reaction, dynamic light scattering, resonance energy transfer, particle size, drug release, cell adhesion, binding, affinity, gene expression, transfection efficiency.

Moreover, a medical applications categorization constructed from visual inspection of the fuzzy clustering categories showed five thematic categories:

- Cancer treatment
- Sensing and detection
- Cells
- Proteins
- DNA

In summary, for medical applications, analysis of nineteen thematic categories obtained from fuzzy clustering of the total 2005 nanotechnology database revealed the following:

- The USA is the publication leader in total health types, and in all the thematic areas as well, most by a wide margin. China was the second most prolific in seven thematic areas, Japan in six, Germany in four, and England in two.
- The University of California system led in five clusters, the Chinese Academy of Science led in four, and the National University of Singapore led in three. The University of California and the Chinese Academy of Science were the most prolific in the non-medical Applications as well, but their orders were reversed. The National University of Singapore is a prolific contributor, especially in pharmaceuticals and biomaterials.
- The journal Langmuir contains the most articles in total health, and is in the top layer of 10 of 19 themes. The only journals in common in the top layers of nonmedical and medical applications and health are *Langmuir* and *Journal of Physical Chemistry B*.
- For total health, the key underlying science areas include cells, proteins, DNA, membranes, binding, drugs, fluorescence, peptides, nanoparticles, detection, lipids, antibodies, immobilization, tissues, receptors, enzymes, genes, drug delivery, self assembly, cell surface, detection limit, escherichia coli, amino acid, molecular weight, particle size, real time, serum albumin, drug release, cell line, cell adhesion, DNA molecules, endothelial cells, surface plasmon resonance, atomic force microscopy, scanning electron microscopy, transmission electron microscopy, differential scanning calorimetry, x-ray photoelectron spectroscopy, bovine serum albumin, poly ethylene glycol, single stranded DNA, double stranded DNA, green fluorescent protein, fourier transform infrared spectroscopy, quartz crystal microbalance, polymerase chain reaction, self assembled monolayer, magnetic resonance imaging, confocal laser scanning, dynamic light scattering, enzyme linked immunosorbent assay, resonance energy transfer, extracellular matrix, laser scanning microscopy, human serum albumin, and poly lactic acid.

REFERENCES

Davidse, R. J. & Van Raan, A. F. J. (1997). Out of particles: Impact of CERN, DESY, and SLAC research to fields other than physics. *Scientometrics, 40*(2), 171-193.

Garfield, E. (1985). History of citation indexes for chemistry - A brief review. *JCICS, 25*(3), 170-174.

Goldman, J. A., Chu, W. W., Parker, D. S., & Goldman, R. M. (1999). Term domain distribution analysis: A data mining tool for text databases. *Methods of Information in Medicine, 38*, 96-101.

Gordon, M. D. & Dumais, S. (1998). Using latent semantic indexing for literature based discovery. *Journal of the American Society for Information Science, 49*(8), 674-685.

Greengrass, E. (1997). Information retrieval: An overview. *National Security Agency*. TR-R52-02-96.

Hearst, M. A. (1999). Untangling text data mining. In *Proceedings of ACL 99, the 37th Annual Meeting of the Association for Computational Linguistics*, University of Maryland.

Karypis, G. (2006). *CLUTO—A clustering toolkit*. Retrieved April 13, 2008, from http://www.cs.umn.edu/~cluto.

Kostoff, R. N., Eberhart, H. J., & Toothman, D. R. (1997a). Database tomography for information retrieval. *Journal of Information Science, 23*(4), 301-311.

Kostoff, R. N. (1997b). Accelerating the conversion of science to technology: Introduction and overview. *Journal of Technology Transfer* [Special Issue on Accelerating the Conversion of Science to Technology], *22*(3) .

Kostoff, R. N., Green, K. A., Toothman, D. R., & Humenik, J. A. (2000). Database tomography applied to an aircraft science and technology investment strategy. *Journal of Aircraft, 37*(4), 727-730.

Kostoff, R. N., Del Rio, J. A., García, E. O., Ramírez, A. M., & Humenik, J. A. (2001). Citation mining: Integrating text mining and bibliometrics for research user profiling. *Journal of the American Society for Information Science and Technology, 52*(13), 1148-1156.

Kostoff, R. N. (2003a). Text mining for global technology watch. In M. Drake (Ed.), *Encyclopedia of library and information science* (2nd ed) (Vol. 4, pp. 2789-2799). New York: Marcel Dekker, Inc.

Kostoff, R. N. (2003b). Stimulating innovation. In L. V. Shavinina (Ed.), *International handbook of innovation* (pp. 388-400). Oxford, UK: Elsevier Social and Behavioral Sciences.

Kostoff, R. N. (2003c). Bilateral asymmetry prediction. *Medical Hypotheses, 61*(2), 265-266.

Kostoff, R. N., Shlesinger, M., & Tshiteya, R. (2004a). Nonlinear dynamics roadmaps using bibliometrics and database tomography. *International Journal of Bifurcation and Chaos, 14*(1), 61-92.

Kostoff, R. N., Shlesinger, M., & Malpohl, G. (2004b). Fractals roadmaps using bibliometrics and database tomography. *Fractals, 12*(1), 1-16.

Kostoff, R. N., Stump, J. A., Johnson, D., Murday, J., Lau, C., & Tolles, W. (2005a). *The structure and infrastructure of the global nanotechnology literature* (DTIC Tech. Rep. No. ADA435984), Defense Technical Information Center, Fort Belvoir, VA. Retrieved April 13, 2008, from http://www.dtic.mil/

Kostoff, R. N., Murday, J., Lau, C., & Tolles, W. (2005b). *The seminal literature of global nanotechnology research* (DTIC Tech. Rep. No. ADA435986), Defense Technical Information Center, Fort Belvoir, VA. Retrieved April 13, 2008, from http://www.dtic.mil/

Kostoff, R. N., Stump, J.A., Johnson, D., Murday, J., Lau, C., & Tolles, W. (2006a). The structure and infrastructure of the global nanotechnology literature. *Journal of Nanoparticle Research, 8*(3-4), 301-321.

Kostoff, R. N., Murday, J., Lau, C., & Tolles, W. (2006b). The seminal literature of global nanotechnology research. *Journal of Nanoparticle Research, 8*(2), 193-213.

Kostoff, R. N. (2006c). Systematic acceleration of radical discovery and innovation in science and technology. *Technological Forecasting and Social Change, 73*(8), 923-936.

Kostoff, R. N., Johnson, D., Bowles, C. A., & Dodbele, S. (2006d). *Assessment of India's research literature* (DTIC Tech. Rep. No. ADA444625), Defense Technical Information Center, Fort

Belvoir, VA. Retrieved April 13, 2008, from http://www.dtic.mil/

Kostoff, R. N,. Koytcheff, R., & Lau, C. G. Y. (2007). *Structure of the global nanoscience and nanotechnology research literature* (DTIC Tech. Rep. No. ADA461930), Defense Technical Information Center, Fort Belvoir, VA. Retrieved April 13, 2008, from http://www.dtic.mil/

Losiewicz, P., Oard, D., & Kostoff, R. N. (2000). Textual data mining to support science and technology management. *Journal of Intelligent Information Systems, 15*, 99-119.

Mukherjee, D. (2006). Promote scientific research. *Central Chronicle.*

Narin, F. (1976). *Evaluative bibliometrics: the use of publication and citation analysis in the evaluation of scientific activity* (monograph). *NSF C-637.* National Science Foundation. Contract NSF C-627. NTIS Accession No. PB252339/AS.

Narin, F., Olivastro, D., & Stevens, K. A. (1994). Bibliometrics theory, practice and problems. *Evaluation Review, 18*(1), 65-76.

Schubert, A., Glanzel, W., & Braun, T. (1987). Subject field characteristic citation scores and scales for assessing research performance. *Scientometrics, 12*(5-6), 267-291.

SCI (2006). Certain data included herein are derived from the *Science Citation Index/Social Science Citation Index* prepared by the THOMSON SCIENTIFIC ®, Inc. (Thomson®), Philadelphia, Pennsylvania, USA: © Copyright THOMSON SCIENTIFIC ® 2006. All rights reserved.

SEARCH (2006). *TechOasis.* Norcross, GA: Search Technology Inc.

Swanson, D. R. (1986). Fish oil, raynauds syndrome, and undiscovered public knowledge. *Perspect Biol Med, 30*(1), 7-18.

Swanson, D. R. & Smalheiser, N. R. (1997). An interactive system for finding complementary literatures: A stimulus to scientific discovery. *Artificial Intelligence, 91*(2), 183-203.

TREC (Text Retrieval Conference) (2004). Retrieved April 13, 2008, from http://trec.nist.gov/.

Viator, J. A. & Pestorius, F. M. (2001). Investigating trends in acoustics research from 1970-1999. *Journal of the Acoustical Society of America, 109*(5), 1779-1783 Part 1.

Zhao, Y. & Karypis, G. (2004). Empirical and theoretical comparisons of selected criterion functions for document clustering. *Machine Learning, 55*(3), 311-331.

Zhu, D. H. & Porter, A. L. (2002). Automated extraction and visualization of information for technological intelligence and forecasting. *Technological Forecasting and Social Change, 69*(5), 495-506.

ENDNOTE

1 The views in this paper are solely those of the authors, and do not represent the views of the Department of the Navy or any of its components, or the Institute for Defense Analyses.

Chapter XII
Early Warning System for SMEs as a Financial Risk Detector

Ali Serhan Koyuncugil
Capital Markets Board of Turkey, Turkey

Nermin Ozgulbas
Baskent University, Turkey

ABSTRACT

This chapter introduces an early warning system for SMEs (SEWS) as a financial risk detector which is based on data mining. In this study, the objective is to compose a system in which qualitative and quantitative data about the requirements of enterprises are taken into consideration, during the development of an early warning system. Furthermore, during the formation of system; an easy to understand, easy to interpret and easy to apply utilitarian model that is far from the requirement of theoretical background is targeted by the discovery of the implicit relationships between the data and the identification of effect level of every factor. Using the system, SME managers could easily reach financial management, risk management knowledge without any prior knowledge and expertise. In other words, experts share their knowledge with the help of data mining based and automated EWS.

INTRODUCTION

The enormous computers of 1950's are now small enough to fit your hand, and are able to assist with the organization of work and daily activities. From the beginning of 1980's, great amounts of data have been accumulated with the usage of the database for computers in everywhere. Information grows when it is shared, therefore, researchers mention information particles before the 1980's, but nowadays they talk about information dews. In another words, the most important contribution of information technology (IT) can be summarized as information accessibility. On the other hand, the prevention of accessibility problem caused another problem—information accuracy. Therefore, the actual problem is accurately reading information from large amounts of information.

In addition, one of the basic insistences of IT is time concept. In the past, time cost meant almost nothing, but today the time is one of the most important factors mostly because of the multispeed processors. At that point, time cost for accessibility of the accurate information became an important factor because of data or information actuality.

Errors, subjectivity, and uncertainty in performance arised from human factors joined with the acceleration arised from IT; it almost took out the human factor in business processes. Intelligent systems began to take part in procedures, transactions, and processes instead of the human factor. As a result, computations done by humans turns into IT-based automated systems.

IT had a rapid improvement in the 1990's and removed almost all borders and distances on the globe in the early 2000's. The concept of "technology" became insufficient to describe that situation. Therefore, the term "knowledge age" was used for description. Another concept associated with knowledge age is "knowledge society." To reach accurate, objective, and useful knowledge in an easy way were it became basic requirements of knowledge society.

Another phenomenon associated with knowledge age is data mining. Towards the end of the 1990's, the idea of strategical usage for great amounts of data led to a fast achievement and popularity in every area that computers have been used for data mining. Data mining is the core of knowledge discovery process, which is mainly based on statistics, machine learning, and artificial intelligence. Generally, data mining discovers hidden and useful patterns in a very large amount of data. But it is difficult to make definitive statements about an evolving area and surely data mining is an area in very quick evolution. Therefore, there is no one single definition of data mining that would be met with universal approval. On the other hand, the following definition is generally acceptable: Data mining is the process of extracting previously unknown, valid and actionable information from large databases and then using the information to make crucial business decisions (Cabena, Hadjinian, Stadler, Verhees & Zanasi, 1997, p. 12).

Data mining is the most realistic method to responds with basic requirements of knowledge society, which are to reach accurate, objective, and useful knowledge in a simplified way. Another concept which is necessary for providing accurate, objective, and useful knowledge is "expertise." It is impossible to provide enough expertise for the entire society, but it is possible to provide "expert knowledge" via IT.

It is possible to provide expert knowledge to nonexperts in every field—business management and economics as well. However, from the business point of view, the firms that mostly need the information are small and medium industrial enterprises (SMEs); they have a great importance with regards to economy. Although SMEs have made an important contribution to the world's rapid economic growth and the fast industrialization process, to enlighten SMEs' managers for overcoming difficulties and improving strategies is critically important. These reasons were what motivated the authors of this chapter to select SMEs as an application area.

SMEs are thrown in financial distress and bankruptcy risk by financial issues. Many SMEs are closed because of this financial distress. These issues of SMEs were grown out of the lack of information and could not use the information in decision-making process. By this approach, SMEs need an early warning system which should give decision support that is easy to understand, easy to interpret, and easy to apply for the decision makers of SMEs. Consequently, the structure of the early warning system:

- Does not require expertise for the calculation and interpretation of the financial and administrative indicators

- Can realize the necessary analysis automatically
- Does not need analytical depth

Therefore, the intention is to bring out the relationship between the financial and administrative variables into the open, to identify the criteria of risk and to use the risk models for decision support. The identification of the criteria of risk by clearifying the relationship between the variables defines the discovery of knowledge from the financial and administrative variables. In this context, automatic and estimation oriented information discovery process coincides the definition of data mining.

This chapter discusses ways of empowering knowledge society from SME viewpoint via designing an early warning system based on data mining. Using the system, SME managers could easily reach financial management, risk management knowledge without any prior knowledge or expertise. They can be a part of the knowledge society via the knowledge, which is provided by early warning system (EWS) based on data mining without any complexity. The people who work for SMEs reach expert knowledge by the way of using EWS. In other words, experts share their knowledge with the help of IT based and automated EWS. Therefore, SMEs can be the part of knowledge society.

The following section provides background about the financial problems of SMEs in some counties and early warning system applications by data mining and other methods.

BACKGROUND

If one looks at the developed countries of the world, such as USA, Japan, and Germany or developing countries such as Thailand, Malaysia, and China, it can be seen that a dynamic and vibrant SME sector is playing a key role in the successful economic growth of these countries. SMEs are defined as nonsubsidiary, independent firms, which employ less than a given number of employees. This number varies across national statistical systems. The most frequent upper limit is 250 employees, as in European Union. However, some countries set the limit at 200 employees, while the United States considers SMEs to include firms with fewer than 500 employees (OECD, http://www.oecd.org). SMEs, which are the vital drivers of the economy, are topic of significant research interest for academics and an issue of great importance for policy makers around the globe. Governments in developing countries, as well as developed countries, started to realize the important role played by SMEs (WIPO, 2002).

Today, globalization is an important factor that has impact on SMEs. SMEs have to be prepared to meet the challenges of the opening markets and the risks associated with it. Opening markets or internationalization of markets provide new opportunities for expansion and growth. But on the other hand it means intensive competition with foreign enterprises, therefore brining threats and challenges. Particularly, after the effects of the globalization have been seen, the financial problems of SMEs have become the subject of several research and reports.

The last five years were evaluated, from the Far East countries to the European countries, and several researches were discovered working on the financial problems of SMEs.

Bukvik and Bartlett (2003) aimed to identify financial problems that prevented the expansion and development of SMEs in Slovenia, Bosnia, and Macedonia. In their studies, they applied a survey on 200 SMEs which were active in Slovenia between 2000-2001. As a result of the study, the major financial problems of SMEs in these countries are defined as high cost of capital, insufficient financial cooperation, the bureaucratic processes of banks, SMEs' lack of information on financial subjects and delay in the collection of the payment.

Sormani (2005) researched the financial problems of small businesses in the UK. The study emphasized the cash flow and unpaid invoices of SMEs. According to this study, there were two main suggestions for SMEs in UK. The first one was managing cash flow for the financial health of SMEs, and the other one was taking into account the time between issuing an invoice and receiving payment in order to run efficiently.

Bitszenis and Nito (2005) determined the financial problems of SMEs while evaluating the obstacles and problems encountered by entrepreneurs in Albania. It was determined that the most important financial problems were lack of financial resources and taxation faced by SMEs in Albania.

Sanchez and Marin (2005) analyzed the management characteristics of Spanish SMEs according to their strategic orientation and the consequences in terms of firm performance and business efficiency. The study was conducted on 1,351 Spanish SMEs. The results confirmed the expected relationship between management characteristics and performance of SMEs in Spain.

Inegbenebor (2006) focused on the role of entrepreneurs and capacity to access and utilize the fund of SMEs in Nigeria. The sample study consisted of 1,255 firms selected to represent 13 identified industrial subsectors. The results of the study showed that the capacities of SMEs to access and utilize the funds were weak in Nigeria.

Kang (2006) analyzed the role of the SMEs in the Korean economy. According to the study, SMEs have been hit hard by the economic slowdown and also face some deep structural problems. The main financial problem of SMEs in Korea was heavy dept financing. Many SMEs are overburdened with debt. Also SMEs are saddled with excess capacity, and they suffer from growing overseas competition. All these factors are affected the profitability of SMEs in Korea.

Nowadays, many entrepreneurs included SMEs are encountered with financial distress and this factor motivated the enterprises for monitoring the financial condition periodically and orderly. Efforts requiring to learn information on firms financial conditions have a long story. The efforts towards the seperation between financially distress and nondistress enterprises started with the z-score that are based on the usage of ratios by Beaver (1996) for single and multiple discriminant analysis of Altman (1968). The examples of other important studies that used multivariable statistical models, are given by Deakin (1972), Altman, Haldeman, and Narayanan (1977), Taffler and Tisshaw (1977) with the usage of multiple discriminant models are also given by Zmijewski (1984), Zavgren (1985), Jones (1987), Pantalone and Platt (1987), with the usage of logit vs. probit models are at the same time given by Meyer and Pifer (1970), with the usage of multiple regression model (Koyuncugil & Ozgulbas, 2006c).

Artificial neural network is used for the identification of problems including financial failure and bankruptcy at 1980's and researchers like Hamer (1983), Coats and Fant (1992), Coats and Fant (1993), Chin-Sheng et al. (1994), Klersey and Dugan (1995), Boritz et al. (1995), Tan and Dihardjo (2001) and Anandarajan et al. (2001) dealt with artificial neural network in their researches (Koyuncugil & Ozgulbas, 2006c).

The basic areas for data mining in financial studies are about equities, exchange rates, estimation on bankruptcy of enterprises, identification and management of financial risk, management of loans, idenfication of customer profiles and the analysis of money laundering (Kovalerchuk & Vityaev, 2000). In addition to this, data mining was used in the studies of Eklund, Back, Vanharanta, and Visa (2003), Hoppszallern (2003), Derby (2003), Chang, Chang, Lin, and Kao (2003), Lansiluoto, Eklund, Barbro, Vanharanta, and Visa (2004), Kloptchenko, Eklund, Karlsson, Back, Vanhatanta, and Visa (2004) and Magnusson, Arppe, Eklund, and Back (2005) for financial performance analysis.

Koyuncugil and Ozgulbas (2006a) emphasized the financial problems of SMEs in Turkey by identifying their financial profiles via data mining. They put forward their suggestions on solutions in addition to the stock markets of SMEs, and expressed that the first step to solve those problems was to identifty the financial profiles of SMEs. Researchers identified the role of SMEs with the method of data mining by the usage of the data from the year 2004 on operations of 135 SMEs in Istanbul Stock Exchange (ISE). As a result of the study, the most important factor that affects the financial performance of SMEs is determined as the strategy of finance. At the end of the study, the basic suggestion indicates that the SMEs who are concentrated on debt financing can increase the financial performance.

Another study by Koyuncugil and Ozgulbas (2006b) on operations of SMEs in İstanbul Stock Exchange (ISE) was a criterion of the financial performance for SMEs that are identified due to factors that affects the financial risk and performance of SMEs.

Once again in another study, that was held by Koyuncugil and Ozgulbas (2006c) on operations of SMEs between the years of 2000-2005 in ISE; the identification of the financial factors that affected the financial failures of SMEs was aimed with the usage of CHAID (chi-square automatic interaction detector) decision tree algorithm. In various studies by Ozgulbas and Koyuncugil (2006) and Ozgulbas, Koyuncugil, and Yilmaz (2006) same set of data showed that the success of the firms was not only based on the financial strength of SMEs, but also based on the scale of the SMEs.

But many companies and their managers have not recognized the symptoms of oncoming financial failure and risk in their business. And when symptoms or signals start occurring, managers do not know what type of action to take first or how to manage the situation. By recognizing some early warning signs of business financial trouble, managers may eliminate, overcome, or at the very least, side step those troubles and risks. Some studies about early warning systems used by data mining and other analytical methods are presented next.

Early warning systems that are used to examine financial failure and risk are investigated for banking sector by Gaytan and Johnson (2002). Collard (2002) emphasized the importance of early warning systems and presented ten early warning signs that pointed business failure and risk to the firm's managers. Mena (2003) mentioned credit card fraud detection via data mining. Gunther and Moore (2003) aimed to develop an early warning model for monitoring the financial condition of bank. They used a statistical models to include 12 financial ratios covering the major categories of financial factors considered under the CAMELS rating system.

Jacops and Kuper (2004) presented an early warning system for six countries in Asia. They used a binomial multivariate qualitative response approach and constructed a model that calculates the probability of a financial crisis.

Apoteker and Barthelemy (2005) focused on financial crises in emerging markets. They used a newly developed nonparametric methodology for country risk signaling. They constructed nine early warning signals to predict financial crises in emerging markets.

Ko and Lin (2005) introduced a modularized financial distress forecasting mechanism based on data mining

Canbas, Onal, Duzakin, and Kilic (2006) aimed to investigate whether or not firms that are taken into the surveillance market in Istanbul Stock Exchange are experiencing financial distress. They developed an integrated early warning model for financial distress prediction by combining principal component analysis and discriminant analysis.

Liu and Lindholm (2006) focused on financial crises that occurred around the world. They showed how the use of fuzzy C-means method can help to identify economic of financial crises as an early warning system.

A novel anomaly detection scheme that uses a data mining to handle computer network security problems is proposed by Shyu, Chen, Sarinnapakorn, and Chang (2006).

Chan and Wong (2007) attempted to find financial stresses and to predict future financial crises for all possible scenarios. To reach this objective they used early warning system to measure the resilience by data mining in their study.

Kamin, Schindler, and Samuel (2007) used early warning systems in emerging markets to identify the roles of domestic and external factors in emerging market crises. Several probit models of currency crises were estimated for 26 emerging market countries. These models were used to identify the separate contributions to the probabilities of crisis of domestic and external variables.

Tan and Quektuan (2007), attemped to use genetic complementary learning (GLC) as a stock market predictor, and bank failure early warning system is investigated. The experimental results show that GCL is a competent computational finance tools for stock market prediction and bank failure early warning system in their study.

Securities exhange markets have early warning or surveillance systems similar to Stock Watch, ASAM, and ATOMS. Stock Watch is the New York Stock Exchange's state-of-the-art computer surveillance unit, which monitors the market in NYSE-listed stocks for aberrant price and volume activity, which may indicate illegal transactions. In addition, automated search and match (ASAM) is another system in which researches and cross-references with publicly available information on individuals, corporations, and service organizations are possibly connected to a particular trading situation (www.nyse.com, 2007).

The Stock Exchange of Thailand (SET) has employed a computerized system for market surveillance. The main tool that handle all market surveillance tasks is called "automated tools for market surveillance" or ATOMS. ATOMS is aimed for monitoring securities trading activities in the SET, analyzing any unusual trading, facilitating any investigation of suspicious cases, and documenting all the tasks to the database (The Stock Exchange of Thailand, 2007).

The most important study towards the purpose of this study is the design of an early warning system based on data mining about the examination of market abuse (manipulation and insider trading) in the stock market was evaluated by Koyuncugil (2006). Koyuncugil determined the success of designed early warning system by testing the system with actual data.

MAIN THRUST

The common result that is gained from these researchs is SMEs' financial problems. The basic reasons of these problems are:

- The economical condition of country
- Underdevelopment of money and capital markets that can provide financial sources to SMEs
- Insufficiency of financial administration and administrators

During a period of time, these problems cause failures and low performance then SMEs recede from the economical environment. As a result of these failures and low performance, only a portion of SMEs can continue their economical activities under difficult conditions. The important role of SMEs for economical development requires guessing the contidition of financial success especially for preventing financial risk under risky conditions. Studies on prediction of financial failures of enterprises and finding the possible reasons of these failures take the attention of administrators, inventors, creditors, inspectors, partners of enterprises, academicians and especially financial administrators.

Actually, it is impossible to solve the problems related to:

- The economical condition of country
- Underdevelopment of money and capital markets

Without country level economical and political policies, although, it is possible to solve problems related to:

- Insufficiency of financial administration and administrators

The basic solution for this insufficiency problem is expert support. However, it is not possible to provide adequate expert for all SMEs and it is not necessary as well. The most rationale approach for solving the problem is to automate the expert knowledge in financial management.

Another development that SMEs are attempting to solve is their problem obtaining financial resources. They would soon face the requirement to comply with and make necessary provisions for the requirements set out in basel-II standard, which is supposed to become effective from 2007.

The basel-II framework describes a more comprehensive measure and minimum standard for capital adequacy that national supervisory authorities are now working to implement through domestic rule-making and adoption procedures. Although, the basel-II is fully based on risk measures, it seeks to improve the existing rules by aligning regulatory capital requirements more closely to the underlying risks that banks face. As a result, it is intended to be more flexible and better able to evolve with advances in markets and risk management practices (http://www.bis.org/publ/bcbsca.htm).

The required support is to provide a tool, which will:

- Provide guidance in financial management
- Provide expert knowledge in more accurate and speedy way without experts
- Discover probable risks and solutions
- Give early warning signs
- Provide roadmaps for prevention financial crisis and distresses
- Easily understand, easily implement and easily interpret, according to the picture given

The tool defined earlier from information technologies viewpoint can be termed as early warning system (EWS). The basic motivation behind the development of EWS for SMEs is to solve financial problems of SMEs in all developed and developing countries. The EWS seeks to identify the risks SMEs may face, today and in the future, and to develop or improve their ability to manage those risks. As a result, it is intended to be more flexible and better able to evolve with advances in markets and risk management practices. Therefore, SMEs reach better governance and financial performance.

Furthermore, operational logic of early warning systems is mainly based on finding unexpected and extraordinary behaviors. Thus, according to Cabena et al. (1997), the definition of data mining in in this aspect: the process of extracting previously unknown, valid and actionable information from large databases and then using the information to make crucial business decisions. From that view point the definitions of EWS and data mining lead to an interesting similarity. Therefore, data mining can be treated as the best analytical approach for early warning systems.

EARLY WARNING SYSTEM FOR SMES BASED ON DATA MINING

Financial Early Warning Systems

Early warning system (EWS) is a technique of analysis that is used to predict the achievement condition of enterprises and to decrease the risk of financial crisis. Applying this technique of analysis, the condition and possible risks of an enterprise can be identified with quantity.

The objectives of EWS are:

- Identification of changes in environment before clarification
- Identification of speed and direction of change for projecting the future
- Identification of the importance in the proportion of change
- Determination of deviations and taking signals
- Determination of possible reactions in direction of privileged deviations
- Investigation of the factors that cause change and the transaction between these factors

However, there is no specific method for total prevention for a financial crisis of enterprises. The important point is to set the factors that cause the condition with calmness, to take corrective precautions for a long term, to make a flexible emergency plan towards the potential future crisis. In essence, the early warning system is a financial analysis technique, and it identifies the achievement analysis of enterprise due to its industry with the help of financial ratios.

Financial early warning systems are grouped under three main categories in the literature (Kutman, 2001):

- The models towards the prediction of profits of enterprise
- The ratio based models towards the prediction of bankrupt/crisis of enterprise
- Economic trend based models towards the prediction of bankrupt/crisis of enterprise

Subsequently, the models that are used by enterprises for early warning model are mostly based on ratio analysis. Some examples of these models are below (Oksay, 2006):

- IRIS (insurance regulatory information system)
- FAST (financial analysis tracking system)
- Neural network systems
- Discriminant models
- Rating systems
- Event history analysis
- Recursive partitioning algorithm

Definition of the Early Warning System for SMEs Based On Data Mining

Nearly all of the financial early warning systems are based on ratio analysis, in other words, based on financial tables. Financial tables are the data sources that reflect the financial truth for early warning system. However, ratio based models ignore the administrative and structural factors and this situation shows that the human factor is not evaluated for the decision mechanism. Therefore, the additional determination of the human factor in every part of the decision process and the characteristics of decision environment to the model will harmonize them with real life and provide an applicable system.

In this study, the objective is to compose a system in which qualitative and quantitative data about the requirements of enterprises are taken into consideration, during the development of an early warning system. Furthermore, during the formation of system; an easy to understand, easy to interpret and easy to apply utilitarian model that is far from the requirement of theoretical background is targeted by the discovery of the implicit relationships between the data and the identification of effect level of every factor. Because of this reason, the ideal method that will help researchers to reach their objective is the data mining method that is started to use frequently nowadays for financial studies.

It is expected that the system will provide benefits to SMEs that have higher proportion of financial risk than larger enterprises, and contributions to the economy and science of country. Some of the contributions of SME early warning system (SEWS) that are expected can be summarized as:

- The financial requirements and weakness of SMEs will be manifested
- Identification of the SMEs' financial strategy with minimum expertise on financial administration will be possible
- The financial risk levels of SMEs will be clearified
- The possibility of happening of a financial crisis will decrease
- The efficient usage of financial resources will be provided
- Loss and gain analysis will be made
- The competition capacity of SMEs against the financial crisis will increase
- The financial comfort will provide opportunities for investments and especially for technological investments
- Financial improvement of SMEs will create a new potential for export
- The decrease of bankruptcy of enterprises, and contribution of employment on economy will create a positive effect
- New enterprises and support of taxes for government will increase
- To provide identification of risk factors and application of risk reducing strategy by SMEs

With the help of a scientifically authentic study, concrete outputs will be offered to the sectoral users, and ultimately, output of the study will lead several new researches.

Method

The main approach for SME early warning system (SEWS) is discovering different risk levels and identifying the factors effected risk levels. Therefore, the SEWS should focus segmentation methods. In the scope of the methods of data mining:

- Logistic regression
- Discriminant analysis
- Cluster analysis
- Hierarchical cluster analysis
- Self organizing maps (SOM)
- Classification and regression trees (C&RT)
- Chi-square automatic interaction detector (CHAID)

can be the principal methods, in addition to other classification/segmentation methods. However, during the preparation of an early warning system for SMEs, one of the basic objectives is to help SME administrators and decision makers, who do not have financial expertise, knowledge of data mining and analytic perspective, to reach easy to understand, easy to interpret, and easy to apply results about the risk condition of their enterprises. Therefore, decision tree algorithms that are one of the segmentation methods can be used because of their easy to understand and easy to apply visualization. Although, several decision tree algorithms have widespread usage today, chi-square automatic interaction detector (CHAID) is separated from other decision tree algorithms because of the number of the branches that are produced by CHAID. Other decision tree algorithms are branched in binary, but CHAID manifests all the different structures in data with its multibranched characteristic. Hence the method of CHAID is used within the scope of this study.

Steps of the SEWS

The SEWS designed similarly with knowledge discovery in databases (KDD) process and has 6 main steps (shown in Figure 1) which are given:

Figure 1. Data flow diagram of the SEWS

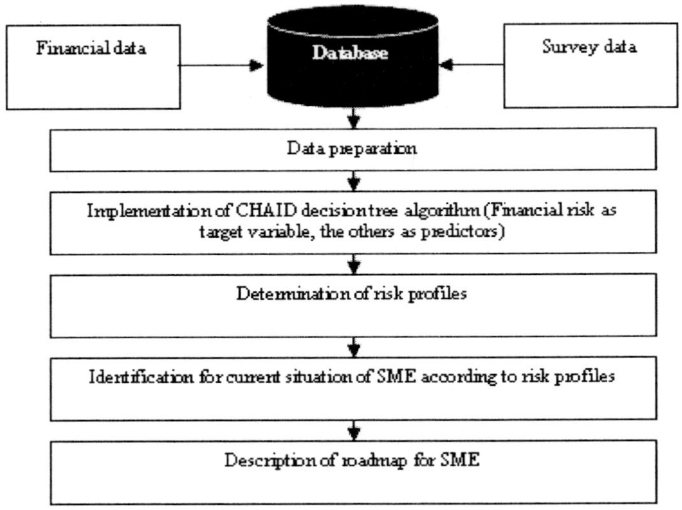

1. Preparation of data collection
2. Organization of data collection
3. Implementation of DM method
4. Determination of risk profiles
5. Identification for current situation of SME from risk profiles
6. Description of roadmap for SME

Preparation of Data Collection

Two data sets can be taken as the foundation for SEWS:

1. Financial data that are gained from balance sheets: Items of balance sheets will be entered as financial data and will be used to calculate financial indicators of system.
2. With the exception of financial data; managerial, demographic, private and structural data that are gained from SMEs by the way of surveys, that include the following parameters:
    - Sector
    - Legal status
    - Number of partners
    - Number of employes
    - Annual turnover
    - Annual balance sheet
    - Financing model
    - The usage situation of alternative financing
    - Technological infrastructure
    - Literacy situation of employes
    - Literacy situation of managers
    - Financial literacy situation of employees
    - Financial literacy situation of managers
    - Financial training need of employes
    - Financial training need of managers
    - Knowledge and ability levels of workers on financial administration
    - Knowledge and ability levels of workers on financial administration
    - Financial problem domains
    - Current financial risk position of SMEs

Organization of Data Collection

1. **Arrangement of data of balance sheets.**
    a. Calculation of financial indicators that are shown in Table 1

b. Reduction of repeating variables in different indicators to solve the problem of Collinearity / multicollinearity
c. Input of missing data
d. Solution of outlier and extreme value problem
2. **Arrangement of survey data.**
a. Input of missing data
b. Solution of outlier and extreme value problem

Implementation of DM Method

Assume that $X_1, X_2,..., X_{N-1}, X_N$ denote discrete or continous independent (predictor) variables and Y denotes dependent variable as target variable in CHAID algorithm where $X_1 \in [a_1,b_1]$, $X_2 \in [a_2,b_2]$,..., $X_N \in [a_N,b_N]$ and $Y \in \{Poor, Good\}$. While "Poor" shows poor financial performance in red bar and "Good" shows good financial performance in green bar in CHAID decision tree in Figure 2.

In Figure 2 we can see that only 3 variables of N have a statistically significant relationship with the target Y:

- X_1 has most statistically significant relation with target Y.
- X_2 has statistically significant relation with X_1 where $X_1 \leq b_1$.
- X_3 has statistically significant relation with X_1 where $b_{11} < X_1 \leq b_{12}$.

Determination of Risk Profiles

Among the target variable and predictor variables, CHAID algorithm organizes Chi-square independency test and starts from branching the variable, which has the strongest relationship, and at the same time arranges statistically significant variables on the branches of the tree in terms of the strength of their relationships. An example of a CHAID decision tree is seen in Figure 2. As it is observed from Figure 2, CHAID has multi-branches, while other decision trees are branched in binary. Thus, all of the important relationships in data can be investigated until the subtle details. In essence, the study identifies all the different risk profiles. Here the term risk means the risk that is caused due to of the financial failures of enterprises.

Table 1. Variables and their definitions

Financial Variables	Definitions
Current Ratio	Current Assets/ Current Liabilities
Quick Ratio (Liquidity Ratio)	(Cash+Marketable Securities+ Accounts Receivable)/ Current Liabilities
Absolute Liquidity	(Cash+Banks+ Marketable Securities+ Accounts Receivable)/ Current Liabilities
Inventories to Current Assets	Total Inventories / Current Assets
Current Liabilities to Total Assets	Current Liabilities / Total Assets
Debt Ratio	Total Dept/Total Assets
Current Liabilities to Total Liabilities	Current Liabilities to Total Liabilities
Long Term Liabilities to Total Liabilities	Long Term Liabilities to Total Liabilities
Equity to Assets Ratio	Total Equity/Total Assets
Current Assets Turnover Rate	Net Revenues/ Current Assets
Fixed Assets Turnover Rate	Net Revenues / Fixed Assets
Days in Accounts Receviable	Net Accounts Receivable/ (Net Revenues /365)
Inventories Turnover Rate	Net Revenues / Average Inventories
AssetsTurnover Rate	Net Revenues / Assets
Equity Turnover Rate	Net Revenues / Equity
Profit Margin	Net Income/ Total Margin
Return on Equity	Net Income/ Total Equity
Return on Assets	Net Income / Total Assets

Figure 2. CHAID decision tree

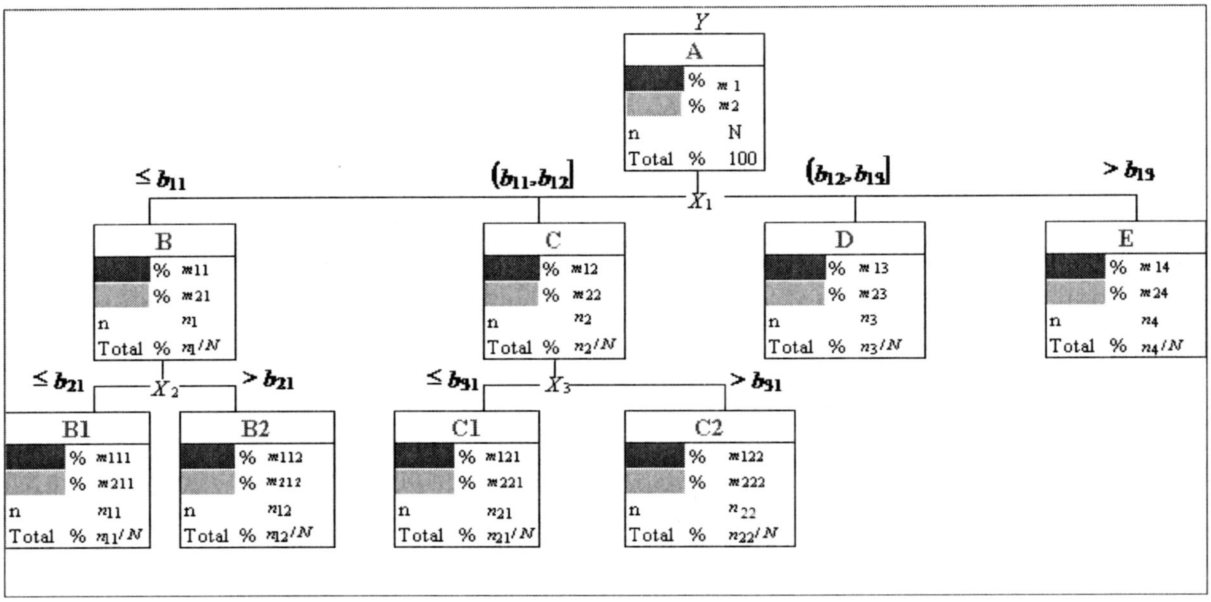

Figure 2 shows that there are six risk profiles:

- **Profile B1** shows that:
  - There are n_{11} samples where $X_1 \le b_{11}$ and $X_2 \le b_{21}$
  - % m_{111} has poor financial performance
  - % m_{211} has good financial performance
- **Profile B2** shows that:
  - There are n_{12} samples where $X_1 \le b_{11}$ and $X_2 \le b_{21}$
  - % m_{112} has poor financial performance
  - % m_{212} has good financial performance
- **Profile C1** shows that:
  - There are n_{21} samples where $b_{11} < X_1 \le b_{12}$ and $X_3 \le b_{31}$
  - % m_{121} has poor financial performance
  - % m_{221} has good financial performance
- **Profile C2** shows that:
  - There are n_{21} samples where $b_{11} < X_1 \le b_{12}$ and $X_3 > b_{31}$
  - % m_{122} has poor financial performance
  - % m_{222} has good financial performance
- **Profile D** shows that
  - There are n_3 samples where $b_{12} < X_1 \le b_{13}$
  - % m_{13} has poor financial performance
  - % m_{23} has good financial performance
- **Profile E** shows that:
  - There are n_4 samples where $X_1 > b_{13}$
  - % m_{14} has poor financial performance

- % m_{24} has good financial performance

If all of the profiles are investigated separately:

- Profile B1 shows that if any firm's variables X_1 and X_2 have values where $X_1 \leq b_{11}$ and $X_2 \leq b_{21}$, poor financial performance rate or in another words risk rate of the firm will be $R_{B1} = m_{111}$.
- Profile B2 shows that if any firm's variables X_1 and X_2 have values where $X_1 \leq b_{11}$ and $X_2 > b_{21}$, poor financial performance rate or in another words risk rate of the firm will be $R_{B2} = m_{112}$.
- Profile C1 shows that if any firm's variables X_1 and X_3 have values where $b_{11} < X_1 \leq b_{12}$ and $X_3 \leq b_{31}$ poor financial performance rate or in another words risk rate of the firm will be $R_{C1} = m_{121}$.
- Profile C2 shows that if any firm's variables X_1 and X_3 have values where $b_{11} < X_1 \leq b_{12}$ and $X_3 > b_{31}$ poor financial performance rate or in another words risk rate of the firm will be $R_{C2} = m_{122}$.
- Profile D shows that if any firm's variable X_1 has values where $b_{12} < X_1 \leq b_{13}$ poor financial performance rate or in another words risk rate of the firm will be $R_D = m_{13}$.
- Profile E shows that if any firm's variable X_1 has values where $X_1 > b_{13}$ poor financial performance rate or in another words risk rate of the firm will be $R_E = m_{14}$.

Identification for Current Situation of SME According to Risk Profiles

Part of this study until this point, is based on the identification of risk profiles from all of the data. In the scope of the data that is about the past of SMEs, the part of the study until this point defines the relationships between financial risk and variables, and also the risk profiles.

At this stage, risk profiles from all of the firms belonging to a bound are identified in the study. This identification is realized by taking the group of variables in the risk profiles into consideration.

All of the firm will look at the values of their own enterprises, in the light of the statistically significant variables in the decision tree. According to Figure 2 these variables are X_1, X_2, and X_3. The firm compares the values of X_1, X_2, and X_3 between decision tree and firms. Then, they can identify their risk profile. For example if any firm has $X_1 > b_{13}$. Therefore, the risk profile of the firm must be Profile E.

Description of Roadmap for SME

According to Figure 2, the risk grades of the firms can easily be determined. Assume that, the risk rates of the firms in the order of E > D > C2 > C1 > B2 > B1. Therefore, the best risk profile will be B1. Then, every firm tries to be in Profile B1. There are two variables X_1 and X_2 related with profile B1. If any firm want to be in Profile B1, the firm must make arrangements to make values $X_1 \leq b_{11}$ and $X_2 \leq b_{21}$.

Enterprise will identify the suitable road map after defining its risk profile. The enterprise can identify the path to reach upper level risk profile and the indicators will require privileged improvement in the light of the priorities of the variables in the roadmap. Furthermore, enterprise can pass to upper level risk profiles step by step and at the same time can reach to a targeted risk profile in the upper levels for improving indicators related to this target. For example, any firm in Profile E has the biggest risk rate. The firm must rehabilitate first the variable X_1 to decrease it between $(b_{12}, b_{13}]$. Therefore, the firm will be in profile D and so on.

THE FUTURE VISION OF EARLY WARNING SYSTEMS BASED ON DATA MINING

Management

To provide information relating to the actions of individual officers, supervisors, and specific units or divisions, early warning management systems should be developed and implemented in every business. In deciding what information to be included in their early warning system, business should balance the need for sufficient information for the system to be comprehensive; with the need for a system that is not too cumbersome to be utilized effectively. The system should provide supervisors and managers with both statistical information and descriptive information about the function of business.

Marketing

Marketing is one of the foremost areas where data mining techniques can be applied. Data mining enables an organization to sort through vast amounts of customer data to target the right customers. This is of vital importance to the marketing department of any organization. Substantial amounts of time and money can be saved if an organization knows who their customers are and are able to predict what their spending patterns will be. Potential uses of data mining in the area of marketing, include:

- **Customer acquisition:** Marketers use data mining methods to discover attributes that can predict customer responses to offers and communications programs. Then the attributes of customers that are found to be most likely to respond are matched to corresponding attributes appended to rented lists of noncustomers. The objective is to select only noncustomer households most likely to respond to a new offer.
- **Customer retention:** Data mining helps to identify customers who contribute to the company's bottom line, but who may be likely to leave and go to a competitor. The company can then target these customers for special offers and other inducements.
- **Customer abandonment:** Customers who cost more than they contribute should be encouraged to take their business elsewhere. Data mining can be used to reveal whether a customer has a negative impact on the company's bottom line.
- **Market basket analysis:** Retailers and direct marketers can spot product affinities and develop focused promotion strategies by identifying the associations between product purchases in point-of-sale transactions.

In this context, early warning systems based on data mining have also been used for identifying dissatisfied customers, customer retention and quality.

Fraud Detection

Fraud detection studies have now become widespread. Fraud detection must also be dealt across all industries, especially the sectors where many transactions are made more vulnerable, such as health care, retail, credit card services, and telecommunications. The pioneers in the use of data mining techniques to prevent fraud were the telephone companies and insurance companies, with banks following close behind. Fraud can result in a business losing substantial amounts of money. Being able to protect a business from the chance of fraud is an important concern for an organization and data mining can help.

Furthermore, early warning systems to detect fraudulent actions can design models and can be built on fraudulent behavior (or potentially fraudulent behavior) done in the past and then use data mining to identify behavior of similar nature.

Manipulation and Insider Trading Detection

Manipulation is intrinsically about making market prices move away from their fair values; manipulators reduce market efficiency. Insider trading is any form of trading based on information that is relevant to the fundamental value of a company, but that is not publicly available. Insider trading will therefore by definition decrease market efficiency. Insider trading is often equated with market manipulation. Market manipulation by contrast takes place whenever nonpublic information is used to push the price of a stock away from its fundamental value. Again, by definition, market manipulation will decrease market efficiency.

Detection of insider trading and manipulation is widely treated as an important function of securities regulation. Early warning systems based on data mining will detect situations that could pose a threat of manipulation or abusive practices.

Health

Today, nearly all processes about patients and hospitals have been done with computers. Unfortunately, data mining applications have not become widespread in health sector, because most of the health and hospital data are not stored by datawarehouse logic. Potential usage of data mining in the area of health sector include the following components:

- **Provider profiling:** Analyze physician practice patterns by measuring clinical, quality, customer satisfaction, and economic indicators. Conduct comparative analysis to identify performance best practices.
- **Clinical decision support:** Measure and view clinical performance across multiple perspectives to optimize resource utilization, cost effectiveness, pathway development, and evidence-based decision-making.
- **Disease/condition management:** Use predictive modeling techniques to identify high-risk patients and to proactively intervene and optimize care across populations;
- **Benchmarking/quality reporting:** Perform necessary data management and analysis to support internal and external comparisons and reporting requirements;
- **Clinical research analysis:** Support the conduct of clinical research and outcomes analysis to generate new knowledge and to optimize clinical care; and
- **Patient safety/error reduction:** Utilize data mining approaches to uncover trends and patterns in clinical errors; identify and investigate key drivers of variation across care settings.

Simultaneously, early warning systems based on data mining have widespread use in fighting against communicable disease and determining pioneer variables and clinical evidence. Early warning systems have been used in the event of epidemics like the global spread of the SARS virus or malaria and early detection of chronic diseases. Early warning systems could also alert health care officials for possible bioterrorist attacks.

Risk Management

Risk management covers not only risks involving insurance, but also business risks from competitive threat, poor product quality, and customer attrition. Customer attrition, the loss of customers, is an increasing problem and data mining is used in the finance, retail, and telecommunications industries to help predict the possible losses of customers.

The key for early warning system is to identify and to manage strategic risk and not to ignore it. Along this perspective, an early warning system may involve three components:

- **Risk identification:** What are the potential market and industry developments to which a company would be vulnerable?
- **Risk monitoring:** What movement exists from competitors or in the business landscape that might indicate these factors are (or will soon be) in play?
- **Management action:** Are executives kept aware of risk dynamics, and are they equipped to launch a swift and aggressive response before their organization is harmed?

Economic Crisis

The risks of financial turmoil and economic instability associated with currency crises have called attention to the importance of monitoring fragility in the foreign exchange market or detecting signs of weakness in the market that may develop into crises. Typically, decision-makers would like to detect the symptoms of a crisis at an early stage so as to adopt preemptive measures. While forecasting the timing of currency crises with a high degree of accuracy remains a difficult task, decision-makers need to develop and improve upon an early warning system that monitors leading indicators of whether the economy is heading to a crisis situation.

Social Risk Detection

Managing social risk is to extend the traditional framework of social policy to the nonmarket based social protection of which its three primary strategies include prevention, mitigation, and coping. Nowadays, it is well understood that social unrest is positively parallel to the poverty and assisting individuals, households and communities to elevate living standard above the poverty level will harmonize global economy and strengthen the social security.

According to the World Bank, the degree of social risks usually varies from idiosyncratic (micro) or regional covariant (meso) to nation-wide covariant (macro). An early warning system based on data mining should give signs about natural risks (rainfall, landslides, volcanic eruption, earthquakes, floods, drought, tornados); lifecycle risks (illness, injury, disability, hungers, food poisoning, pan epidemics, old ages and death); social risks (crimes, domestic violences, drug addiction, terrorism, gangs, civil strife, war, social upheaval, child abuses); economic risks (unemployment, harvest failure, resettlement, financial or currency crisis, market trading shocks); administrative and political risks (ethnic discrimination, ethnic conflict, riots, chemical and biological mass destruction, administrative induced accidents and disasters, political induced malfunction on social programs, coup); and environmental risks (pollution, deforestation, nuclear disasters, soil salinities, acid rains, global warming).

CONCLUSION

At the present time, competitive conditions are an increasing threat to the enterprises day by day. If the importance of SMEs for the economies of countries is taken into consideration, economic fragility of SMEs, means the economic fragility of countries in other words. After the investigation of SMEs, the financial administration is seen as one of the biggest problem of SMEs. Finding practical solutions to these problems will not only help to SMEs but also to the economies of countries. Therefore, having information about their financial risk, monitoring this financial risk and knowing the required roadmap for the improvement of financial risk are very important for SMEs to take the required precautions.

However, the main problem preventing the SMEs on the stage of taking required precautions is insufficient administrators. Administrators' insufficiency mostly based on inadequate financial knowledge. The most practical way closing this knowledge gap is designing a tool which will present expert knowledge to nonexperts. One of the most important contribution of knowledge age is giving the chance to share knowledge in different ways via IT.

Actually, there are a lot of ways for empowering knowledge society. One of these ways is adding knowledge products as a part of daily life or working life of society. Put some facilities in society's working life is increasing transition speed to become a part of knowledge society. In case of showing the society utilities of Information Age, it will make them aware about Information Age and finaly make them a part of knowledge society.

Main motivation of this chapter was structured with the lights of picture in the earlier sections. Designing a useful tool for SMEs, which will make SMEs' administrators aware about the utilities to become a part of knowledge society. In addition, change their working process that they were used to do, with new habits. The best approaches to put some new elements in their usual working process are showing the way and making them stronger against their weaknesses. In this context, when the lack of financial management knowledge is taken into the consideration, it can be seen that the most appropriate tool must be to act as a financial advisor. Moreover, it should be felt that the power of knowledge is greater than financial advisory. Therefore, the designed system warns SMEs by taking differences among the SMEs into the consideration. In a way, this process is treated as knowledge discovery by data mining.

Furthermore, data mining is the reflection of information technologies in the area of strategical decision support system. A system can be developed based on data mining for finding solutions to the financial administration as one of the most suitable application area for SMEs as the vital point of economy.

In this chapter, an easy to understand and easy to use system is designed to observe the financial risk condition of SMEs and provide financial risk reduction system for SMEs. The system that provides the determination of financial risk also provides a roadmap for the risk reduction of SMEs, and gives opportunity to SMEs to be proactive.

SEWS, is an early warning system that provides the administration of financial risks to the individuals who do not have expert knowledge, for a subject like financial administration that requires expertise. SEWS, offers a prototype for an early warning system that is based on data mining for the every area that requires strategical decision making and proactivation.

It is apprehended further that, administrators of SMEs will easily be able to use the most up to date IT methods, most complicated algorithms and the most depth expert knowledge via SEWS. In addition, this tool will provide sustainability for their works, and ultimately they will become an efficient member of the emerging knowledge society.

REFERENCES

Altman, E., Haldeman, G., & Narayanan, P. (1977). Zeta analysis: A new model to identify bankruptcy risk of corporations. *Journal of Banking and Finance*, June, 29-54.

Anandarajan, M., Picheng, L. & Anandarajan, M. (2001). Bankruptcy prediction of financially stressed firms: An examination of the predictive accuracy of artificial neural networks. *International Journal of Intelligent Systems in Accounting, Finance and Management*, *10*, 69-81.

Apoteker, T. & Barthelemythierry, S. (2005). Predicting financial crises in emerging markets using a composite non-parametric model. *Emerging Markets Review, 6*(4), 363-375.

Beaver, W. (1966). Financial ratios as predictors of failure. *Journal of Accounting Research*, pp. 71-111.

BIS, The Bank for International Settlements. (2006). *Basel II: Revised international capital framework*. Retrieved April 13, 2008, from http://www.bis.org/publ/bcbsca.htm.

Bitzenis, A. & Nito, E. (2005). Obstacles to entrepreneurship in a transition business environment: The case of Albania. *Journal of Small Business and Enterprise Development, 12*(4), 564-578.

Boritz, E. J., & Kennery, D. (1995). Effectiveness of neural network types for predicition of business failure. *Expert Systems with Applications, 9*,503-512.

Bukvic, V. & Bartlett, W. (2003). Financial barriers to SME growth in Slovenia. *Economic and Business Review, 5*(3), 161-181.

Cabena, P., Hadjinian, P., Stadler, R., Verhees, J., & Zanasi, A. (1997). *Discovering data mining: from concept to implementation*. Upper Saddle River, NJ: Prentice Hall PTR.

Canbas, S., Onal, B. Y., *Duzakin*, H. G., & *Kilic*, S. B. (2006). Prediction of financial distress by multivariate statistical analysis: The case of firms taken into the surveillance market in the Istanbul Stock Exchange. *International Journal of Theoretical & Applied Finance, 9*(1), 133.

Chan, N. H. & Wong, H. Y. (2007). Data mining of resilience indicators. *IIE Transactions, 39*, 617–627.

Chang, S., Chang, H., Lin, C., & Kao, S. (2003). The effect of organizational attributes on the adoption of data mining techniques in the financial service industry: An empirical study in Taiwan. *International Journal of Management, 20*, 497-503.

Chin-Sheng, H., Dorsey, R. E., & Boose, M.A. (1994). Life insurer financial distress prediction: A neural network model. *Journal of Insurance Regulation, 13*(2), 131-168.

Coats, P. K., & Frant, F. L. (1992). A neural network approach to forecasting financial distress. *The Journal of Business Forecasting Methods & Systems, 10*, 9-12.

Coats, P. K., & Frant, F. L. (1993). Recognizing financial distress patterns using a neural network tool. *Financial Management, 22*(3), 142-155.

Collard, J. M. (2002). Is your company at risk? *Strategic Finance, 84*(1), 37-39.

Deakin, E. B. (1972). A discriminant analysis of predictors of business failure. *Journal of Accounting Research, 10*(1), 167-179.

Derby, B. L. (2003). Data mining for improper payments. *The Journal of Government Financial Management, 52*, 10-13.

Eklund, T., Back, B., Vanharanta, H., & Visa, A. (2003). Using the self- organizing map as a visualization tool in financial benchmarking. *Information Visualization, 2*, 171-181.

European Commission (2003). *2003 observatory of European SMEs: SMEs in Europe* (Tech. Pep. No.7). European Commission.

Gaytan, A. & Johnson, A. J. (2002). *A review of the literature on early warning systems for banking crises* (Working papers No: 183). Central Bank of Chile.

Gunther, J. W. & Moore, R. R. (2003). Early warning models in real time. *Journal of Banking, 27*(10), 1979-2001.

Hamer, M. (1983). Failure prediction: Sensitivity of classification accuracy to alternative statistical method and variable sets. *Journal of Accounting and Public Policy, 2*, 289-307.

Hoppszallern, S. (2003). Healthcare benchmarking. *Hospitals & Health Networks, 77*, 37-44.

Inegbenebor, A. U. (2006). Financing small and medium industries in Nigeria-case study of the small and medium industries equity investment scheme: Emprical research finding. *Journal of Financial Management & Analysis, 19*(1), 71-80.

Jacobs, L. J. & Kuper, G. H. (2004). Indicators of financial crises do work! An early-warning system for six Asian countries. *International Finance, 0409001*, 39.

Jones, F. (1987). Current techniques in bankruptcy prediction. *Journal of Accounting Literature, 6*, 131-164.

Kamin, S. B., Schindler, J., & Samuel, S. (2007). The contribution of domestic and external factors to emerging market currency crises: An early warning system. *International Journal of Finance and Economics, 12*(3), 317-322.

Kang, K. (2006) *Outlook and reforms for the Korean economy in 2006*. Retrieved April 13, 2008, from http://www.keia.org/2-Publications/2-2-Economy/Economy2006/01cover.pdf

Klersey, G. F. & Dugan, M.T. (1995). Substantial doubts: Using artificial neural networks to evaluate going concern. In *Advanced in Accounting Information Systems*. Greenwich: JAI Press.

Kloptchenko, A., Eklund, T., Karlsson. J., Back, B., Vanhatanta, H., & Visa, A. (2004). Combinig data and text mining techniques for analyzing financial reports. *Intelligent Systems in Accounting Finance and Management, 12*, 29-41.

Ko, P. C. & Lin, P. C. (2005). An evolutionary modularized data mining mechanism for financial distress forecasts. In A. Ghosh, & L.C. Jain (Eds.), *Evolutionary Computation in Data Mining* (pp. 249-263). Berlin Heidelberg, Germany: Springer-Verlag.

Kovalerchuk, B. & Vityaev, E. (2000). *Data mining in finance*. Hingham MA: Kluwer Academic Publisher.

Koyuncugil, A. S. (2006). *Fuzzy data mining and its application to capital markets*. Unpublished doctoral dissertation, Ankara University, Ankara.

Koyuncugil, A. S. & Ozgulbas, N. (2006a). Financial profiling of SMEs: An application by data mining. *The European Applied Business Research (EABR) Conference*, Clute Institute for Academic Research.

Koyuncugil, A. S. & Ozgulbas, N. (2006b). Is there a specific measure for financial performance of SMEs? *The Business Review, 5*(2), 314-319.

Koyuncugil, A. S. & Ozgulbas, N. (2006c). Determination of factors affected financial distress of SMEs listed in ISE by data mining. In *Proceedings of the 3rd Congress of SMEs and Productivity*, KOSGEB and Istanbul Kultur University, Istanbul.

Kutman, O. (2001). Researching the early warning signals for the enterprises in Turkey. *Journal of Dogus University, 4*, 59-70.

Lansiluoto, A., Eklund, T., Barbro, B., Vanharanta, H., & Visa, A. (2004). Industry-specific cycles and companies' financial performance comparison using self-organising maps. *Benchmarking, 11*, 267-286.

Liu, S. & Lindholm, C. K. (2006), Assessing early warning signals of currency crises: A fuzzy clustering approach. *Intelligent Systems in Accounting, Finance and Management, 14*(4), 179-184.

Magnusson, C., Arppe, A., Eklund, T., & Back, B. (2005). The language of quarterly reports as an indicator of change in the company's financial staus. *Information & Management, 42*, 561-570.

Mena, J. (2003). *Investigative data mining for security and criminal detection*. USA: Elsevier Science.

Meyer, P. A., & Pifer, W. H. (1970). Prediction of bank failures. *The Journal of Finance, 25*(4), 853-868.

OECD (2000). *Policy briefs small and medium-sized enterprises: Local strength, global reach.* Retrieved May 9, 2008, from www.oecd.org/dataoecd/3/30/1918307.pdf

Oksay, S. (2006). *Publication of insurance research and analysis.* Turkey: TSRSB.

Ozgulbas, N. & Koyuncugil, A. S. (2006). Profiling and determining the strengths and weaknesses of SMEs listed in ISE by the data mining decision trees algorithm CHAID. In *Proceedings of the 10th National Finance Symposium,* Izmir.

Ozgulbas, N., Koyuncugil, A. S., & Yılmaz, F. (2006). Identifying the effect of firm size on financial performance of SMEs. *The Business Review, 5*(2), 162-167.

Pantalone, C., & Platt, M. (1987). Predicting failures of savings and loan associations. *AREUEA Journal, 15*, 46-64.

Sanchez, A. & Marin, G. S. (2005). Strategic orientation, management characteristics, and performance: A study of Spanish SMEs. *Journal of Small Business Management, 43*(3), 287-309.

Shyu, M. L., Chen, S.C., Sarinnapakorn, K., and Chang, L. (2006). Principal component-based anomaly detection scheme. In T.S. Lin, S. Ohsuga, J. Liau, & X. Hu (Eds.), *Foundations and Novel Approaches in Data Mining* (pp. 311-329) Springer-Verlag.

Sormani, A. (2005). Debt causes problems for SMEs. *European Venture Capital & Capital Equity Journal, 1*, 1.

Taffler, R. & Tisshaw, H. (1977). Going, going gone - four factors which predict. *Accountancy*, March, 50-54.

Tan, C. N., & Dihardjo, H. (2001). A study on using artificial neural networks to develop an early warning predictor for credit union financial distress with comparison to the probit model. *Managerial Finance*, *27*(4), 56-78.

Tan, Z. & Quektuan, C. (2007). Biological brain-inspired genetic complementary learning for stock market and bank failure prediction. *Computational Intelligence, 23*(2), 236-242.

The New York Stock Exchange (2007). Retrieved April 13, 2008, from www.nyse.com.

The Stock Exchange of Thailand (2007). Retrieved April 13, 2008, from www.set.or.th/en/index.html.

WIPO, World Intellectual Property Organization (2002). *Interregional forum on small and medium-sized enterprises (SMEs) and intellectual property* (Tech. Rep. No. 02/01). Moscow: Document of WIPO.

Zavgren, C. (1985). Assessing the vulnerability to failure of American industrial firms: A logistics analysis. *Journal of Accounting Research, 22*, 59-82.

Zmijewski, M. E. (1984). Methodological issues related to the estimation of financial distress prediction models. *Journal of Accounting Research*, (Supplement), 59-82.

Chapter XIII
What Role is "Business Intelligence" Playing in Developing Countries? A Picture of Brazilian Companies

Maira Petrini
Fundação Getulio Vargas, Brazil

Marlei Pozzebon
HEC Montreal, Canada

ABSTRACT

Constant technological innovation and increasing competitiveness make the management of information a considerable challenge, requiring decision-making processes built on reliable and timely information from internal and external sources. Although available information increases, this does not mean that people automatically derive value from it. After years of significant investment to establish a technological platform that supports all business processes and strengthens the operational structure's efficiency, most organizations are supposed to have reached a point where the implementation of information technology (IT) solutions for strategic purposes becomes possible and necessary. This explains the emergence of "business intelligence" (BI); a response to information needs for decision-making through intensive IT use. This chapter looks at BI projects in developing countries—specifically, in Brazil. If the management of IT is a challenge for companies in developed countries, what can be said about organizations struggling in unstable contexts such as those often prevailing in developing countries?

INTRODUCTION

The final decades of the 20th century and the beginning of the 21st have been marked by a staggering proliferation of information and communication technologies throughout the industrialized world (Steinmueller, 2001). Not only do globalization trends bring a turbulent and most often unequal competitive environment, they also propagate waves of "managerial imperatives"—such as total quality; reengineering and integrated systems—that exert tremendous pressure on organizations wanting not only to survive, but to succeed. In addition to performance and effectiveness, global organizations are asked to display ethical, social and environmental responsibility. This entire context makes the task of managing information a formidable challenge.

Information management is seen as one of the biggest challenges characterizing today's corporate context. A combination of constant technological innovation and increasing competitiveness makes the management of information a difficult task, one which requires decision-making processes that are built on reliable and timely information gathered from internal and external sources. Although the volume of information available is increasing, this does not automatically mean that people are able to derive value from it (Burn & Loch, 2001). In the IT field, after years of significant investments to create technological platforms that support all business processes (processes that are "reengineered" and "integrated") and that strengthen the efficiency of the operational structure (after undergoing "quality" programs), organizations are supposed to have reached a point where the implementation of IT solutions for strategic decision-making processes becomes possible and necessary. This context explains the emergence of the area generally known as "business intelligence" (BI), seen as an answer to current needs in terms of information for strategic decision-making through intensive use of information technology (IT).

This perception of IT as a strategic resource is not exclusive to developed countries. IT is expected to play a key role in developing countries as well. Because IT offers significant potential benefits for socioeconomic development, the likely gains in efficiency of production and services are at least as relevant in developing countries as in advanced economies (Avgerou, 2002). The possibility of technology transfer is seen as an opportunity for organizations in developing countries to bypass stages of growth in their programs for industrialization and advancement (Steinmueller, 2001). However, very often the resulting IT-based solutions these companies deploy have had little impact in terms of the goals they were intended to reach (Sahay & Avgerou, 2002). One can argue that this is partly due to the fact that IT solutions developed in certain contexts—the "developed" world, for instance—are not necessarily translated beneficially into other contexts, such as those of countries considered "in development." Such considerations have motivated the authors to put forward a research project aiming to investigate the status and role, if any, of BI projects in the context of developing countries, more specifically, in Brazil.

The conventional definition of BI refers to the consolidation and analysis of internal data (e.g., transactional POS (point of sales system) data) and/or external data (e.g., purchased consumer demographics) for the purpose of effective decision-making. Several reasons for BI's relevance are generally put forward. Historically (and continuing today in the vast majority of firms), companies have spent too much time closing their books and preparing data and financial reports, and too little time on analysis and review. This causes a gap between analysis and action (decision-making) (Rasmussen, Goldy & Solli, 2002). In addition, a BI initiative includes objectives like creating a vision for the organization, coaching the organization to set realistic goals, and supporting optimal decision-making. Although the current push, promoted by IT vendors arguing for the importance of BI applications, may be seen as simply one more IT-driven management fashion, it is difficult to deny the benefits of deep and meaningful insights that easy and rapid access to relevant and consolidated data can provide to all organizational members, particularly decision makers.

Within a broad enquiry—"What role is BI playing in developing countries?"—two specific research questions are explored in this chapter. First, *what approaches, models or frameworks have been adopted to implement BI projects in Brazilian companies?* The purpose is to determine whether those approaches, models or frameworks are tailored for particularities and the contextually situated business strategy of each company, or if they are "standard" and imported from "developed" contexts. In addition, the authors want to verify whether the temporal dimension affects the degree of sophistication of a firm's approach, that is, if more mature projects have improved their methodologies and use of performance indicators.

Second, *what is the perceived "value" of BI to strategic management of Brazilian companies?* The purpose here is to analyze: what type of information is being considered for incorporation by BI systems; whether they are formal or informal in nature; whether they are gathered from internal or external sources; whether there is a trend that favors some areas, like finance or marketing, over others, or if there is a concern with maintaining multiple perspectives; who in the firms is using BI systems, and so forth.

Considering that information technology use takes place within a context of "globalization," and being aware that companies that participate in such a globalized process do not compete under equal conditions, the hypothesis is that BI applications can help firms in developing countries to improve their competitive advantage. Exploring these questions and discussing them with Brazilian entrepreneurs, the purpose is two-pronged: to sketch the nature and quantity of BI projects implemented and used by Brazilian companies, and to indicate their perceived "value" in terms of gaining competitive advantage in a globalized world.

BACKGROUND

IT and Developing Countries

It is often assumed that the impact and implementation of IT will be uniform, with little regard to particular social or cultural contexts. Drawing on experience and research in different parts of the world, including Europe and Latin America, Avgerou (2002) holds a different view. She developed a conceptual approach to account for the organizational diversity in which IT innovation takes place, showing how the processes of IT innovation and organizational change reflect local aspirations, concerns and action, as well as the multiple institutional influences of globalization.

Such a perspective raises issues about whether IT implementation can be handled similarly in developing and industrialized countries. For example, there are special requirements that should be taken into consideration by ISD (information systems development) methodologies in Africa (Mursu, Soriyan, Olufokunbi & Korpela, 2000). These special requirements are based on the local socioeconomic conditions as well as on wider sociopolitical issues including sustainability, affordability and community identity. Although these issues are also relevant to industrialized countries, they are more critical for developing countries and are not sufficiently addressed by existing ISD methodologies.

Indeed, a number of estimates suggest that a significant majority of IS projects in developing countries fail in some way. Why should this be? According to Heeks (2002), central to developing countries' IS success and failure is the amount of change between "where we are now" and "where the information system wants to get us." The former will be represented by the current reality of the particular context (part of which may encompass subjective perceptions of reality). The latter will be represented by the model or conceptions, requirements and assumptions that have been incorporated into the new information system's design. "Design conceptions" derive largely from the worldview of the stakeholders who dominate the IS design process. Putting this a little more precisely, it can be said that the likelihood of success or failure depends on the size of the gap that exists between "current realities" and "design conceptions" of the information system (Heeks, 2002).

Those gaps arise especially when designs and dominant design stakeholders are remote (physically or symbolically) from the context of IS implementation and use. This can occur in a number of ways, but approaches to IS projects in developing countries are particularly dominated by the mechanistic transfer of Northern designs to Southern realities (Heeks, 2002). An example of country-context gaps can be drawn from an experience in the Philippines, where an aid-funded project to introduce a field health information system was designed according to a Northern model that assumed the presence of "skilled" programmers, "skilled" project managers, a "sound" technological infrastructure, and a need for information outputs like those used in an American health care organization (Heeks, 2002). In reality, none of these elements was present in the Philippine context, and the IS project failed.

Globalization is a contradictory process, implying increased interconnectedness of local actors along with "globalism" in the form of increased trans-national uniformity (Beck, 2000). IT is a powerful tool which, if well adapted, can help countries promote their development (Meier, 2000). Implementing technologies across locations represents a huge challenge as "global" principles and multiple choices have to be translated into "local" contexts and requirements (Williams, 1997). People need to improve their capacity to address contextual characteristics and particular requirements in order to better implement and manage an IT application conceived elsewhere (Avgerou, 2002).

Recent studies in IS show the importance of the local context, particularly the importance of adapting global practices based on IT when implementing them in developing countries (Pozzebon, 2003). However, the nature of these adaptations and the factors that create them are poorly understood (O'Bada, 2002). It is believed that an emergent and important role for IT research is to study *particular* individuals, groups, organizations or societies in detail, and *in context.* In this way, studies of IS projects from all parts of the world might form the basis for comparisons and inferences from a global viewpoint (Avgerou, 2002). In this vein, this work on BI in developing countries seeks to contribute to advancing such knowledge of local/global gaps in IS implementation and use. By outlining the use of a particular type of IT application—namely, BI - by particular firms—Brazilian companies—and the particularities of these Brazilian projects, the authors hope to improve knowledge of this important area.

Business Intelligence

The literature review of BI reveals few studies. Most of the articles are conceptual. What's more, throughout the literature one encounters the traditional "separation" between technical and managerial aspects, outlining two broad patterns (Table 1).

The technological approach, which prevails in most studies, presents BI as a *set of tools* that support the storage and analysis of information. This encompasses a broad category of applications and technologies for gathering, storing, analyzing and providing access to data to help users in the enterprise make better business decisions. Those BI tools include decision support systems, query and reporting, online analytical process-

Table 1. Two approaches to BI

	Managerial Approach	Technological Approach
Main focus	Focus on the *process* of gathering data from internal and external sources and of analyzing it in order to generate relevant information	Focus on the *technological tools* that support the process
References	(Kalakota & Robinson, 2001; Liautaud, 2000; Schonberg, Cofino, Hoch, Podlaseck & Spraragen, 2000; Vitt, Luckevich & Misner, 2002)	(Dhar & Stein, 2006; Giovinazzo, 2002, Hackathorn, 1998; Kudyba & Hoptroff, 2001; Scoggins, 1999; Watson, Goodhue, & Wixon, 2002)

ing (OLAP), statistical analysis, forecasting and data mining.

The focus is not on the process itself but on the technologies that allow the recording, recovery, manipulation and analysis of information. For instance, Kudyba and Hoptroff (2001) conceive of BI as data warehousing: technology allows users to extract data (demographic and transactional) in structured reports that can be distributed within companies through Intranets. Having determined that some organizations get greater returns on implementation of data warehousing than others, Watson, Goodhue, and Wixon (2002) developed research showing how data warehousing can change an organization, what its impact on the organization is, and how such impacts can be quantified and measured. Sophisticated use of warehoused data occurs when advanced *data mining* techniques are applied to change data into information (Scoggins, 1999).

Data mining is the utilization of mathematical and statistical applications that process and analyze data. Mathematics refers to equations or algorithms that process data to discover patterns and relationships among variables. Statistics generally shed light on the robustness and validity of the relationships that exist in the data-mining model. Leading methods of data mining include regression, segmentation classification, neural networks, clustering, and affinity analysis.

The synergy created between data warehousing and data mining allows knowledge seekers to leverage their massive data assets, thus improving the quality and effectiveness of their decisions. The growing requirements for data mining and real time analysis of information will be a driving force in the development of new data warehouse architectures and methods and, conversely, the development of new data mining methods and applications (Kudyba & Hoptroff, 2001). In this vein, Hackathorn (1998) approaches the convergence of technologies of data warehousing, data mining, hypertext analysis and Web information resources as a major challenge in creating architecture for all these technologies in an organizational BI platform.

In short, BI is a wide set of tools and applications for collection, consolidation, analysis and dissemination aiming to improve the decision-making process. The components of BI that focus on collection and consolidation can involve data management software to access data variables, extract, transform, and load tools that also enhance data access and storage in a data warehouse or data mart. In the analysis and distribution phases, each time more different products are launched and integrated with attention paid to the different uses of the information. These products can include: creation of reports, fine-tuned dashboards containing customized performance indicators, visually rich presentations that use gauges, maps, charts and other graphical elements to juxtapose multiple results; generation of OLAP cubes; and data mining software that reveal information hidden within valuable data assets through use of advanced mathematical and statistical techniques, making it possible to uncover veins of surprising, golden insights in a mountain of factual data.

Figure 1 proposes an overview of BI architecture, distributing each different technology and application in terms of its main contribution in each step in the BI process.

The managerial approach sees BI as a *process* in which data from inside and outside the company are integrated in order to generate information relevant to the decision-making process. The role of BI here is related to the whole informational environment and process by which operational data gathered from transactional systems and external sources can be analyzed to reveal the "strategic" business dimensions.

From this perspective emerge concepts such as the "intelligent company": one that uses BI to make faster and smarter decisions than its competitors (Liautaud, 2000). Put simply, "intelligence" entails the distillation of a huge volume of data into knowledge through a process of filtering, analyzing and reporting information (Kalakota & Robinson, 2001). The explanation of how companies acquire "intelligence" would lie in the data-information-intelligence transformation. Traditional wisdom emerges here: data is raw and mirrors the operations and daily transactions of a company; information is the data that has passed through filtering and aggregation processes and acquired a certain level of contextual meaning; intelligence elevates the information to the highest level, as the result of a complete understanding of actions, contexts and choices.

Both approaches—technical and managerial—rely on an objective and positive view that "strategic decisions based on accurate and usable information lead to an intelligent company." All the subjectivism inherent in social interactions is evacuated, and cultural and political issues are not evoked. In addition, this literature requires some facility with managerial and IT-driven "language games," where the use of buzzwords—like data warehousing, data mining and OLAP—and jargon—like "intelligent enterprise" and "strategic dimensions"—is ubiquitous.

Whether the reviewed studies are managerial or technological, they share a common idea: (1) the core of BI (process or tool) is information *gathering, analysis and use*, and (2) the goal is to support the strategic *decision-making process*. Taking into account the relative scarcity of literature, the authors looked for other areas that could help in reaching a more comprehensive understanding of BI. They revisited three distinct but interrelated areas: information planning, balanced scorecard and competitive intelligence.

Figure 1. BI architecture (authors' proposal)

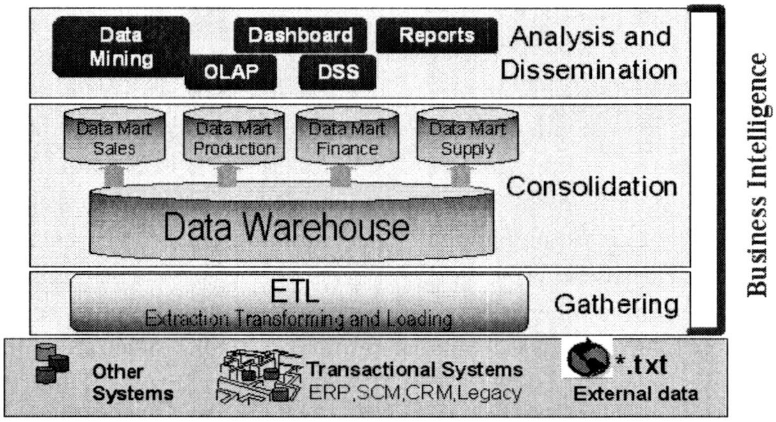

The Contribution of Information Planning Literature: Limited and Strategic Information, Collectively Identified

An important question that has been widely debated and must be considered in an extended definition of BI involves the relationship between strategic planning and IT. The harmony or alignment of organizational strategy and IT strategy seems to be increasingly identified as a key factor in the success or failure of IT implementations, especially for BI projects. A survey of 67 IT executives on three different continents has shown that they perceive the alignment between IT and corporate objectives as the most important task (Van Der Zee & De Jong, 1999).

This means that the chances of success in the application of any technology are directly related to how it is articulated in terms of organizational strategy and of the characteristics of each industry. Reich and Benbasat (2000) define the alignment of IT and organizational strategy as "the level in which IT mission, objectives and plans support and are supported by the mission, objectives and business plans" (p. 82). When talking about the information planning phase, they emphasize the importance of identifying *limited but strategic information* (Reich & Benbasat, 2000). Eisenhardt and Sull (2001) also reinforce the need for only limited but direct information and rules for establishing organizational strategies. Such information and rules should clearly indicate how processes are executed and what the business focus is.

According to Connelly, McNeill, and Mosimann (1998), relevant information for decision-making is likely to already exist within the company or to be clearly defined in managers' minds. They also consider that the information most valuable for decision-making in a company is likely to be concentrated in a relatively small number of points (*sweet spots)* that already exist in the information that flows along the value chain of organizations, categorized according to the area to which these spots refer, and which should be monitored. In order to make the decision-making process more effective, they suggest that the information should be presented in a manner compatible with the managers' way of thinking, respecting the corollary that different managers analyze information from different points of view.

In short, authors concerned with information planning assume that information relevant to decision-making is *limited but strategic* and that it should be *collectively identified*, but that often it is already identified within the company or clearly defined in managers' minds. This sets up a paradox counterpoising the need for limited, thus standardized, information, with respect to individual decision-making styles and points of view, thus personalized information. Achieving such a balance between standardization and customization is one of the biggest challenges in contemporary IS projects, particularly for BI.

The Contribution of the Balanced Scorecard Approach: Multidimensional Information

Additional support for this study of BI has been found in the *balanced scorecard approach*, as it associates indicators and measures with the monitoring of the company's strategic objectives (Kaplan, 1996). The concept of *balanced scorecard* encompasses a set of measurements that provide high-level executives with a quick and understandable view of the business (Kaplan & Norton, 1992). Its development was motivated by dissatisfaction with traditional performance measurements that were concerned only with financial metrics and that focused on the past rather than the future. As a result, financial measures are complemented with measurements related to internal process, clients and organizational learning perspectives. Niven (2002) explored the limitations of exclusive reliance on financial measures of performance and explained how the balanced scorecard could overcome them.

Balanced scorecard, like value-based management, is a component of a broader concept: strategic performance measurement systems (SPMS). SPMS are described as powerful tools for executing strategy. Value-based management is the term given to a process used to determine the drivers of a particular strategy, to understand how those drivers link to value creation, and then to break those drivers down into steps for action and activities that can be pushed throughout an organization all the way to the shop floor. Value-based management shouldn't be confused with the actual design of strategy—it represents a vehicle and process for strategy execution translated into the specific value drivers of that particular organization (Frigo, 2002).

The distinctive feature of balanced scorecards is that they are designed to present managers with financial and nonfinancial measures spanning different perspectives which, in combination, provide a way of translating strategy into a coherent set of performance measures (Chenhall, 2005). Chenhall (2005) defined a key dimension of balanced scorecards: integrative information. Three interrelated dimensions of integrative information were identified in his study. The first, strategic and operational linkages, is a generic factor that captures the overall extent to which the systems provide for integration across elements of the value chain. The second, customer orientation, focuses on customer linkages and includes financial and customer measures. The third, supplier orientation, is based on linkages to suppliers and includes business process and innovation measures.

The balanced scorecard's main contribution to this study is its idea of *multiple perspectives.* In a world that is continually becoming more and more globalized, and in which management strategies are constantly being revised, the idea of taking into account multiple perspectives when analyzing firms' performance seems worthy of attention. For instance, the "internal processes" dimension or the "learning and growing" dimension invite the firm to give greater consideration to their employees' motivation and training. BSC can lead to learning organizations. In addition to these four classic perspectives proposed by Kaplan and Norton (1992)—financial, customer, internal processes and learning and growing—others can be conceived. The firm's social and environmental roles, regarding its local community and country, are also increasing in visibility and importance. Interorganizational processes and social responsibility could be considered examples of new perspectives that should be articulated by balanced scorecard approaches.

It is our view that such multidimensionality should be integrated into BI processes.

The Contribution of the Competitive Intelligence Area: Contextualized Information

Finally, the authors have borrowed insights from an adjacent discipline, *competitive intelligence.* The Society of Competitive Intelligence Professionals (SCIP) defines intelligence as the process of collection, analysis and dissemination of accurate, relevant, specific, current and visionary intelligence related to the company, to the business environment, and to competitors (Miller, 2002). Gilad and Gilad (1988) defined *competitive intelligence* as the activity of monitoring the firm's external environment to gather information that is relevant to the decision-making process. Indeed, some authors apply competitive intelligence as a kind of synonym for BI. Despite their common concern with data collection and analysis, and with the conceptual distinction between information and intelligence, the focus of competitive intelligence is external information about competitors and markets, and it deals with information that is essentially qualitative and textual, informal, and ambiguous.

The objective of *competitive intelligence* is not to "steal" competitors' trade secrets or other private information, but rather to gather, in a systematic and open manner (i.e., legally), a wide range of information that, once put together and analyzed (i.e., in context), provides a fuller understanding of a competing firm's structure, culture, behavior, capabilities, and weaknesses (Sammon, Kurland, & Spitalnic, 1984). CI uses public sources to locate and develop information on competition and competitors (McGonagle & Vella, 1990).

The main contribution of competitive intelligence is the focus on *contextualized information*. For example, in a manufacturing business, the level of "waste" constitutes information that can be analyzed over time and according to product line (*the context*). These analyses might indicate, for example, that waste levels are higher during a specific period, and in one specific product line. Deeper analyses, including *external information*, may show that such increments coincide with the increase in air humidity. The intelligence which can be acted upon is the fact that the materials used in that specific product line are more sensitive to air humidity than other materials. Although information is factual and intelligence is something that can be more purposively acted upon, both are *contextual*. Competitive intelligence makes us more conscious of the contextual nature of information, and of the value of aggregating more external information that is qualitative and informal in nature.

The Meaning of Business Intelligence: Proposing an Extended Definition

All the approaches reviewed are clearly dominated by an objectivist mindset that disregards the socially constructed and political process of information production in any organization. Aiming to develop a more critical appreciation, the authors put forward a distinct definition of BI. Table 2 summarizes the contribution of each area to building an extended definition.

First of all, the authors have favored the notion of *process* over that of *tool*. The "tool" view of technology is still predominant in IT literature, with its assumption that technology is an engineering artifact, expected to do what its designers intend it to do (Orlikowski & Iacono, 2001). The "tool" view tends to see IT independently of the social or organizational arrangements within which ICT is developed and used. It black-boxes technologies and assumes that they are stable, settled artifacts that can be passed from hand to hand and used as is, by anyone, anytime, anywhere. In contrast, BI can be seen as an organizational process, composed of information-related activities carried out by people in particular settings, and whose perceptions and cultural background will influence the way they interpret and use information and technology.

Second, they outlined the collective and socially constructed nature of the process of defining, collecting, transforming, analyzing and sharing relevant information for decision-making purposes. It is collective because these activities depend on people interacting within the firm. People should acknowledge their active role as information producers and consumers, and the role of their company (political and institutional constraints), in that context.

The process is socially constructed because information is usually understood as a type of commodity that can be unambiguously defined and mechanistically treated and transferred, eliminating the subjectivity inherent in any information-related process and, most importantly, the intersubjectivity inherent in human interactions when dealing with "information" (Easterby-Smith, Araujo & Burgoyne, 1999). If information is a collective social construction, tone must focus on the particular activities people engage in when producing and managing it, and these are likely to change across different cultures, industries and even organizations. The

Table 2. The contribution of each area in building an extended definition of BI

Literature revisited	Main concepts retained	Our elaboration from these concepts
"Pure" BI literature: managerial and technical approaches (see also Table 1)	The core of BI (process or tool) is information gathering, analysis and use, and the goal is to support the strategic decision-making.	BI is an organizational *process* consisting of *information-related activities.*
Information planning literature (Connelly, McNeill, & Mosimann, 1998; Eisehhardt & Sull, 2001; Reich & Benbasat, 2000; Van Der Zee & De Jong, 1999)	Relevant information for decision-making can be collectively identified in the company while respecting individual decision-making styles and points of view.	BI is a *collective* process; information-related activities (definition, collection, transformation, analysis and distribution of few but strategic indicators) are inherently collective
Balanced scorecard approach (Chenhall, 2005; Frigo, 2002; Kaplan & Norton, 1992; Kaplan, 1996; Niven, 2002)	The need for multiple perspectives, from traditional "BSC" dimensions (learning and growing, internal process, customers, financial) to emergent like inter-organizational processes and social responsibility.	BI is a *multidimensional* process; information-related activities require multiple perspectives
Competitive intelligence literature (Gilad & Gilad, 1988; McGonagle & Vella, 1990; Miller, 2002;Sammon, Kurland, & Spitalnic, 1984)	Valuable information, both internal and eternal, is always contextual.	BI is a *contextual and culturally situated* process; information-related activities are essentially contextual and culturally situated.
An extended definition of BI	BI is an organizational process consisting of a range of activities and interactions wherein organizational members define, collect, transform, analyzes, and share information for decision-making purposes. This process is likely to be collective, socially constructed, multidimensional, contextual and culturally situated.	

contextual and culturally situated character of IT projects is supported by abundant literature on IT and development, including efforts at learning more about the history, culture, social relations and local competencies (Avgerou, 2002).

Finally, the authors outlined the multidimensional nature of information and BI processes. They believe that relevant information and decision-making processes go beyond the financial dimension and that all the proactive and creative aspects of information production depend on multiple perspectives. In a nutshell, BI is seen as an organizational process composed of a range of activities and interactions wherein organizational members define, collect, transform, analyze, and share information for decision-making purposes.

This process is likely to be collective, socially constructed, multidimensional, contextual and culturally situated. The implications of this extended definition for the present study are that, regarding firms located in Brazil, this chapter focuses on the particularities of BI projects' adoption and use, and examines the consequences of these particularities in terms of their "impact" on perceived organizational benefits. In addition, this view goes beyond a "technical" appreciation of functionalities and technological features. In the rest of this chapter, although how BI is referred to may vary, the greatest importance is placed on the processual notion—BI is a process—and expressions such as *BI applications*, *BI tools*, *BI solutions* and *BI systems* are merely used to refer to the software components of *BI projects*.

MAIN THRUST

Research Methods

The present research has been conceived as a qualitative study aimed at describing and understanding complex phenomena whose contextual factors must be deeply analyzed (Stake, 1998). To date, no academic study aimed at researching BI projects from a social or "developmental" viewpoint has been carried out in Brazil. Therefore, this research fills this gap by investigating the implementation and use of Brazilian BI projects, that is, what approaches to BI implementation are being applied by Brazilian companies and what the perceived "value" or benefits of these projects are.

Sample and Data Collection

This research employs criterion sampling, a strategy that should include cases that meet a set criterion that is useful for quality assurance (Miles & Huberman, 1990, p. 28). The logic of *criterion sampling* is to review and study cases that meet a predetermined criterion of importance. In this study, this means that all selected cases should meet the same criterion: medium-to-large companies that have implemented a BI project for more than one year and are currently, and effectively, using the BI system. This strategy can add an important qualitative component to a quantitative analysis of an information system: all cases that exhibit certain pre-determined criterion characteristics are routinely identified for in-depth qualitative analysis. Criterion sampling can also be used to identify cases from standardized questionnaires for in-depth follow-up. This strategy can only be used where respondents have willingly supplied contact information (Dubé & Paré, 2003).

The unit of analysis is BI projects and the first endeavor was to canvass all firms belonging to the FIESP (Federation of Industries of the State of São Paulo). This entity's goal is to lead Brazilian industry to a high rank among the world's most industrially advanced countries, and to support its associated companies and trade unions. One hundred and twenty-nine (129) trade unions are represented, and FIESP serves as a reference in the search for solutions, helping businesspeople manage their firms through strategies, orientation and information.

However, after dozens of calls that yielded no uniquely BI projects, authors decided to change their strategy. In order to identify BI projects implemented by Brazilian firms, the four major vendors of BI tools in Brazil (Business Objects, Cognos, Hyperion and Oracle) were contacted and asked to provide a list of clients with qualifications potentially meeting the sample selection criteria. In this way, 30 companies were contacted initially, only 15 agreed to participate.

Those companies that declined to participate cited commercial confidentiality or lack of interest as reasons for their refusal. This lack of interest is not related to the role of BI in developing economies per se. Rather, it reflects a common picture in Brazil: in general, companies do not see value in academic research. Although the final sample, 15 firms, might seem extremely small, it actually is not since, at the time of this research, the number of Brazilian companies using BI systems was limited.

Research by International Data Center ("Business Intelligence: Aspectos," n.d.) about BI scenarios and tendencies in Brazilian companies surveyed 250 Brazilian firms and showed that only 12% of them (30 firms) had already invested in some BI project (which does not imply that the implementation had succeeded). The main barrier to Brazilian firms' adoption of a BI application is the classic decisional criterion of real return on investment (ROI) ("Ferramentas de Business," n.d.). Authors believe that their 15-firm sample allows them to sketch an initial picture of the situation. Table 3 presents a list of the companies studied, their industry, and which BI vendor had provided the software application. These companies operate in different industries: manufacturing, financial, insurance, consumer goods, chemical, health, and technology. In size, they range from medium to large, having between 500 and 1,000 employees.

The methodological strategy is based on recent studies showing the value of telephone interviews when the main purpose is to outline an initial picture of a given phenomenon (Hannula & Pirttimaki, 2003; Harvey, 1988; Miller, 1995; Robey, Ross, & Boudreau, 2002). In order to cover a large number of Brazilian companies and to assure a comprehensive view, semi-structured phone interviews seemed the best strategy. Sturges and Hanrahan (2004) conducted research that reports the results of a comparison between face-to-face interviewing and telephone interviewing in a qualitative study and concluded that telephone interviews can be used productively in qualitative research.

Data collection consisted of telephone interviews with one person at each company. The organizational member identified as being in charge of BI projects was contacted and interviewed by phone. Authors used an interview protocol in Portuguese (which is available upon request) to conduct the interviews. They started each interview by explaining the traditional concept of BI and verifying whether it corresponded to the interviewee's conception. Concrete examples of

Table 3. Summary of data collection

Company*	Industry	Type	Interviews	*Vendor*
CoServ	Service	Mutinational	1	*Cognos*
CoChem	Chemical	Multinational	1	*Business Objects*
CoFo	Food and Drugs	Multinational	1	*Cognos*
CoBank	Bank	Multinational	1	*Oracle*
CoBanka	Bank	National	1	*Hyperion*
CoInsur	Insurance	National	1	*Business Objects*
CoPaper	Paper and Cellulose	Mutinational	1	*Cognos*
CoHosp	Hospital	National	1	*Cognos*
CoSid	Siderurgy	Mutinational	1	*Oracle*
CoTele	Telecommunications	Mutinational	1	*Business Objects*
CoBankb	Bank	Multinational	1	*Hyperion*
CoPapera	Paper and Cellulose	Multinational	1	*Cognos*
CoChema	Chemical	Multinational	1	*Cognos*
CoFoa	Food	Mutinational	1	*Business Objects*
CoInsrua	Insurance	National	1	*Hyperion*
Total			*15*	

**All company names are pseudonyms*

the implementation and use of BI activities and applications were requested.

The interviews, although semistructured, allowed respondents to present particular aspects of their BI projects, to describe, in varying degrees of detail, the nature of their decision-making processes, and to report events, constraints, interpretations, and insights that could be seen as unique to each organizational experience. The interviews took place between January and March 2003. Each interview lasted from 30 to 60 minutes and was not tape-recorded. Immediately following each interview, the interviewer wrote a detailed summary from notes taken during the interview.

Data Analysis

The analysis was conducted in two main ways. First, descriptive analysis was applied in examining the answers to interview questions. Because the purpose of this chapter is not to test a theory, but to draw an initial picture in terms of the nature and amount of BI projects implemented and used by Brazilian companies, authors believe that statistical description is an appropriate method of data analysis in an initial phase.

However, in addition to simple descriptive analysis, they coded the interviews according to different categories, like "IT jargon," local or cultural expressions, and particular interpretations and insights that respondents expressed during the interviews. Although these grounding categories were not sufficient for building initial concepts or relationships between concepts, they were helpful for developing an insightful discussion concerning the perceived value of BI projects from the perspective of Brazilian managers.

RESULTS

This section presents the main results of the interviews, according to the questions included in the interview protocol. Figure 2 shows these results graphically, in a presentation of the seven most important questions. These questions are closely related to the first research question (*What approaches, models or frameworks have been adopted to implement BI projects in Brazilian companies*?) but only tangential to the second research question (*What is the perceived "value" of BI to strategic management of Brazilian companies?)*. As previously described, the data collected were not restricted to the interview protocol questions, due to the semistructured character of the interviews, additional comments and questions were posed, depending on the course of each conversation, so that the two research questions could be explored in different ways.

For How Many Years Have These Brazilian Companies Been Engaged in a BI Process?

BI projects are relatively recent arrivals to Brazilian business. Of the companies interviewed, 73% had begun the operation of a BI system within the last three years (Chart 1). The oldest BI application was six years old. Because BI projects were called EIS (Executive Information Systems) in the past, we asked if they had had an EIS before implementing their BI project, and the answer was no.

Did These Brazilian Companies Use a Specific Methodology to Identify Their BI Indicators?

One of the main concerns relates to what approaches to BI implementation (in terms of identification of performance indicators) are being applied by Brazilian companies. Surprisingly, 73% of the interviewed companies were not using a specific methodology for developing their BI (Chart 2). Instead of following a given methodology, they had simply replicated the existing indicators or

Figure 2. A picture of the use of BI projects by Brazilian companies

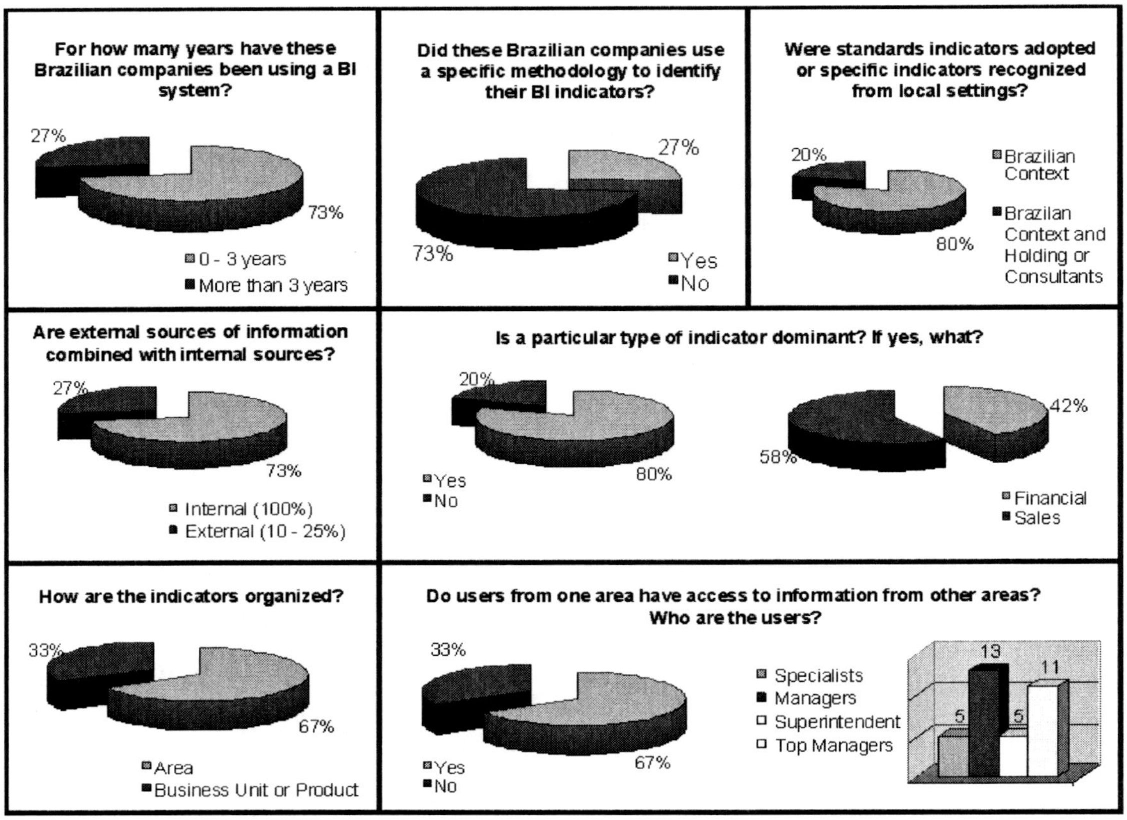

measures already being used in traditional MIS reports and spreadsheets.

Actually, they seemed to pay much greater attention to building and managing the data warehouse from a technical perspective than to thinking about its content. In sum, there was no collective process for identifying key indicators that could effectively help the decision-making process. Among the remaining 27% of companies actually using a specific BI methodology, the preferred method was the balanced scorecard. It was observed that firms having started their BI projects during the previous two years were those using specific methodologies to identify their BI indicators, but insufficient information was gathered to conclude that there exists an association between these two facts.

Were Standard Indicators Adopted or Were Specific Indicators Drawn from Local Settings?

The question regarding how indicators were identified concerns the verification of whether the collection of indicators integrated into the BI system reflects local aspirations, concerns and actions, and respects the idea that different countries have different requirements. Results show that 87% of the companies defined their indicators from a Brazilian context while 13% used indicators suggested by international headquarters or by external consultants (and even in those cases, they tried to take advantage of and add indicators from the local context) (Chart 3).

The implications of these findings are discussed in the next section.

Are External Sources of Information Combined with Internal Sources?

Regarding sources of information, results show that the focus is on information produced from operational or transactional systems. Few companies were concerned with external information. Only 27% of the companies had external information in their BI. In these cases, external information amounted to around 10-25% of the total information used. Among the main sources of external information, market institutes (i.e., market share), governmental institutes (i.e., demographic information) were identified and market research for a specific proposal were customized (Chart 4).

Is a Particular Type of Indicator Dominant?

Recent approaches, like the balanced scorecard, warn of the danger of performance measurement that essentially reflects a financial dimension and a type of monitoring which is perceived as reactive, rather than proactive. Beyond seeking to avoid dominance by a financial perspective, these approaches argue that indicators should be balanced, that is, with no single dominant perspective but rather a mix of several perspectives. This study reveals that in 80% of the companies, some kind of indicator was predominant. In 58% of them, sales indicators were the most powerful while financial indicators dominated in 42%. This is understandable because, in general, BI systems had first been implemented in the commercial or financial areas of the companies and had been restricted to those areas (Chart 5).

How Are the Indicators Organized?

67% of the firms organized the indicators by area, such as financial, sales, supply, human resources, while 33% organized them by product, such as credit card, leasing, investments, car insurance and life insurance (Chart 6). (Actually, all the companies that presented indicators organized by product were in the banking and insurance industries). Companies using a balanced scorecard approach based their indicators on the four well-known perspectives.

Do Users from One Area Have Access to Information from Other Areas?

Finally, questions were asked concerning organization and access to information. The users were managers (87%), top managers (73%), superintendents (33%), and specialists (33%) from different areas. On the one hand, all the specialists (33%) used data mining tools; on the other hand, neither managers and top managers nor superintendents used these tools. Authors believe that this finding reflects the fact that the users who use data mining tools have specific skills, with expertise in math, statistics, and analysis, and also speak the business language.

In 67% of these companies, users from one area could access information from other areas, but no company shared any information with its suppliers or clients (Chart 7). Access control was based on hierarchical levels—that is, top managers and managers could access information from all areas, but superintendents or specialists could only access their area—or on specific roles, that is, users from one branch could access all information, but only in their branch.

RECOMMENDATIONS AND FUTURE DIRECTIONS

In assembling this first collection of semi-structured interviews, authors observed an interesting picture of BI use in Brazil. Some of the results were expected but many brought surprises, such as the discrepancy between what was seen in advertising and consulting claims and what was found in practice. IT consultants and vendors try to convince public opinion that BI is already a "reality" for most companies, especially among the leaders in each segment or industry. But, this field study shows that although many firms "intend" to start a BI process, few have already embarked on such projects. Is this gap between advertising and practice specific to the Brazilian context, or can a similar scenario be found worldwide?

A very similar situation prevails in Canada, based on the experience of one of the researchers who teaches at a Canadian university: there is a lot of interest in BI as a new IT trend, but there are relatively few projects that have reached maturity. For these reasons, the actual benefit of investing in BI processes still remains to be proven. Authors believe that the scenario is slightly different in the US, if one takes into account Kaplan and Norton's claims (1992) that 90% of the Fortune 1000 firms are using a balanced scorecard to monitor their business (usually, a balanced scorecard is the interface of a BI system). Aside from the US, the supposed absence of mature BI projects worldwide is something that Brazilian companies can take advantage of, by starting to use BI systems before or at the same moment as most of their competitors elsewhere in the world.

A second interesting insight concerns the methodological approach adopted by BI users. This answered this study's first research question: *What approaches, models or frameworks have been adopted in a Brazilian context in the implementation of BI projects*? Surprisingly, the majority of companies did not use any "standard" model or methodology imported from "developed" contexts, which could be seen as a positive sign in itself. The results show that although Brazilian firms have adopted "globalized" products (like Cognos and Business Objects), they have not adopted imposed or imported *methodologies* for implementing their BI applications.

In other words, the collection of indicators integrated into the BI system reflects local aspirations, concerns and actions, and respects the idea that different countries have different requirements. It suggests that these Brazilian firms have resisted against the imposition of key performance indicators pushed by "globalized" companies, especially when dealing with global corporations such as the major IT vendors. The experience of these firms provides a portrait of local and cultural factors influencing the adoption and impact of BI projects.

The Brazilian situation combines an inward-oriented economy with strong linkages to international sources of technology. Part of the reason for Brazil's inward orientation is its size and its distance from major markets and global production networks. While Brazil's economy has become somewhat more globally oriented during the last 10 years, local rather than global forces still drive IT adoption in this country (Tigre & Dedrick, 2004). This national feature helps explain the Brazilian pattern of importing technologies but resisting imported performance indicators when using these technologies.

It is important to recognize the difference between importing key performance indicators that reflect contexts other than local, and adopting approaches to BI implementation that help in the identification of key performance indicators. Nonetheless, this study's findings suggest that the absence of any specific or well-defined approach when engaging in a BI project is dangerous. The lack of any methodology to improve the capacity to identify performance indicators can jeopardize the very existence of a BI process. In addition, although they paid attention to local and contextualized information, BI systems in the companies studied tended to favor formal and internal information over that which is informal and external (as suggested by "competitive intelligence"). What's more, the focus on financial and commercial areas is also problematic because important areas like innovation, employee motivation and collective learning may be neglected, and these represent some of the strongest sources of competitive advantage (Kaplan & Norton, 1992).

The previous consideration is directly related to the second research question: *What is the perceived "value" of BI to strategic management of Brazilian companies?* Considering that IT use takes place within a context of "globalization," and being aware that companies participating in such a globalized process do not compete under equal conditions, the authors expected that BI processes would help companies in developing countries to find competitive advantage. As previously discussed, this research suggests that few Brazilian companies are, in fact, using BI processes, and even among those using them, there is an absence of well-defined methodologies. This tendency must be revisited. Seeking the reasons for such a situation, an attempt was made to determine whether it represented a well-known phenomenon in IT: the adoption of an IT innovation without really understanding its nature or value.

In the Brazilian context, IT consultants and vendors have exercised a strong influence and have pushed new IT solutions on companies, even when the nature or value of such innovations is not really understood (Pozzebon, 2003). This means that a BI system can be adopted as a "strategic" project, but end up being used as a technical solution for operational and tactical problems. This belief is reinforced by Brazilian managers' current emphasis on BI projects in "data warehousing."

However, a BI process needs alignment with organizational strategy in order to produce the expected benefits, and the lack of understanding of such a strategic role for BI should be overcome. The word "strategic" is often used to increase the perceived value of a BI project or a vendor's offerings, but this element is not always integrated into the business process during implementation. It is worth recalling Buytendijk's words (2001) regarding the meaning of "strategic," one of the most abused terms in BI projects. What happens when BI is applied from a tactical/operational but not a strategic point of view?

A *strategic deployment of BI* means the BI application becomes embedded in the systems and processes of the business to build a more agile enterprise that can anticipate and react faster than its competitors to changing business conditions and new profit opportunities. On the other hand, a *tactical deployment of BI* aims at making a current process more efficient, usually the existing management reporting process.

A *strategic use of information* focuses on how well the organization is meeting predefined goals and objectives. Furthermore, this use provides perspective on, and direct support for, how the organization is able to change its ways, going beyond improving current operations. When a *tactical use of information* prevails, it provides insight into the status of current and day-to-day processes, or insight into how to improve current processes.

Failure to understand these differences usually leads to the BI/data warehousing "graveyard." This can entail overspending on a data warehouse infrastructure that serves only a few BI applications or underspending that causes huge project delays and failure because the infrastructure components have been underestimated. Similarly, the use of BI may be tactical rather than strategic, in that the preoccupation is not with defining strategic information aligned to strategic objectives, but with recovering indicators or measures that already exist in spreadsheets and are already being used in traditional managerial reports. The BI ends up being used as a traditional MIS that is simply more flexible and has graphical functionalities, but not as a "business intelligence process" for strategic decision-making.

A BI system's worth lies in the value of the indicators and information dealt with by the people that are interacting. If there is no awareness of how to conceptualize, produce, analyze and share such information, and of what strategic insights are likely to be triggered, the benefit from a BI process is likely to decrease or disappear. This research suggests that the strategic and social role of IT is not always perceived. Behind any IT application there lie social and political choices.

To trigger a BI process is much more of an organizational or management issue than a technological one. Much of the potential benefit of a BI project disappears when firms pay more attention to how to technically build and effectively manage a centralized data repository/data warehousing than to how to collectively and socially build a mechanism that produces and disseminates useful and timely information for decision-making.

This encourages the promotion of a less technical view of BI, as reflected in the authors' extended definition. They propose that BI should be seen as organizational processes instead of as simple BI tools or applications. These processes are likely to be collective, socially constructed, multidimensional, contextual and culturally situated. When a technical approach dominates, the potential benefits of BI processes tend to disappear.

The exceptions to this pessimistic scenario are those Brazilian firms which have decided to use a balanced scorecard as their methodological approach. By its very nature, the balanced scorecard approach calls for rethinking strategic goals, requiring the alignment of key indicators with top goals and functional objectives. For these reasons, the authors suggest that firms using adapted balanced scorecards are perhaps the only ones using their BI processes for truly strategic purposes.

On the other hand, precisely because Brazilian firms do not have mechanistically "imported" models and approaches from other contexts, the authors find here an opportunity to stimulate these companies to adopt a framework that is intrinsic to Brazilian social, economical and cultural contexts. The indicators have been identified according to the Brazilian context, and this fact can represent an important element to be explored by researchers and practitioners aiming to use IT as a vector for development.

CONCLUSION

Given the exploratory character of this study—it is, in fact, the first phase of a research project of this nature—the authors think that they have shed some light on a subject not yet investigated which could be further explored through additional theoretical lenses. Most people managing BI systems in the Brazilian companies investigated were more concerned with technology than with business. In other words, the companies implemented their systems with a technological focus, that is, how to structure data warehousing, which technology vendor is better, and so forth.

Furthermore, there is a lack of attention to determining what information is most relevant to business, or aligning indicators with strategic objectives. In the companies where balanced scorecards were used to drive the development of the BI system, a greater alignment between indicators and strategic objectives was found. That most of the Brazilian companies investigated have paid close attention to the Brazilian context in defining their indicators seems a good sign, as recent IT studies suggest the danger of mechanistically transposing global principles to local contexts, especially in developing countries.

However, the fact that most of the companies did not employ any specific methodology seems to interfere with the creation of value or competitive advantage from their BI projects. The lack of methodology is a weakness and invites future research into locally conceived approaches to BI. The authors believe that their extended definition of BI, although provisional and in progress, could be helpful in developing a framework that adheres to the company's particular and contextually situated business strategy, with a greater likelihood of obtaining "value" and benefit from these projects.

This research has revisited existing views of BI and proposes a reformulated definition. Its authors believe that a collective, contextualized and critical process of information management may help companies in developing countries derive value from their BI projects. IT can be a powerful tool that helps countries promote their own development and emphasize the local context within which the IT-based solutions are implemented. However, the nature of these "adaptations" and the factors that influence them are poorly understood, and this chapter's main contribution has been to shed some light on these elements regarding BI processes.

REFERENCES

Avgerou, C. (2002). *Information systems and global diversity.* London, UK: Oxford University Press.

Beck, U. (2000). *What is globalization?* Cambridge, UK: Polity Press.

Burn, J. M. & Loch, K. D. (2001). The societal impact of the World Wide Web—Key challenges for the 21st century. *Information Resources Management Journal, 14*(4), 4-14.

Business intelligence: Aspectos e tendências do uso de ferramentas de análise corporativa. Retrieved March 12, 2003, from www.idcbrasil.com.br.

Buytendijk, F. (2001). *Strategic BI: Its definition and effect on infrastructure.* Gartner Group.

Chenhall, R. H. (2005). Integrative strategic performance measurement systems, strategic alignment of manufacturing, learning and strategic outcomes: An exploratory study. *Accounting, Organizations and Society, 30*(5), 395-423.

Connelly, R., McNeil, R., & Mosimann, R. (1998). *The multidimensional manager - 24 ways to impact your bottom line in 90 days.* Ottawa, ON: Cognos Incorporated.

Dhar, V. & Stein, R. (1996). *Seven methods for transforming corporate data into business intelligence.* Upper Saddle River, NJ: Prentice Hall.

Dubé, L. & Paré, G. (2003). Rigor in information systems positivist case research: Current practices, trends and recommendations. *MIS Quarterly, 27*(4), 597-635.

Easterby-Smith, M., Araujo, L., & Burgoyne, J. (1999). *Organizational learning and the learning organization: Developments in theory and practice.* London, UK: Sage Publications.

Eisenhardt, K. M. & Sull, D. N. (2001). Strategy as simple rules. *Harvard Business Review, 79*(1), 106-117.

Ferramentas de business intelligence no Brasil (2003). Retrieved March 12, 2003, from www.idcbrasil.com.br.

Frigo, M. L. (2002). Strategy-focused performance measures. *Strategic Finance, 84*(3), 10-13.

Gilad, B. & Gilad, T. (1988). *The business intelligence system: A new tool for competitive advantage.* New York: Amacom.

Giovinazzo, W. A., (2002). *Internet-enabled business intelligence.* Upper Saddle River, NJ: Prentice Hall.

Hackathorn, R. D. (1998). *Web farming for the data warehouse: Exploiting business intelligence and knowledge management.* San Francisco: Morgan Kaufmann Publishers.

Hannula, M. & Pirttimaki, V. (2003). Business intelligence empirical study on the top 50 Finnish companies. *Journal of American Academy of Business, 2*(2), 593-599.

Harvey, C. D. (1988). Telephone survey techniques. *Canadian Home Economics Journal, 38*(1), 30-35

Heeks, R. (2002). Information systems and developing countries: Failure, success and local improvisation. *Information Society, 18*(2), 101-112.

Kalakota, R. & Robinson, M. (2001). *E-business 2.0—Roadmap for success*. New York: Addison-Wesley.

Kaplan, R. & Norton, D. (1992). The balanced scorecard—Measures that drive performance. *Harvard Business Review, 70*(1), 71-79.

Kaplan, R. (1996). *The balanced scorecard: Translating strategy into action*. Boston: Harvard Business School Press.

Kudyba, S. & Hoptroff, R. (2001). *Data mining and business intelligence: A guide to productivity*. Hershey, PA: Idea Group Publishing.

Liautaud, B. (2000). *E-business intelligence: turning information into knowledge into profit*. New York: McGraw-Hill.

McGonagle, J. J. & Vella, C. M. (1990). *Outsmarting the competition*. Naperville, IL: Sourcebooks.

Meier, R. L. (2000). Late-blooming societies can be stimulated by information technology. *Futures, 32*(2), 163.

Miles, M. B. & Huberman, A. M. (1990). *Qualitative data analysis*. London: Sage Publications.

Miller, C. (1995). In-depth interviewing by telephone: Some practical considerations. *Evaluation and Research in Education, 9*(1), 29-38.

Miller, J. (2002). *O milênio da inteligência competitiva*, Brazil: Bookman.

Mursu, A., Soriyan, H. A., Olufokunbi, K., & Korpela, M. (2000). Information systems development in a developing country: Theoretical analysis of special requirements in Nigeria and Africa. In *Proceedings of the 33rd Hawaii International Conference on System Sciences*. Maui, Hawaii: IEEE.

Niven, P. R. (2002). *Balanced scorecard step-by-step: Maximizing performance and maintaining results*. New York: J. Wiley & Sons.

O'Bada, A. (2002). Local adaptations to global trends: A study of an IT-based organizational change program in a Nigerian bank. *Information Society, 18*(2), 77.

Orlikowski, W. J. & Iacono, C. S. (2001). Research commentary: Desperately seeking "IT" in IT research—A call to theorizing the IT artifact. *Information Systems Research, 12*(2) 121-156.

Pozzebon, M. (2003). *The implementation of configurable technologies: Negotiations between global principles and local contexts*. Unpublished doctoral dissertation, McGill University, Montreal, Canada.

Rasmussen, N., Goldy, P. S., & Solli, P. O. (2002). *Financial business intelligence—Trends, technology, software selection, and implementation*. New York: John Wiley and Sons.

Reich, B. & Benbasat, I. (2000). Factors that influence the social dimension of alignment between business and information technology objectives. *MIS Quarterly, 24*(1), 81-113.

Robey, D., Ross, J., & Boudreau, M. (2002). Learning to implement enterprise systems: An exploratory study of the dialectics of change. *Journal of Management Information Systems, 19*(1), 17.

Sahay, S. & Avgerou, C. (2002). Information and communication technologies in developing countries. *Information Society, 18*(2), 1-5.

Sammon, W. L., Kurland, M. A., & Spitalnic, R. (1984). *Business competitor intelligence: Methods for collecting, organizing, and using information*. New York: John Wiley & Sons.

Schonberg, E., Cofino, T., Hoch, R., Podlaseck, M., & Spraragen, S. (2000). Measuring success. *Communications of the ACM, 43*(8), 53-57.

Scoggins, J. (1999). A practitioner's view of techniques used in data warehousing for sifting through data to provide information. In *Proceedings of The Eight International Conference on Information and Knowledge Management*, Kansas City, MI.

Stake, R. E. (1998). Case studies. In N. K. Denzin & Y. S. Lincoln (Eds.), *Strategies of qualitative inquiry* (pp. 86-109). Thousand Oaks, CA: Sage Publications.

Steinmueller, W. E. (2001). ICTs and the possibilities for leapfrogging by developing countries. *International Labour Review, 140*(2), 193-210.

Sturges, J. & Hanrahan, K. (2004). Comparing telephone and face-to-face qualitative interviewing: a research note. *Qualitative Research, 4*(1) 107-118.

Tigre, P. B. & Dedrick, J. (2004). E-commerce in Brazil: local adaptation of a global technology. *Electronic Markets, 14*(1) 36-40.

Van Der Zee, J. T. M. & De Jong, B. (1999). Alignment is not enough: Integrating business and information technology management with the balanced score card. *Journal of Management Information Systems, 16*(2), 137-158.

Vitt, E., Luckevich, M., & Misner, S. (2002). *Business intelligence*. Microsoft Press.

Watson, H., Goodhue, D., & Wixon, B. (2002). The benefits of data warehousing: Why some organizations realize exceptional payoffs. *Information & Management, 39*(6), 491-502.

Williams, R. (1997). Universal solutions or local contingencies? Tensions and contradictions in the mutual shaping of technology and work organization. In I. McLoughlin & M. Harris (Eds), *Innovation, organizational change and technology*. London, UK: International Thomson Business Press.

Chapter XIV
Building an Environmental GIS Knowledge Infrastructure

Inya Nlenanya
Center for Transportation Research and Education, Iowa State University, USA

ABSTRACT

Technologies such as geographic information systems (GIS) enable geospatial information to be captured, updated, integrated, and mapped easily and economically. These technologies create both opportunities and challenges for achieving wider and more effective use of geospatial information in stimulating and sustaining sustainable development through smart policy making. This chapter proposes a simple and accessible conceptual knowledge discovery interface that can be used as a tool to accomplish that. In addition, it addresses some issues that might make this knowledge infrastructure stimulate sustainable development with emphasis on sub-Saharan Africa.

INTRODUCTION

Technologies such as geographic information systems (GIS) enable geographic information to be captured, updated, integrated, and mapped easily and economically. These technologies create both opportunities and challenges for achieving wider and more effective use of geoinformation in stimulating and sustaining sustainable development through smart policy making. With the start of a new millennium humankind faces environmental changes greater in magnitude than ever before as the scale of the problem shifts from local to regional and to global. Environmental problems such as global climate change and unsustainable developments in many parts of the world are evolving as major issues for the future of the planet and of mankind. Acidification of lakes and rivers, destruction of vital natural wetlands, loss of biotic integrity and habitat fragmentation, eutrophication of surface waters, bioaccumulation of toxic pollutants in the food web, and degradation of air quality contribute some of the many examples of how human-induced changes have impacted the Earth system. These human induced changes are stressing natural systems and reducing biological diversity at a rate and magnitude not experienced for millions of years (Speth, 2004). Also, anthropogenic stresses such as those associated with population growth, dwindling resources, chemical and biological pollution of water resources are expected to become more acute and costly.

The approach in dealing with these environmental issues requires a balanced response in the form of an environmental management strategy. Such a response must utilize the best available scientific understanding and data in addition to an infrastructure that combines both in order to deliver sound science-based solutions to the myriad of environmental problems. In the Fall 2003 edition of the Battelle Environmental Updates, it was argued that such a response would result in a complex decision network. This argument must have inspired the National Science Foundation (NSF) in 2004 to propose a network of infrastructure called National Ecological Observatory Network (NEON). NEON supports continental-scale research targeted to address the environmental challenges by facilitating the study of common themes and the transfer of data and information across research sites (NAS, 2004). This creates a platform that enables easy and quick access to the environmental data needed to tackle the environmental challenges.

NEON is based on the same concept as grid computing. Grid computing eliminates the need to have all data in one place through on-demand aggregation of resources at multiple sites (Chetty & Buyya, 2002). This creates an enabling platform for the collection of more specialized data with the hope of integrating them with data from other related areas. This has particular benefit in environmental data management and analysis since both data and specific processing methods are frequently exchanged and used within various organizations (Vckovski & Bucher, 1996). Together, NEON and grid computing form the enabler for the construction of an environmental cyberinfrastructure that will permit the transfer of data, the specific processing methods and the interoperability of these methods so as to reduce the time wasted in duplication of resources. This infrastructure is necessary especially in the face of unprecedented data availability.

During the last decade, the society has witnessed a tremendous development in the capabilities to generate and acquire environmental data to support policy and decision-making. Furthermore, the rapid and exploding growth of online environmental data due to the Internet and the widespread use of ecological and biological databases have created an immense need for intuitive knowledge discovery and data mining methodologies and tools.

However, in Africa, where according to Song (2005) the bandwidth speed of an average university has the same aggregate bandwidth as a single home user in North America or Europe and costs more than 50 times for this bandwidth than its counterparts in Europe or North America deserves special attention while establishing such networks. This statistics is from a continent where the major issues include hunger, poverty, AIDS, and political instability and these summarizes why sub-Saharan Africa in this knowledge age is still undeveloped and unable to tackle her own environmental problems. Clearly, a survey of the wealthiest nations in the world would quickly reveal that GDP is directly proportional to the volume of digital information exchange. Technology transfer has not been able to make a mark in Africa simply because the proponents ignored the social and economic questions of access to markets, fair wages, water, land rights, and so forth, in favor of purely technical questions and rejecting the indigenous knowledge in the process. Hence with all the progress made in cutting edge technology for data acquisition, there is still a dearth of geographic information exchange in sub-Saharan Africa.

Sobeih (2005) argues that, "GIS is considered to be one of the most important tools for increased public participation and development that offers new socio-economic development opportunities. It can encourage human resource development within the country, facilitate the participation of youth in public life, help provide an analytic and scientific understanding of development issues, and much more" (p. 190). Evidently, in a region of the world marked with political instability, the role of the private sector and the ordinary citizen has become elevated. Hence, the need to increase capacity for handling GIS tools in environmental policy making. All the more important is this environmental GIS knowledge interface as the wealth of the continent lies in the environment. The participants at the AFRICAGIS 2005 Conference which held at South Africa concluded deliberations by recognizing the opportunity provided by geospatial information for use in the development in Africa. Consequently, the specific objectives of this chapter are:

1. To develop a simple and accessible conceptual knowledge infrastructure that can be used as a tool to introduce GIS into the education curriculum in sub-Saharan Africa
2. To adapt (1) to the current context of sub-Saharan Africa taking into effect the prevailing social and economic questions
3. To proffer policies for development in sub-Saharan Africa

BACKGROUND

From the history of GIS, it is without doubt that environmental application has been one of the motivating factors that led to the development of GIS in the mid-1960's (Longley, Goodchild, Maguire & Rhind, 2001). This is due to the fact that environmental issues arise as a result of human activities and almost all human activities involve a geographic component (Blaschke, 2001; Longley et al., 2001; Rautenstrauch & Page, 2001). From infancy in land use applications in Canada, GIS has evolved to an all enveloping technology that has found useful applications in every facet of human enterprise. Technologies such as global positioning systems (GPSs) and remote sensing satellites have been largely responsible for the GIS evolution complemented with reductions in the cost of computer hardware, electronic storage media, etc (Chainey & Ratcliffe, 2005; Longley et al., 2001). Ratcliffe (2004) believes that in addition to the technology aspect of GIS evolution, the discipline has also benefited immensely from what he refers to as the scientific development of the discipline, an angle developed by Goodchild (1991) and Longley et al. (2001). As a result, GIS has seen the adaptation of analytical methods, techniques and processes to problems with a spatial component—and every human activity has a spatial axis, thereby making GIS omnipresent in modern life (Chainey & Ratcliffe, 2005) and a partner in development. As a partner in development, there is need to leverage all the utility of GIS to increase the environmental knowledge base in sub-Saharan Africa.

In the global economy, knowledge is everything, which is one thing that industrialized countries have in common (Mokyr, 2002). But before one gets to knowledge, data is needed. There is a dearth of environmental data in developing countries (Kufoniyi, Huurneman & Horn, 2005; Rüther, 2001). And where they are available, they are not in digital format (Dunn, Atkins & Townsend, 1997). Organizations such as Environmental Information Systems-Africa (EIS-Africa), USAID and other notable international organizations have been in the forefront of the campaign to bridge the environmental knowledge gap by concentrating on human and institutional capacity building in the GIS sector and in encouraging the integration of GIS into policy making. As a way of strengthening these efforts, this chapter proposes a knowledge discovery interface.

KNOWLEDGE DISCOVERY INTERFACE

A knowledge discovery interface (KDI) is a type of interface that provides the means by which users can connect the suite of data mining tools to communicate with each other irrespective of their implementation and at the same time communicate with the data. KDI defines the range of permissible inputs, outputs and controls between the elements that make up the knowledge discovery process in order to encourage more participation from various fields of study which may not be part of the traditional data mining research catchment's area.

The knowledge discovery process is a computationally-intensive task consisting of complex interactions between a human and a large database, supported by heterogeneous suite of tools (Brachman & Anand, 1996). Consequently a knowledge discovery interface defines the rules for the complex interactions between not just the user and a large database but most importantly between the heterogeneous suite of tools and a large assortment of databases. It is very important that this suite of data mining tools sees the assortment of databases as a *whole* and not just as a *sum of the parts* since the best picture is being looked for. In this case, the best picture is one that takes from all sources and presents an output that is unique to all its sources. This is very significant because in knowledge discovery the object is not to look for the obvious but for some interesting pattern (Fayyad, Piatetsky-Shapiro & Smyth, 1996) that can be used for decision making.

To further understand the concept of the KDI, the author is going to look at some of the definitions of knowledge discovery in database in order to get a better understanding of the knowledge discovery process. Koua and Kraak (2004) defines knowledge discovery as a higher level process using information from the data mining process to turn it into knowledge or integrate it with prior knowledge. They went on to present a more general definition borrowing from Miller and Han (2001) and Fayyad et al. (1996), as "discovering and visualizing the regularities, structures and rules from data, discovering useful knowledge from data and for finding new knowledge." This definition takes into account the research area of data visualization, which hitherto has been largely ignored in knowledge discovery research (Lee, Ong & Quek, 1995). From this definition, a KDI provides the means of integrating the discovering of information from a database via statistical techniques and machine learning with visualization techniques so that the two work seamlessly to extract new knowledge or add to existing.

A KDI provides the means for controlling the complex processes of extraction, organization and presentation of discovered information (Brachman & Anand, 1996) from a database. This definition encompasses the various steps in the knowledge discovery process. In Miller (in press) the knowledge discovery process is grouped into the following steps as shown in Figure 1:

- **Data preprocessing** (data selection, data cleaning and data reduction)
- **Data mining** (choosing the data mining task, choosing the data mining technique and data mining)
- **Knowledge construction** (interpreting the mined patterns and consolidating the discovered knowledge)

The steps are independent and, therefore, a KDI provides the protocol that connects the steps. Mitra and Acharya (2003) add a new dimension in their assessment of the knowledge discovery process as involving all that have been mentioned above plus the modeling and the support of the overall human machine interaction. A KDI, in short, is the connection between the user, the knowledge discovery tools and the data.

Figure 1. Overview of the knowledge discovery process

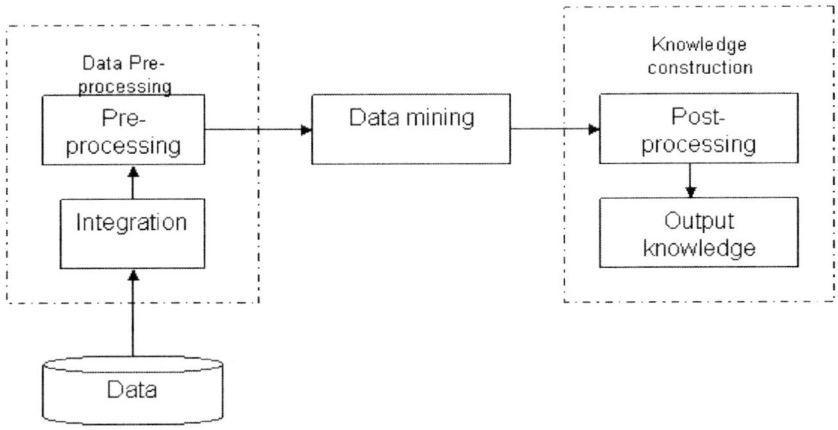

Why a Knowledge Discovery Interface is Needed

A KDI simplifies the process of knowledge discovery by making it easy for the user to interact with the wealth of environmental data and the suite of data mining tools available. This has the potential to encourage more participants and to expand the knowledge input into the research field. It provides the key to a human-centered knowledge discovery process that Brachman and Anand (1996) emphasize since it gives the user control over the tools. This control is very important as advances in knowledge discovery technologies are yielding more tools than the user can grasp without the help of a KDI.

As a result of the breakthroughs in data storage and data collection technologies, datasets for environmental studies now come in tera- or gigabytes of memory space. This factor is responsible for influencing the advances in artificial neural networks that have enhanced the analysis and visual mining of large volumes of data. Keim (2002), in his assessment of visual data mining, argued that it gives an overview of the data and it is particularly good for noisy, inhomogeneous and large datasets which are a few of the characteristics of the data available for environmental modeling. He continued to present that visual data mining can be seen as a hypotheses generation process. It can also be used for hypotheses verification, but in union with automatic techniques from statistics or machine learning (Lee, Ong, Toh & Chan, 1996). Additionally, visual data mining can help the user to determine whether a certain data is the best choice for the targeted learning process. This makes visual data mining an important member of the suite of knowledge discovery tools. Hence, the need for KDI to integrate the visual data mining tools with nonvisual data mining tools.

With a repertoire of already existing tools for handling spatial data, spatiotemporal data, and nonspatial data, a knowledge discovery interface will eliminate the need to create a holistic system that handles all the forms of data from scratch. Instead, through a well-developed interface, these existing tools can be integrated for the benefit of knowledge extraction from all kinds of data models available for environmental applications. Examples of these existing tools include ArcGIS and spatial OLAP or SOLAP (which is an integration of geographic information system (GIS) and OLAP (Bedard, Gosselin, Rivest, Proulx, Nadeau & Lebel, 2003). Accordingly, the KDI reduces the time required for the deployment of a state-of-art knowledge discovery infrastructure.

The iterative nature of the knowledge discovery process which is highlighted in Fayyad et al. (1996), Han (1999), NAS (2003), and Mitra et al. (2003) suggests that the process of applying tools and transformations in the task of knowledge discovery is repeated until the analyst discovers some striking regularities that were not known. This iterative character has the advantage of allowing the entire process to be broken into modules. KDI is very useful where modules exist because it defines the rules for inter-modular interaction. As a result, KDI enables a platform that leads to specialized stand-alone applications such that modifications can be made to one part without affecting the entire system. This is a view that Thuraisingham (1999) shares by recommending the development of data mining modules as a collection of generic reusable tools.

The contribution of grid computing to the knowledge discovery process comes with its own attendant problem. With the availability of data in intranet repository and geodata on the Internet, the problem arises of what kind of data would be best for a particular learning process. Albertoni, Bertone, and De Martino (2003) captures this by acknowledging the urgent need for methods and tools to support the user in data exploration. He proposed a solution based on the integration of different techniques including data mining, visualization and graphical interaction techniques. His approach aims to aid the user in making the right choice of data by offering both an automated presentation of data to dynamically visualize the metadata and interactive functionalities to discover the relationship among the different metadata attributes. This approach is hinting at creating a common control platform for these interactive functionalities to be integrated so that the user can manage them. KDI provides that common control. Metadata is mentioned here to underscore its prime place in data mining (Thuraisingham, 1999).

The bulk of the knowledge discovery process is in the data preprocessing stage. Miller and Han (2001) describes the pre-processing of data which is partly accomplished in data warehouses as fundamental to the knowledge discovery process because it integrates and represents data in a manner that supports very efficient querying and processing. Zaiane, Han, Li, and Hou (1998) highlights its importance by observing that most of the studies done in knowledge discovery are confined to the data filtering step which is part of the data preprocessing stage. This presumes that the success of the overall process centers on how well the data is prepared before mining since the data preparation process has the power to bias the knowledge that can be extracted. Thuraisingham (1999) makes his own case for the importance of data warehousing in these words, "good data is the key to good mining." As a result, advances in data warehousing and database integration would play a very important role in enhancing the knowledge discovery process. Database integration plays a role here because it provides the input to the data warehousing stage. For environmental applications, the data of choice is geo-spatial. Currently, conventional conceptual database models do not provide a straightforward mechanism to explicitly capture the semantics related to space (Khatri, Ram & Snodgrass, 2004). However, research is underway to develop tools for automatic integration of geo-spatial data (from well-structured vector and raster models to unstructured models such as georeferenced multimedia data) from heterogeneous sources into one coherent source (NAS, 2003). This will enable applications to be designed that integrate geospatial data from different sources. The next logical step would be to provide a KDI that will integrate these applications into the overall knowledge discovery process.

Another argument for the need for a KDI is the fact that data mining, and consequently the overall process of knowledge discovery, is a relatively young and interdisciplinary field, drawing from such areas as database management systems, data warehousing, data visualization, information retrieval, performance computing, and so forth. It needs the integration of approaches from multiple disciplines (Han, 1999). These fields that support knowledge discovery all have their own standards which creates the need for integration. In addition, research is currently underway to address the development of formalized platforms to enhance multidisciplinary research investments (NAS, 2003). A KDI would be advantageous to fully utilize the results of this research.

KDI System Conceptualization

The KDI system consists of GIS components, data mining components and the interactions between the two. The degree of integration which is a measure of the interaction between the components would be loose coupling as opposed to tight coupling. In loose coupling, the interaction is limited in time and space, that is, control and data can be transferred directly between the components. Nevertheless the interaction is always explicitly initiated by one of the components or by an external agent. In tight coupling, knowledge and data are not only transferred, they can be shared by the components via common *internal* structures (Alexandre, 1997). A comparison of the two degrees of integration would show that tightly coupled systems would definitely be difficult to upgrade without tearing down everything. Also scalability and reusability problems would arise. It would be difficult to integrate such a system outside of the application domain that warranted its design. Longley et al. (2001) believes that as standards for software development become more widely adopted, software developers or users would prefer software systems whose components are reusable. This would give them the choice of building from scratch or building by components (Longley et al., 2001). From a purely financial standpoint, choice is everything. Consequently, the three main components of the KDI are geospatial component, spatial analysis component, and the knowledge component as shown in Figure 2.

Figure 2. KDI architecture

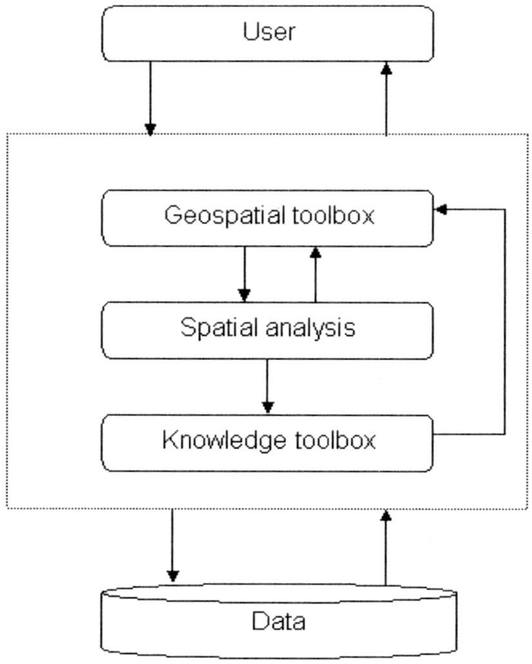

The Geospatial Component

The geospatial component is purely a GIS-based tool or assortment of tools. Blaschke (2001) differentiates GIS from other spatial environmental information systems on the basis of its data linkages. Rautenstrauch and Page (2001) in making a case for environmental informatics, argue that environmental studies should not be limited to just ecological data, a view which Groot and McLaughlin (2000) already saw the need for by opining that the pendulum is moving in favor of a geospatial data infrastructure (GDI). GDI has been described by Coleman and McLaughlin (1997) as an information system linking environmental, socioeconomic and institutional databases. A key characteristic of geospatial data is its potential for multiple applications (Groot & McLaughlin, 2000) which is a reflection of the technologies, including GPS and remote sensing, that are used to collect such data. In other words, a better understanding of environmental issues lies in integrating purely environmental subjects like ecology, land use, and so forth, with other factors that influence them, for example, the economy. A typical example of this application would be in the area of sustainable development which is the ability to simultaneously tackle the economic and environmental proportions of resource distribution and administration (Groot & McLaughlin, 2000).

GIS plays the role of presenting these linkages in such a way as to force an environmental view of reality. The idea of the geospatial component is to provide a toolbox in the KDI that handles data in a way that bridges the gap between data on paper and reality on ground. According to Thurston, Poiker, and Moore (2003), the accuracy of a model is in its ability to recreate reality accurately. They emphasize that accuracy is a function of the quality of information included in the model. Looking from another perspective, given that the quality of the information (data) is acceptable both in content and in how well the contents are integrated; the value of the knowledge extracted from this information would be a function of how the data is encapsulated and manipulated.

Figure 3 shows the processing that goes on within the geospatial component. The query connector acts as a kind of filter for selecting the data specified by the user. The selected data is promptly represented using either of the GIS data models- vectors and raster. This ensures consistency between how the features are stored in the database and how they are represented in visual form. The vector/raster block represents the encapsulation of the data for presentation in a visual product or map. The cyclic route linking the components shows a tightly coupled connection which ensures that at each point on the route the data will remain the same giving no loophole for data corruption. The two-sided arrow connecting the query to the visual product represents the connection between the data and visual product

Figure 3. Schematics of the interactions within the geospatial component

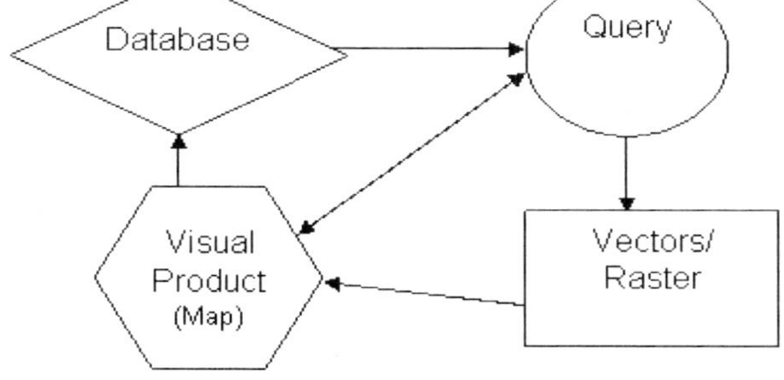

which makes it easy to query the database via the visual product.

Against this background, the geospatial component provides the tools that enable the user to edit, display, customize, and output spatial data. It acts as the display unit of the KDI making use of the geovisualization functionality of the GIS. As a result, researchers are adopting the geovisualization view of GIS in the conceptualization of the geospatial component. This geovisualization view basically sees GIS as a set of intelligent maps or views that show features and feature relationships on the earth (Zeiler, 1999). In this setup, the map acts as a window into the geodatabase (ESRI, 2004) which is at the heart of GIS architecture. Also from a knowledge discovery point of view, the geospatial component acts as the exploratory data analysis tool that gives the user a summary of the problem at hand ensuring a greater grasp of environmental issues.

The Spatial Component

Vital to understanding the need for a spatial component is the fact that the map or any other visual product is merely a representation of information stored in a database. The database is the depository of spatial information and not the map (Thurston et al., 2003). Kraak (2000) argues that a map has three major functions in the manipulation of geospatial data:

1. It can function as a catalog of the data available on the database.
2. It can be used to preview available data.
3. It can form part of a database search engine.

In a sense, the map is a guide to other information and as a result can be used to direct the extraction of information from the database. The information so extracted is then subjected to spatial analysis for the purpose of extracting knowledge, which can form the basis for updating the database in terms of reorganizing the way data is integrated or linked.

Spatial analysis refers to the ability to manipulate spatial data into different forms and extract additional information as a result (Bailey, 1994). Combining spatial analysis and GIS has been a study area many researchers have been interested in. Wise and Haining (1991) identified the three categories of spatial analysis as statistical spatial data analysis (SDA), map-based analysis and mathematical modeling. Haining (1994) believes that for GIS to attain its full measure, it needs to incorporate SDA techniques.

The nature of this link between spatial analysis and GIS is the subject of the spatial component implementation. Based on the study of the linkage between GIS and spatial analysis, Goodwill et al. (1991) distinguished between four scenarios:

1. Free standing spatial analysis software
2. Loose coupling of proprietary GIS software with statistical software
3. Close coupling of GIS and statistical software
4. Complete integration of statistical spatial analysis in GIS

Of all the four, most attention is on close coupling or loose coupling (Gatrell & Rowlingson, 1994) mostly because both options give the developers/users freedom in implementing the linkage in the way that will best accomplish their task. Also it makes it easy to integrate other components as the need arises.

The spatial component is the integration of GIS tools and statistical tools. While the geospatial component seeks to encapsulate the data in a way that will enhance knowledge discovery, the spatial component deals with manipulating the raw data in a way that will enhance application of the appropriate levels of theory and modeling capability in real problem solving situations

(O'Kelly, 1994). To this end, the spatial component provides the tools for analysis and transformation of spatial data for environmental studies. To be able to perform the analysis, the spatial component must be able to extract the data first. To extract the data, it needs access to the GIS tools for data integration, filtering, cleaning and all the necessary data preprocessing tasks. The spatial component must possess tools that will allow the results of the spatial analysis to be used to update the database in addition to the ability to view the results. Figure 4 provides the schematics of how the spatial component works.

Knowledge Component

Spatial decision support systems (SDSS) are very important tools for planning and decision making for environmental management. Normally, SDSS combine spatially explicit observational data and simulation of physical process with a representation that is suited for nonspecialist decision makers and other stakeholders (Taylor, Walker & Abel, 1998). It also provides the users and decision-makers with the tools for dealing with the ill- or semistructured spatial problems in addition to providing an adequate level of performance (Abiteboul, 1997; Hopkins, 1984; Stefanakis et al., 1998; Taylor et al., 1998). According to Ting (2003), sustainable development demands complex decision making that combines environmental, social and economic consequences of the choices made with regards to resource management. Such decision making, she continues, requires ready access to current, relevant and accurate spatial information by decision makers and stakeholders. Feeney (2003) argues that spatial information is one of the most critical elements underpinning decision making for many disciplines. She went on to define decision support as the automation, modeling and/or analysis that enables information to be shaped from data. The task of the knowledge component is to transform the information extracted into knowledge thereby improving the quality of the decision making process. It accomplishes this task by providing the necessary input for creating new environmental models or validating/updating existing ones.

As a result, the knowledge component is made up of a collection of learning algorithms. The knowledge component acts as the decision support of the entire system. As the decision support component, it can be used to structure, filter and integrate information, model information where gaps occur in data, produce alternative solution scenarios as well as weight these according to priorities, and most importantly facilitate group as well as distributed participation in decision making (Feeney, 2003). The interactions taking place in the component are shown diagrammatically in Figure 5. The interactions form the basis of the implementation. The figure shows a tight-coupled

Figure 4. Schematics of the interactions within the spatial component

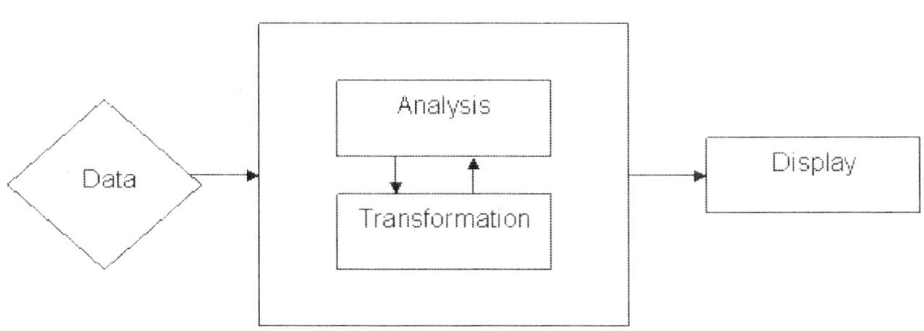

integration of the components—ensuring that the data and knowledge extracted are tied together.

POTENTIAL APPLICATIONS OF THE KDI

The possibility of packaging data mining methods as re-useable software applications objects have opened up the whole realm of knowledge discovery to people outside the traditional usage base.

The KDI enables a knowledge discovery platform optimized for environmental applications by integrating state of the art standalone GIS application and data mining functionality in a closely coupled, open, and extensible system architecture. The data mining functionality can be used to validate models for decision support systems used in generating environmental policies. For example, Mishra Ray, and Kolpin (in press) conducted a research on the neural network analysis of agrichemical occurrence in drinking water wells to predict the vulnerability of rural domestic wells in several Midwestern states of USA to determine agrichemical contamination. The research objectives included studying the correctness of results from the neural network analysis in estimating the level of contamination with known set of data and to show the impact of input parameters and methods to interpret the results. Also, Hsu et al. (1995), Shamseldin (1997), Shukla, Kok, Prasher, Clark, and Lacroix (1996), Kao (1996), Yang, Prasher, and Lacroix (1996), Maier and Dandy, 1996), Schaap and Bouten (1996) and Schaap and Linhart (1998), and Basheer, Reddi, and Najjar (1996) have all applied the neural network analysis to various problems in agriculture, water resources and environmental domains. The KDI is a great resource for applying data mining applications to water and environment–related problems. It also provides the platform for performing machine learning analysis on the magnitude of environmental data collected in order to keep up with the pace at which they are being collected.

The KDI can be used as a teaching tool for an introductory course for students interested in the partnership between GIS, data mining and environmental management without overloading them with advanced GIS. It will also widen the scope of environmental students in the area of programming languages, by challenging them to design models that are portable and reuseable. It can also serve as a teaching aid in creating the need for more public participation in environmental resource management within the framework of the long distance education paradigm.

The KDI is a prototype for using object oriented programming platform to enable the design of environmental modeling systems that are reuseable with well-defined inputs, outputs and controls for easy integration.

Figure 5. Interactions within the knowledge component

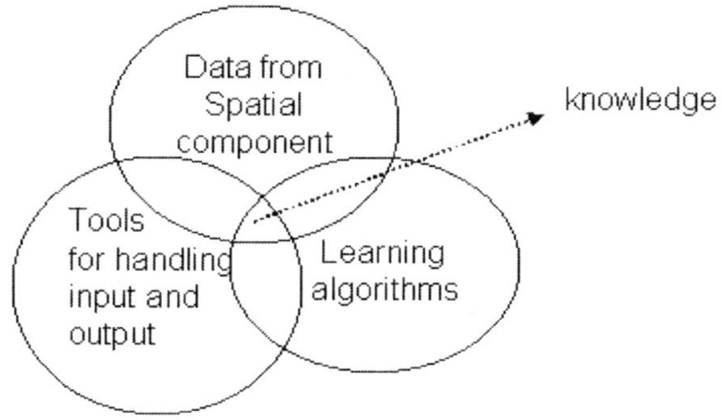

KDI IMPLEMENTATION

In the implementation, a GIS-based system which integrates logic programming and relational database techniques has been adopted. It is well documented that geographic analysis and spatial visualization improve operational efficiency, decision making, and problem solving. Software developers need the flexibility to build domain specific, easy-to-use applications that incorporate the power of GIS technology into a focused, user-friendly application (ESRI, 2004). The KDI implementation consists of a GIS component and Java Foundation Classes, JFC (see www.java.sun.com). JFC encompass a group of features for building graphical user interfaces (GUIs) and adding rich graphics functionality and interactivity to applications. The KDI application consists of a GIS component, the data mining application, and JFC that provide the interface for connecting them (Onwu, 2005). The GIS component is implemented using the ArcGIS® engine which is an integrated family of GIS software products from ESRI® that delivers complete, scalable GIS at the project, work group, and enterprise levels (ESRI, 2004). It contains developer application programming interfaces (APIs) that embed GIS logic in nonGIS-centric applications and efficiently build and deploy custom ArcGIS applications on the desktop. The data mining application can be implemented with any third party data mining application. Basically it provides the collection of machine learning algorithms for data mining tasks. This includes algorithms for classification and regression, dependency modeling/link analysis and clustering. The algorithms can either be applied directly to a dataset or called from user-defined Java code.

The choice of ESRI® products is informed by the fact that they are one of the few GIS companies with a commitment to human and institutional capacity building in Africa. JFC is open-source and will not cost the users anything to implement. The portability of the application ensures that it can be used in any operating systems –environment- which is an important consideration in an environment where there are not a lot of choices. In addition, JFC ensures that the implementation can be done in any other language besides English. This makes it possible for the end users to be able to take ownership of the application.

FUTURE ISSUES

This section discusses some of the emerging technologies and how they are redefining the need for KDI and the future of the synergy of GIS and data mining.

GeoSensors

GIS has been evolving over the years in response to the advances in database management technologies. Currently, advances in sensor technology and deployment strategies are transforming the way geospatial data is collected and analyzed and the quality with which they are delivered (NAS, 2003). This presupposes that the current methods of storing geospatial information are bound to change. The change is predicated on the fact that homogenous collection of data is now being replaced by heterogeneous collection of data for an area of interest, for example, video and temperature feeds. The nature of these feeds or data will warrant that pieces of information will vary in content, resolution and accuracy in addition to having a spatiotemporal component (Nittel & Stefanidis, 2005). The current trend of geosensor technology is also going to affect the time for data analysis. Usually, data is analyzed after it has been downloaded but with regards to energy considerations for the sensors, there might be the need to do real-time analysis of data being collected so that the sensor can discard unnecessary data and transmit useful data in accordance to the study requirements. This implies having an on-chip data mining capability. As a result, the geospatial nature of the data being collected would necessitate the call for GIS functionality tightly integrated with the on-chip data mining application. This is targeted to minimizing the energy consumption of the sensors by reducing amount of data to transmit. Nittel and Stefanidis (2005) suggested a minimization of data acquisition time as a solution to energy optimization.

In response to the ongoing development of sensor technology, Tao, Liang, Croitoru, Haider, and Wang (2005) is proposing the Sensor Web which would be the sensor-equivalent of the Internet. According to them, the Sensor Web would be a global sensor that connects all sensors or sensor databases. The Sensor Web would be interoperable, intelligent, dynamic, scalable, and mobile. This is certainly going to revolutionize the concept of GPS systems. With these sensors connecting wirelessly via Internet and possibly by satellite linkages, the possibility increases of having live feed for each location on the surface of the earth complete with video, audio and other parameters of interest at a particular time. It is going to be more like having a live Webcam with the added benefit of knowing the current wind speed, temperature, humidity, etc. all wrapped up within the framework of a GIS so that the geospatial component is not lost. This will obviously warrant a multimedia data mining application to tap into the vast knowledge trapped in the video images. The knowledge from the video feeds is then integrated with the knowledge from the nonspatial data in order to get a perfect or approximate picture. Although the emphasis would not be on *perfect,* but on approximate because as Evan Vlachos framed it in his opening address to GIS 1994, "it is better to be approximately right rather than precisely wrong," (Vlachos, 1994).

The success of the scenario painted in the foregoing paragraphs can only be accomplished in a closely coupled working multidisciplinary partnership. All the stakeholders involved must be accommodated at the outset to offset the possibility of creating integration problems later down the road. With object oriented programming platforms, each solution would be implemented as reuseable software application object.

Geographic Data Mining

The new ArcGIS 9 from ESRI is revolutionizing the concept of a geodatabase. The new ArcSDE has the capability for storing and managing vector, raster and survey dataset within the framework of the relational database management system (RDBMS) (ESRI, 2004b). This implies that not only is the geospatial data linked to nonspatial dataset; it is also linked to images. Hence the user has the choice of what kind of map to view-vector maps or satellite imagery. This creates an enabling environment for geographic data mining (GDM).

Geographic data mining is at best a knowledge discovery process within the context of a map instead of a database. Miller and Han (2001) define it as the application of computational tools to reveal interesting patterns in objects or events distributed in geographic space and across time. The time component specifically refers to the satellite images which represent pictures taken over time. GDM is closely related to geographic visualization (GV) which is the integration of cartography, GIS, and scientific visualization for the purpose of exploring geographic data in order to communicate geographic information to end-users (MacEachren & Kraak, 1997). With these developments, the possibility of performing machine learning analysis on a map object will greatly increase the knowledge available for environmental management as this will reduce the level of abstraction of spatial data and preserve the loss of spatial information. Also GDM would be the best way to capture the contribution of the time component in the knowledge extracted. There is still the problem of how to incorporate a time component in the RDBMS (NAS, 2003). But GDM of satellite maps would remove the need to abstract the time component making satellite images a repository of spatial-temporal data. The next task will be to encapsulate these developments in reuseable application objects with well defined user interfaces in order to make it accessible to the managers of environmental resources.

CONCLUSION

GIS started as a technology for data creation and has now evolved into one for data management. This research focused on the development and implementation of a prototype KDI for environmental science applications. This was predicated on the need to help policy makers to grasp with the current environmental challenges. The extensible nature of the KDI makes it a dynamic tool since it allows for integration with other tools. The challenge is now to package this concept in a cost effective way as a tool to introduce GIS in the educational curriculum.

In all, what GIS does is very simple. It makes a point aware of its position vis-à-vis other points. Stretching this understanding, the concept of a network becomes obvious. The challenge before sub-Saharan African countries becomes how to create a social infrastructure that will connect these points so that they can work for a common goal and avoid duplication of resources. That is the first step in taking the initiative to bridge the knowledge gap with the rest of the world. In 2003, USAID Success Stories captured the current state of Africa's efforts in bridging this gap in the following lines:

We find currently that a chasm exists, separating the users of environmental information, policymakers, and scientists from one another. We often think of this as a divide between continents, but more importantly, it is also a divide between islands of expertise. There is a divide between highly dedicated and competent analysts in Africa from the state of the art in the rest of the world, but also between the analysts and decision-makers, and between the scientific expertise of metropolitan centers and the innate local knowledge of the environment in rural areas (USAID/AFR, 2003).

Juma (2006) believes that African universities should take the initiative in community development by developing an educational curriculum that addresses the needs of the community. African universities should re-align themselves so that they became active participants with the international organizations in institutional capacity building. African universities should provide the leverage needed to bring the expertise together.

The road to sustainable development in sub-Saharan Africa will not be complete without addressing the role of governments. Juma (2006) proposes the role of governments as a facilitator. With government as facilitator, this creates a level field for public-private partnerships in the form of nongovernmental organizations to step in and get the knowledge to the rural communities by creating urban-rural partnerships and investing in youths as the harbinger of rural development. As a facilitator, African government should be committed to the fact that knowledge is the currency of development and if the developing countries must join their developed counterparts in providing basic services to their citizens, there is the need to create a unified system of tracking the vast potentials in Africa and organizing it in such a way that it can provide insights that would produce policies that would bring about development in Africa. The main benefit for the establishment of a GIS based system is to stimulate and assist development activities in the region. One way of doing this is by creating a commission tasked with the creation of baseline geographic data at the local government level and converting existing data into digital format. The funding for this commission can be sourced from private companies, or international agencies/foreign aid. The availability of baseline data makes it easy for international development agencies to track the progress of development in a region.

In the face of the failure of technology transfer in the developing countries, there is need for a GIS system that answers the more fundamental social and economic questions as well as the technical ones, an opinion exemplified by Ficenec (2003). GIS is very important for stimulating community development by providing a way for policy makers to match resources with potentials available in a community. This leads to grassroot development, poverty reduction, job opportunities, and overall, an economically viable state.

REFERENCES

Abiteboul, S. (1997). Querying semi-structured data. In *Proceedings of the International Conference on Database Theory*, Delphi, Greece.

AfricaGIS (2005). *Conference resolutions draft.* Retrieved April 13, 2008, from http://www.africagis2005.org.za/agp/africagispapers/AfricaGIS 2005 Resolutions draft 041105.doc

Albertoni, R., Bertone, A., & De Martino, M. A. (2003). Visualization-based approach to explore geographic metadata. In *Proceedings of the 11th International Conference in Central Europe on Computer Graphics, Visualization and Computer Vision, WSCG 2003*, Plzen-Bory, Czech Republic.

Alexandre, F. (1997). *Connectionist-symbolic integration: From unified to hybrid approaches.* Mahwah, NJ: Lawrence Erlbaum Associates.

Bailey, T. C. (1994). A review of statistical spatial analysis in geographical information systems. In A. S. Fotheringham & P. A. Rogerson (Ed.), *Spatial analysis and GIS* (pp. 14-44). London, UK: Taylor and Francis.

Basheer, I. A., Reddi, L. N., & Najjar, Y. M. (1996). Site characterization by NeuroNets: An application to the landfill siting problem. *Ground Water, 34*, 610-617.

Bedard, Y., Gosselin, P., Rivest, S., Proulx, M., Nadeau, M., Lebel, G., & Gagnon, M., (2003). Integrating GIS components with knowledge discovery technology for environmental health decision support. *International Journal of Medical Informatics, 70*, 79-94.

Blaschke, A. (2001). Environmental monitoring and management of protected areas through integrated ecological information systems- An EU perspective. In C. Rautenstrauch & S. Patig (Ed.), *Environmental information systems in industry and public administration* (pp. 75-100). Hershey, PA: Idea Group Publishing.

Brachman, R. J. & Anand, T. (1996). The process of knowledge discovery in databases. In U. Fayyad, G. Piatetsky-Shapiro, P. Smyth & R. Uthurusamy (Ed.), *Advances in knowledge discovery and data mining* (pp. 37-57). Cambridge, MA: AAAI/MIT Press.

Chainey, S. & Ratcliffe, J. (2005). *GIS and crime mapping.* Chichester, West Sussex: John Wiley and Sons.

Chetty, M. & Buyya, R. (2002). Weaving computational crids: How analogous are they with electrical grids? *IEEE Computing in Science and Engineering,* July/August, 61-71.

Coleman, D. J. & McLaughlin, J. D. (1997). Information access and network usage in the emerging spatial information marketplace. *Journal of Urban and Regional Information Systems Association, 9*, 8-19.

Dunn, C. E., Atkins, P. J., & Townsend, J. G. (1997). GIS for development: A contradiction in terms? *Area, 29*(2), 151-159.

ESRI (2004a). *ArcGIS 9: What is ArcGIS?* A White Paper. Redlands, CA: Environmental Systems Research Institute.

ESRI (2004b). *ArcSDE: Advanced spatial data server.* White Paper. Retrieved May 8, 2008 from http://esri.com/library/whitepapers/pdfs/arcgis_spatial_analyst.pdf

Fayyad, U., Piatetsky-Shapiro, G., & Smyth, P. (1996). *From data mining to knowledge discovery in databases. AI Magazine, 17*, 37-54.

Feeney, M. F. (2003). SDIs and decision support. In I. Williamson, A. Rajabifard, & M. F. Feeney (Ed.), *Developing spatial data infrastructures: From concept to reality* (pp. 195-210). London, UK: Taylor & Francis.

Ficenec, C. (2003, June). Explorations of participatory GIS in three Andean watersheds. *Paper presented at the University Consortium of Geographic Information Science (UCGIS) Summer Assembly 2003*, Pacific Grove, CA.

Gatrell, A. & Rowlingson, B. (1994). Spatial point process modeling in a GIS environment. In A.S. Fotheringham & P.A. Rogerson (Ed.), *Spatial analysis and GIS* (pp. 148-163). London, UK: Taylor and Francis.

Goodchild, M. F., Haining, R., & Wise, S. M. (1991). Integrating GIS and spatial data analysis: Problems and possibilities. *International Journal of Geographic Information Systems, 6*, 407-423.

Groot, R. & McLaughlin, J. (2000). Introduction. In R. Groot & J. McLaughlin (Eds.), *Geospatial data infrastructure: Concepts, cases and good practice* (pp. 1-12). Oxford, UK: Oxford University Press.

Haining, R. (1994). Designing spatial data analysis modules for GIS. In A.S. Fotheringham & P.A. Rogerson (Eds.), *Spatial analysis and GIS* (pp. 46-63). London, UK: Taylor and Francis.

Han, J. (1999). Data mining. In J. Urban & P. Dasgupta (Eds.), *Encyclopedia of distributed computing*. Kluwer Academic Publishers.

Hopkins, L. D. (1984). Evaluation of methods for exploring Ill-defined problems. *Environmental Planning B: Planning and Design, 11*, 339-348.

Hsu, K-L., Gupta, H. V., & Soroosian, S. (1995). Artificial neural network modeling of the rainfall-runoff process. *Water Resour. Res., 31*, 2517-2530.

Juma, C. (2006, April). *Reinventing African economies: Technological innovation and the sustainability tansition.* Paper presented at The John Pesek Colloquium on Sustainable Agriculture, Ames, Iowa

Kao, J-J. (1996). Neural net for determining DEM-based model drainage pattern. *Journal of Irrigation and Drainage Engineering, 122*, 112-121.

Keim, D. A. (2002). Information visualization and visual data mining. *IEEE Transactions on Visualization and Computer Graphics, 7*, 100-107.

Khatri, V., Ram, S., & Snodgrass, R. T. (2004). Augmenting a conceptual model with geospatiotemporal annotations. *IEEE Transactions on Knowledge And Data Engineering, 16*, 1324-1338.

Koua, E. L. & Kraak, M. J. (2004). Geovisualization to support the exploration of large health and demographic survey data. *International Journal of Health Geographics, 3*,12.

Kraak, M.-J. (2000). Access to GDI and the function of visualization tools. In R. Groot & J. McLaughlin (Eds.), *Geospatial data infrastructure: Concepts, cases and good practice* (pp. 217-321). Oxford, UK: Oxford University Press.

Kufoniyi, O., Huurneman, G., & Horn, J. (2005, April). *Human and institutional capacity building in geoinformatics through educational networking.* Paper presented at the International Federation of Surveyors Working Week 2005, Cairo, Egypt.

Lee, H. Y., Ong, H. L., & Quek, L. H. (1995). *Exploiting visualization in knowledge discovery.* In *Proceedings of the 1st International Conference on Knowledge Discovery and Data Mining* (pp. 198 – 201), Montreal, Canada.

Lee, H. Y., Ong, H. L., Toh, E. W., & Chan, S. K. (1996). A multi-dimensional data visualization tool for knowledge discovery in databases. In *Proceedings of IEEE Conference on Visualization*, pp. 26–31.

Longley, P. A., Goodchild, M. F., Maguire, D.J., & Rhind, D. W. (2001). *Geographic information systems and science*. West Sussex, England: John Wiley and Son, Ltd.

Maier, H. R. & Dandy, G. C. (1996). The use of artificial neural networks for the prediction of water quality parameters. *Water Resour. Res., 32*, 1013-1022.

MacEachren, A. M. & Kraak, M.-J. (1997). Exploratory carthographic visualization: Advancing the agenda. *Computer and Geosciences, 23*, 335-343.

Miller, H. J. (in press). Geographic data mining and knowledge discovery. In J.P. Wilson & A. S. Fotheringham (Eds.), *Handbook of geographic information science*. Blackwell.

Miller, H. J. & Han, J. (2001). *Geographic data mining and knowledge discovery*. London: Taylor and Francis.

Mishra, A., Ray, C., & Kolpin, D. W. (in press). Use of qualitative and quantitative information in neural networks for assessing agricultural chemical contamination of domestic wells. *Journal of Hydrological Engineering.*

Mitra, S. & Acharya, T. (2003). *Data mining: Multimedia, soft computing and bioinformatics*. Hoboken, NJ: John Wiley and Sons, Inc.

Mokyr, J. (2002). *The gifts of Athena: Historical origins of the knowledge economy.* New Haven: Princeton University Press.

National Academy of Sciences (NAS) (2003). *IT roadmap to a geospatial future*., Washington, D.C.: The National Academies Press.

National Academy of Sciences (NAS) (2003). *IT roadmap to a geospatial future.* Washington, D.C.: The National Academies Press.

Nittel, S. & Stefanidis, A. (2005). GeoSensor networks and virtual GeoReality. In S. Nittel & A. Stefanidis (Eds.), *GeoSensor networks* (pp. 1-9). Boca Raton, FL: CRC Press.

O'Kelly, M. E. (1994). *Spatial analysis and GIS.* In A.S. Fotheringham & P.A. Rogerson (Eds.), *Spatial analysis and GIS* (pp. 66-79). London, UK: Taylor and Francis.

Onwu, I. (2005). *Knowledge discovery interface for environmental applications.* Unpublished master's thesis, Iowa State University, Ames.

Ratcliffe, J. (2004). *Strategic thinking in criminal intelligence*. Sydney: Federation Press.

Rautenstrauch, C. & Page, B. (2001). *Environmental informatics-methods, tools and applications in environmental information processing.* In C. Rautenstrauch & S. Patig (Eds.), *Environmental information systems in industry and public administration* (pp. 2-11). Hershey, PA: Idea Group Publishing.

Rüther, H. (2001, October). EIS education in Africa – The geomatics perspective. *Paper presented at the International Conference on Spatial Information for Sustainable Development*, Nairobi, Kenya

Schaap, B. D. & Linhart, S.M. (1998). *Quality of ground water used for selected municipal water supplies in Iowa, 1982-96 water years* (p. 67). Iowa City, IA: U.S. Geological Survey Open File Report 98-3.

Schaap, M. G. & Bouten, W. (1996). Modeling water retention curves of sandy soils using neural networks. *Water Resour. Res., 32*, 3033-3040.

Shamseldin, A. Y. (1997). Application of a neural network technique to rainfall-runoff modeling. *Journal of Hydrology, 199*, 272-294.

Shukla, M. B., Kok, R., Prasher, S. O., Clark, G., & Lacroix, R. (1996). Use of artificial neural networks in transient drainage design. *Transactions of the ASAE, 39*, 119-124.

Sobeih, A. (2005). *Supporting natural resource management and local development in a developing connection: Bridging the policy gap between the information society and sustainable development.* A publication of the International Institute for Sustainable Development (IISD), pp. 186-210.

Song, S. (2005). Viewpoint: Bandwidth can bring African universities up to speed. *Science in Africa,* September 2005. Retrieved April 13, 2008, from http://www.scienceinafrica.co.za/2005/september/bandwidth.htm

Speth, J. G. (2004). *Red sky at morning: America and the crisis of the global environment.* Yale University Press.

Stefanakis, E., Vazirgiannis, M., & Sellis, T. (1999). Incorporating fuzzy set methodologies in a DBMS repository for the application domain of GIS. *International Journal of Geographic Information Science, 13*, 657-675.

Taylor, K., Walker, G., & Abel, D. (1999). A framework for model integration in spatial decision support systems. *International Journal of Geographic Information Science, 13*, 533-555.

Tao, V., Liang, S., Croitoru, A., Haider, Z. M., & Wang, C. (2005). GeoSwift: Open geospatial sensing services for sensor web. In S. Nittel & A. Stefanidis (Eds.), *GeoSensor Networks* (pp. 267-274). Boca Raton, FL: CRC Press.

Thuraisingham, B. M. (1999). *Data mining: Technologies, techniques, tools and trends.* Boca Raton, FL: CRC Press.

Thurston, J., Poiker, T. K., & Moore, J. P. (2003). *Integrated geospatial technologies: A guide to GPS, GIS, and data logging.* Hoboken, NJ: John Wiley & Sons.

Ting, L. (2003). Sustainable development, the place for SDIs, and the potential of e-governance. In I. Williamson, A. Rajabifard & M. F. Feeney (Eds.), *Developing spatial data infrastructures: From concept to reality* (pp. 183-194). London, UK: Taylor & Francis.

USAID (2003). *USAID Africa success stories.* Retrieved April 13, 2008, from http://africastories.usaid.gov:80/print_story.cfm?storyID=23

Vckovski, A. & Bucher, F. (1996). Virtual data sets - Smart data for environmental applications. In *Proceedings of the Third International Conference/Workshop on Integrating GIS and Environmental Modeling*, Santa Fe, NM.

Vlachos, E. (1994). GIS, DSS and the future. In *Proceedings of the 8th Annual Symposium on Geographic Information Systems in Forestry, Environmental and Natural Resources Management*, Vancouver, Canada.

Wise, S. M. & Haining, R. P. (1991). The role of spatial analysis in geographical information systems. *Westrade Fairs, 3*, 1-8.

Yang, C.-C., Prasher, S. O., & Lacroix, R. (1996). Application of artificial neural networks to land drainage engineering. *Trans. ASAE, 39*, 525-533.

Zaïane, O. R., Han, J., Li, Z.-N., & Hou, J. (1998). Mining Multimedia Data. In *Proceedings of the CASCON'98: Meeting of Minds* (pp. 83-96), Toronto, Canada.

Zeiler, M. (1999). *Modeling our world: The ESRI guide to Geodatabase design.* Redlands, CA: ESRI Press.

Chapter XV
The Application of Data Mining for Drought Monitoring and Prediction

Tsegaye Tadesse
National Drought Mitigation Center, University of Nebraska, USA

Brian Wardlow
National Drought Mitigation Center, University of Nebraska, USA

Michael J. Hayes
National Drought Mitigation Center, University of Nebraska, USA

ABSTRACT

This chapter discusses the application of data mining to develop drought monitoring tools that enable monitoring and prediction of drought's impact on vegetation conditions. These monitoring tools help decision makers to assess the current levels of drought-related vegetation stress and provide insight into the possible future trends in vegetation conditions at local and regional scales, which can be used to make knowledge-based decisions. The chapter summarizes current research using data mining approaches (e.g., association rules and decision-tree methods) to develop these types of drought monitoring tools and briefly explains how they are being integrated with decision support systems. Future direction in data mining techniques and drought research is also discussed. This chapter is intended to introduce how data mining is be used to enhance drought monitoring and prediction in the United States and assist others to understand how similar tools might be developed in other parts of the world.

INTRODUCTION

Over the past few decades, many parts of the world have experienced devastating impacts from the frequent occurrences of both short- and long-term droughts, and decision makers such as policy makers and farmers are faced with the difficult challenge of dealing with these natural disasters. Although drought characteristics are complex and the prediction of such events is difficult, decisions must still be made to manage and mitigate drought impacts whenever this natural disaster occurs. With an increase in population growth and the resultant demand for natural resources (e.g., food and water), the vulnerability of people to natural disasters such as drought has dramatically increased. As a result, droughts of identical magnitude and spatial coverage will incur more damages and greater impacts today than they would a few decades ago.

There is a growing need for improved drought monitoring tools to assist people in making more informed drought risk management decisions. Such tools would help decision makers to implement effective responses (crisis management) that include technical, financial, and humanitarian assistance to drought-affected areas. Improved drought-related information is needed to make more efficient and effective planning and mitigation decisions. This requires new tools that can deliver more accurate and detailed drought information in a timely and reliable fashion.

Many studies have focused on developing improved drought monitoring tools that can assist in the decision-making process (Goddard, Harms, Reichenbach, Tadesse & Waltman, 2003). Most of these studies have relied on traditional statistical methods to build models based on the relationships of atmospheric, climatic, and oceanic variables to drought events. However, traditional statistical techniques are often insufficient for identifying drought and its characteristics (e.g., intensity) because of the complex interplay of these variables, which affect the occurrence, geographic extent, intensity, and duration of drought. As a result, researchers are focusing on developing drought monitoring tools using new analytical techniques that can explore these complex relationships. Recently, data mining techniques were used to develop improved drought monitoring tools and better understanding of drought characteristics (Harms, Deogun& Tadesse, 2002; Tadesse, Wilhite, Harms, Hayes & Goddard, 2004).

The primary strength of data mining techniques is their capability to search databases for hidden patterns and find predictive information that experts may miss because it lies outside their expectations (Berry & Linoff, 2000; Cabena, Stadler, R., Verhees & Zanasi, 1998; Groth, 1998). In addition, data mining can be used to answer difficult questions or problems that would be too time-consuming and/or complex to resolve using traditional methods. The automated, prospective analyses offered by data mining move beyond the analyses of past climatic events commonly used for drought monitoring and allow complex relationships between many diverse variables (or indictors) to explore for this application. Data mining tools also have the potential to predict future trends and behaviors, and this information could allow decision makers to make proactive, knowledge-driven decisions (Tadesse, Brown & Hayes, 2005a).

This chapter reviews the use of data mining techniques for drought monitoring in the United States and highlights the challenges facing this application. The chapter briefly explains the potential of data mining techniques for drought monitoring and the current research activities in developing drought monitoring tools and integration systems to enhance drought assessment and prediction for the continental United States. The chapter also presents examples of the results of this ongoing collaborative research by computer scientists, remote sensing specialists, water resources specialists, and climatologists in the central United States.

BACKGROUND

Drought Monitoring

Drought is characterized by its intensity, spatial extent, and duration (Wilhite, 2000). The determination of these characteristics in real time is often complicated (Kottegoda, Natale & Raiteri, 2004; Svoboda, LeComte, Hayes, Heim, Gleason, Angel, 2002; Wilhelmi & Wilhite, 2002). The capability to characterize these different dimensions of drought is important because effective drought planning and mitigation actions require drought indicators based on sound science that provide useful information about a drought event and its impacts. Drought indicators can be based on atmospheric, hydrologic, and/or satellite observations that either directly or indirectly influence the occurrence of drought in a specific area or region. Modern technical capabilities, which include the development of computer algorithms to identify hidden patterns within multiple datasets, could help improve drought indices that are often used to make planning decisions and trigger mitigation actions.

Because of the varied and potentially catastrophic losses resulting from drought in many parts of the world, both governmental and nongovernmental decision makers need improved access to accurate and timely monitoring and prediction tools to assist them in dealing more effectively and efficiently. Better early warning and prediction of drought is the foundation of the new paradigm for risk-based drought management. Technological advances (e.g., computing capabilities, algorithms, and Web-based services) will allow improved and enhanced drought monitoring tools to be developed, which will improve the ability to more effectively manage water and other shared natural resources during periods of drought.

Data Mining to Identify Drought

Large historical data sets are needed to identify relationships between different climatic, oceanic, and biophysical parameters and to distinguish patterns that may be used to predict drought. In light of this, it is essential to have an efficient way to extract information from large databases and to deliver relevant and actionable information for drought mitigation. One of the recently developed techniques relevant for such purposes is *data mining*.

Data mining is a process that uses a variety of data analysis tools to discover patterns and relationships among a number of variables in different data sets. This approach integrates techniques from machine learning, pattern recognition, statistics, databases, and visualization and has been used by numerous disciplines (e.g., science, business, and medicine) to address the issue of information extraction from large databases.

The data mining approach is commonly used in the commercial sector by companies to design strategies to increase profitability. For example, data mining is used to predict (or identify) consumers that are the most likely to buy certain products. Based on this information, companies can effectively identify the market demand. Data mining can also be used by businesses to understand trends in the marketplace to reduce costs and improve the timeliness of products reaching the market. Recent studies found this method to be one of the best tools to identify the patterns of supply and demand for specific products; this type of information is necessary to be profitable in a competitive market (Berry & Linoff, 2000; Cabena et al., 1998; Larose, 2005).

Data mining is also being increasingly used for environmental applications (De'ath & Fabricus, 2000; White, Kumar, & Tcheng, 2005) and holds considerable potential for identifying complex relationships among atmospheric, oceanic, and other environmental variables as they relate to droughts. The most common data mining algorithms and models, which include decision trees, associations, clustering, classification, multiple linear regression, sequential patterns, and time-series forecasting, have the potential to identify drought patterns and characteristics. Association, clustering, and sequence discovery approaches may be useful tools for investigating and describing the occurrence and intensity of drought, while classification, regression, and time-series analyses may be appropriate for mapping and monitoring drought patterns.

One of the main challenges of data mining in drought research is interpreting model results. Models from decision trees are easier to interpret because their classification and rules structure is transparent to the user, while the results from neural networks are the least comprehensible and most difficult to understand because of the non-linear combination of many parameters in their models and their relatively 'black box' modeling environment. Using a combination of different models and comparing the model outputs may provide a better understanding of the results.

Data mining techniques can identify "local" patterns better than traditional time-series analysis techniques, which largely focus on global models such as statistical correlations. The infrequent and complex nature of drought requires alternative analysis techniques that emphasize the discovery of local patterns of climate and oceanic data. For example, one may consider the occurrence of drought and its association with climatic and oceanic parameters instead of all precipitation patterns that include both dry and wet periods. In other words, since drought monitoring is particularly concerned with dry episodes, the data-mining algorithm is needed to discover the associations between oceanic and/or atmospheric conditions/patterns and the resulting drought event(s).

Other techniques can be utilized to maximize the results acquired from the data mining algorithms to identify drought characteristics. Among these techniques are geographic information systems (GIS), which have the capability to integrate geospatial data sets of different types and spatial scales, analyze this information, and present the results of the analysis in a geospatial format (Aguilar, 2002). The integration of data mining and GIS techniques into a drought monitoring tool is useful for fully exploring large, diverse databases; developing predictive drought models based on historical climate-ocean-biophysical relationships and occurrences; and applying the models in a geospatial environment to map and monitor drought patterns.

Drought characteristics can be better monitored after their patterns have been recognized using data mining algorithms. Models that are based on historical data and their patterns can be applied on near-real time geospatial data. This capability allows more informed decisions to be made at the earliest stages of drought onset and intensification. Also, some studies have recently used data mining to build predictive models that provide outlooks of drought conditions and patterns (Tadesse et al., 2004). This information can be used for proactive drought management

CURRENT RESEARCH USING DATA MINING

In the United States, recent research has developed drought assessment and predictive models using data mining techniques that include association rules, regression-trees, and neural networks (Brown, Tadesse & Reed, 2002; Goddard et al., 2003; Harms et al., 2002; Tadesse et al., 2004; Tadesse, Wilhite, Hayes, Harms & Goddard, 2005b). To enhance the efficiency and accuracy of drought monitoring for agricultural and water resources management, these models and algorithms used databases containing oceanic, climate, biophysical (e.g., land cover, soil, and irrigation data), and satellite-based vegetation condition information. These databases can be efficiently accessed, manipulated, and integrated with data mining techniques to develop improved drought monitoring tools. In the following sections, some examples of these current research activities are presented to demonstrate the utility of data mining techniques for identifying and monitoring drought.

Drought Monitoring Tool Integrating Climate and Satellite Data

A collaborative research effort between the National Drought Mitigation Center (NDMC) and the U. S. Geological Survey's (USGS) National Center for Earth Resources Observation Science (EROS) was recently undertaken in the United States to improve the country's national drought monitoring capabilities. The objective of this research was to develop and implement a new drought monitoring indicator called the Vegetation Drought Response Index (VegDRI) across the conterminous United States. The VegDRI integrates climate, satellite, and other biophysical information (e.g., land cover, percentage of irrigated agriculture, soil available water capacity, and ecosystem type) in a data mining environment to produce a 1-kilometer (km) resolution drought indicator.

The VegDRI is expected to provide improved, more spatially precise information regarding drought-induced vegetation stress than traditional drought indicators that are based solely on climate indices or satellite-based vegetation condition observations. Drought indices based on meteorological station observations have a coarse spatial resolution. These stations are not uniformly distributed and, as a result, the extrapolated data for areas between the stations does not necessarily represent accurate spatial information of drought patterns. In contrast, the satellite data have uniform and continuous coverage over large areas and contain spatially detailed information. However, the satellite data require ground-truth (e.g., climate and biophysical evidence) to validate the environmental factors influencing changes in the vegetation condition observations over time (Ji & Peters, 2003; McVicar & Bierwirth, 2001). Satellite observations highlight areas with anomalous vegetation conditions that might be caused by drought, flooding, pest infestations, or late green-up. The integration of the satellite and climate data allows the VegDRI model to utilize the climate information to identify the vegetation condition anomalies related to drought.

The VegDRI model is built using a regression-tree data mining technique that incorporates information from climate-based drought indices (i.e., Standardized Precipitation Index [SPI] and Palmer Drought Severity Index [PDSI]), 1-kilometer satellite-based observations of general vegetation conditions (derived from a time series of normalized difference vegetation index [NDVI] data from the advanced very high resolution radiometer [AVHRR]), and other environmental data sets that summarize land use/land cover (LULC), soil characteristics, and the ecological setting (Tadesse et al., 2005a). Figure 1 summarizes the specific inputs and processing steps for the development of the VegDRI and the dissemination of the VegDRI information.

Currently, a semioperational, biweekly VegDRI product is being generated for the U.S. Northern Great Plains with plans to further expand coverage to the conterminous United States (http://edc2.usgs.gov/phenological/drought/index.html). A 1-km VegDRI map is produced at 2-week intervals to provide timely information of drought effects on vegetation during the growing season. Figure 2 shows an example of the VegDRI map over fifteen states in the central and mid-western United States. The VegDRI map in Figure 2 shows the large areas of Wyoming, South Dakota, Colorado, Nebraska, and New Mexico that experienced severe to extreme drought in 2002. The high 1-km resolution of the VegDRI allows users to zoom in on the map to a more localized level (e.g., county) and identify more specific areas that are affected by drought. Moreover, overlaying the land cover map on the VegDRI map, it is possible to calculate the percentage of a land cover type (e.g., grassland or cropland) affected during a drought event. Such information can be utilized by agricultural producers and other

Figure 1. This flow chart shows the data inputs and model building processes for VegDRI and VegOut, as well as the dissemination mechanism for the information.

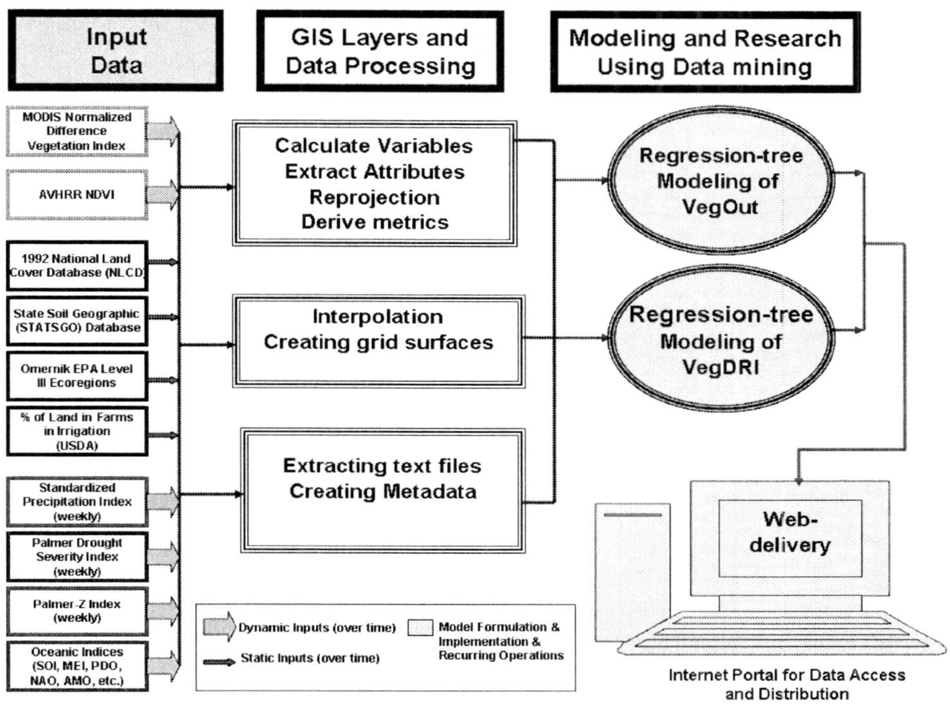

decision makers for a variety of drought planning and mitigation activities.

Vegetation Outlook: Integrating Climate and Satellite Data Using Data Mining

Research has also been undertaken to use the predictive capability of data mining techniques to develop a tool called the Vegetation Outlook (VegOut), which provides an outlook of the vegetation conditions a few weeks in advance during the growing season. The VegOut map is similar to the VegDRI map but shows a general outlook of the vegetation conditions in the upcoming weeks. Inputs, processing steps, and dissemination mechanisms similar to those shown in Figure 1 are also used for VegOut. The main objective of the VegOut research is to develop a tool through the use of data mining and knowledge discovery techniques that will enable the anticipation of drought conditions and the assessment of consequential landscape and vegetation response at local scales based on ocean-atmosphere-land interactions and their relationships with drought. Figure 3 (a) shows an experimental VegOut map that provides a 2-week outlook of vegetation conditions expressed as Standardized Seasonal Greenness (SSG) for the central United States for July25, 2006. The actual SSG observed by satellite on July 25 is shown in Figure 3 (b). The SSG patterns predicted 2 weeks in advance in the VegOut map are in strong agreement with the SSG patterns observed by satellite on July 26

Figure 2. (a) Vegetation Drought Response Index (VegDRI) map for July 25, 2002. VegDRI shows the severe drought conditions that plagued most of Colorado, New Mexico, and Wyoming and the western parts of Kansas, Nebraska, and South Dakota in 2002. The favorable vegetation conditions that occurred in the eastern part of this area (Illinois, Iowa, Minnesota, Missouri, and Wisconsin) during that same time were also represented in the VegDRI map. (b) The 15-states study area within the United States is highlighted in grey.

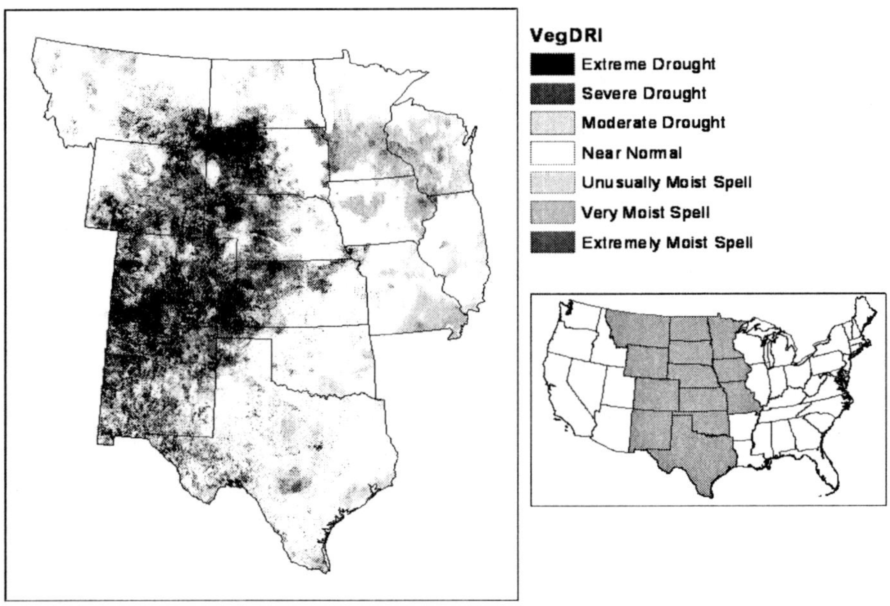

for most areas in the central United States. The correlation (r^2) between the actual SSG and the predicted SSG from this experimental VegOut model was 0.98.

In the VegOut research, Tadesse et al. (2005a) showed that future outlooks of drought-induced vegetation stress patterns could be delivered up to six weeks in advance and at a high spatial resolution (1-km^2). According to Tadesse et al. (2005a), the accuracy of the prediction is lower (r^2=0.44) for the early phenological phases because of instability of vegetation condition close to the start of the growing season. However, the correlation (r^2) between the predicted VegOut model results and the actual satellite-observed SSG values ranged from 0.67 to 0.85 during the remainder of the growing season. VegOut's predictive capability improved during this period because the vegetation activity is usually more stable during the maturity and senescence (i.e., desiccation and leaf-drop) phases of the seasonal growth cycle.

The VegOUT is designed to complement the "current" drought condition information represented in the VegDRI products by providing drought conditions and patterns into the future, which could assist decision makers in both their short- and long-term planning.

Research is currently underway to build on the initial VegOut research by Tadesse et al. (2005a) and test the hypothesis that global oceanic conditions (e.g., El Niño or La Niña) are an important precursor to drought conditions over the continental United States. A better understanding and

Figure 3. (a) The Vegetation Outlook (VegOut) map that predicted general vegetation conditions (expressed as standardized seasonal greenness – SSG) 2 weeks prior to the biweekly period ending on July 25, 2006. (b) The observed SSG for the biweekly period ending 25 July 2006 that was derived from the satellite data. The maps show general agreement in the intensity and pattern of vegetation conditions except for some localized areas of severe to extreme vegetation stress over northeastern Montana, southern North Dakota, western South Dakota, and central Nebraska.

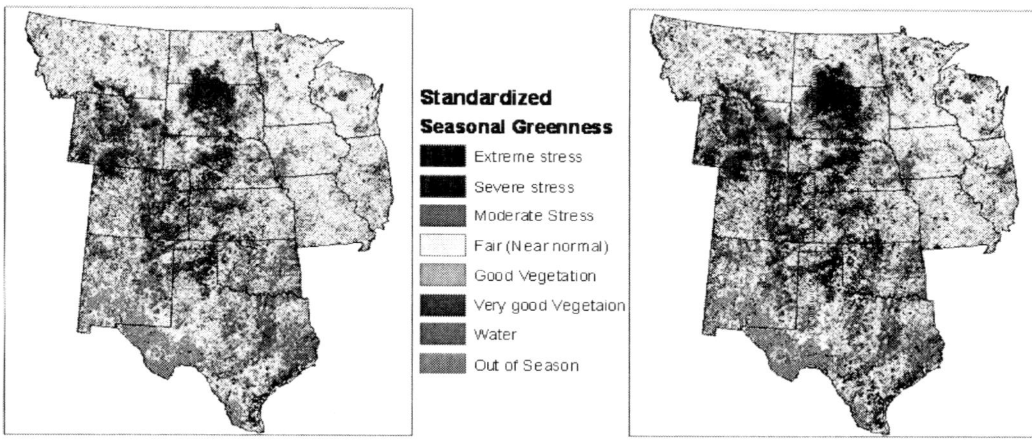

representation of the oceanic-climate-drought relationship in the VegOut model will improve our predictive capabilities for providing useful short-term outlooks (i.e., 2-, 4-, and 6-week) of drought-induced vegetation stress. Different approaches and techniques including association rules, regression decision-tree method, and Case Based Reasoning (CBR) methods are being tested to build better predictive models. To enhance the existing VegOut model, a regression-tree method is being tested with scenarios that are based on the probability of occurrence of precipitation, and with El Niño/La Niña conditions (teleconnection) that are based on changes in ocean-atmosphere dynamics of the Pacific and Atlantic oceans. For this reason, oceanic variables based on Pacific and Atlantic observations are also being tested to identify patterns that may be used for predicting drought.

This research is built on previous studies (Brown et al., 2002; Goddard et al., 2003; Harms et al., 2002; Tadesse et al., 2004; Tadesse et al., 2005a) that have shown the importance of data mining in exploring and discovering the relationships between ground-based climatic observations and oceanic observations to enable a better understanding of their causes and effects as they pertain to drought.

This current VegOut research is expected to lead to a greater understanding of the drought-related relationships among remote sensing, climatic, oceanic, and biophysical data, which can be used to improve the spatial and temporal resolutions of our drought monitoring and prediction capabilities.

Identifying Drought Using Association Rules

Another approach uses association rules to identify drought at a specific weather station or geographic location. Tadesse et al. (2005b) developed a new data mining algorithm called the minimal occurrences with constraints and time lags (MOWCATL) to identify relationships between oceanic parameters and drought indices at a specific location. Rather than using traditional global statistical associations, the MOWCATL algorithm identifies historical drought episodes from periods of normal and wet conditions and then uses these drought episodes to find time-lagged relationships with oceanic parameters. As with all association-based data mining algorithms, MOWCATL is used to find existing relationships between historical drought events and the corresponding oceanic signals in the data, and is not by itself a prediction tool.

Using the MOWCATL algorithm (Tadesse et al., 2005b), the analyses of the rules generated for selected stations and state-averaged precipitation and temperature data for Nebraska from 1950 to 1999 indicated that most occurrences of drought were preceded by positive values of the Southern Oscillation Index (SOI), negative values of the Multivariate ENSO Index (MEI), negative values of the Pacific/North American (PNA) Index, negative values of the Pacific Decadal Oscillation (PDO), and negative values of the North Atlantic Oscillation (NAO). The frequency and confidence of the time-lagged relationships found between these five oceanic indices and drought at the selected stations in Nebraska indicates that oceanic parameters can be used as indicators of drought in the central United States.

FUTURE TRENDS

As has been shown, identifying patterns of drought characteristics using climatic, oceanic, and satellite data and finding their associations with vegetation conditions is of great importance for drought monitoring. New drought monitoring tools that integrate these diverse types of information should also be utilized in early warning systems (EWS) in an effort to more effectively and efficiently use available resources to reduce the impacts of drought.

In an effort to help build an integrated EWS, researchers are continuing to: (1) investigate the relationships between drought and oceanic parameters over the continental United States and use this information to identify triggering mechanisms for the onset, continuance, and end of drought at regional, state, and local levels; (2) build predictive models to assess and predict drought using an integrated database of satellite, climate, oceanic, and biophysical information; and (3) evaluate the results from these predictive models using crop yield data in an effort to identify and predict the impacts of drought on agriculture. These types of studies will continue to improve drought risk analysis through the creation of knowledge-based decision-making tools that can be integrated with other drought risk management systems (both implemented and in development).

Progress in solving these drought risk management challenges has reached a stage where the value of data mining approaches has been established, but further research is needed to fully explore the utility of these methods. The computer science community is also challenged to develop new, alternative data mining techniques that can be used to better characterize the complex environmental relationships associated with drought and produce improved drought indicators. In the future, researchers need to continue to work on bringing data mining and GIS together for meteorological and environmental data exploration, analysis, and visualization. This combination will allow a much higher level of interaction between users and database systems. As the integration of data mining and GIS technologies improves, so will our capabilities to solve more complex environmental problems and provide more effective management and monitoring tools.

At present, the VegDRI and VegOut models are experimental and have only been implemented over the central United States. Research is currently underway to expand the models to the other parts of the United States. These models may also be tested and applied internationally with considerable potential to improve researchers' monitoring capabilities in developing parts of the world. However, international expansion of both models will be heavily dependent on the availability of similar climatic, satellite, oceanic, and biophysical data that are needed as input variables into both models. The models' input variables may need to be modified or substituted by other variables depending on data availability and their relationship(s) with vegetation conditions in a specific country or region in the world.

CONCLUSION

Accurate and timely assessment of the onset, spatial extent, and severity of drought is critical for responding to a multitude of environmental and socio-economic impacts. Identifying the triggering mechanisms and associations of various parameters related to drought occurrences are important tasks in improving drought monitoring and prediction and the resultant tools and information that are available to decision makers. Currently, existing tools are limited in their ability to predict drought, and development of such tools is a research priority.

Increased understanding of drought patterns and characteristics can improve the development and implementation of planning and mitigation actions. One of the challenges in understanding drought is the difficulty of extracting meaningful information from large volumes of data for numerous climate and hydrologic variables and indices, which are produced at a variety of spatial (e.g., station, climate division, or regional) and temporal (e.g., weekly, biweekly, or monthly) scales. As mentioned in this chapter, data mining has proven useful in analyzing these large collections of diverse data sets and gleaning improved drought-related information, as compared to traditional drought indicators and tools.

Data mining techniques have demonstrated great potential in identifying drought characteristics and their spatial and temporal patterns as well as finding the association of these characteristics with oceanic processes, which can be used to improve both drought monitoring and prediction. The use of such techniques will help drought researchers and policy makers to: (1) develop effective drought monitoring capabilities, (2) improve water management based on area-specific drought predictions, (3) improve the allocation of human and financial resources during drought events, (4) improve financial protection strategies such as crop insurance, (5) implement effective drought policies to reduce vulnerability to drought, and (6) develop alternative food supply options in relation to drought hazards.

The VegDRI is an example of a current drought monitoring approach that uses data mining techniques. By collectively considering a number of climate and biophysical variables, this tool identifies drought-induced vegetation stress that cannot be identified solely by traditional climate drought indices and satellite vegetation indices. Data mining has proven to be an effective and efficient means of integrating and analyzing this diverse collection of variables for drought characterization. Data mining techniques also offer the potential to move beyond monitoring and begin to predict drought conditions, as was demonstrated in this chapter by the VegOut and MOWCATL tools that have been recently developed. The availability of these advanced analytical tools allows researchers to enhance the drought characterization capabilities by exploring and analyzing diverse, and often complex, sets of information in ways that were not possible with more traditional analysis techniques.

These new and improved drought monitoring tools can be used to provide improved information for more effective drought planning, management, and risk analysis. They can also be used to develop a better understanding of the impacts of drought on available resources, which assist decision makers in taking appropriate and timely mitigation actions. Lastly, these tools can be utilized for policy decisions related to sustainable development and preparation for future challenges in resource-limited areas.

REFERENCES

Aguilar, A. M. (2002). Integrating GIS, circular statistics and KDSD for modelling spatial data: A case study. *Geographical and Environmental Modelling, 6*(1), 5-25.

Berry, J. A. & Linoff, G. (2000). *Mastering data mining: The art and science of customer relationship management.* New York: John Wiley & Sons, Inc.

Brown, J. F., Tadesse, T., & Reed, B. C. (2002). Integrating satellite data and climate data for US drought mapping and monitoring. In *Proceedings of the 15th Conference on Biometeorology and Aerobiology joint with 16th International Congress on Biometeorology*, (pp. 147-150). Kansas City, Missouri.

Cabena, P. H., Stadler, R., Verhees, J., & Zanasi, A. (1998). *Discovering data mining: From concept to implementation.* New Jersey: IBM.

De'ath, G. & Fabricus, K. E. (2000). Classification and regression trees – A powerful yet simple technique for ecological data analysis. *Ecology, 8*(11), 3178-3192.

Goddard, S., Harms, S. K., Reichenbach, S. E., Tadesse, T., & Waltman, W. J. (2003). Geospatial decision support for drought risk management. *Communication of the ACM, 46*(1), 35-37.

Groth, R. (1998). *Data mining: A hands-on approach for business professionals.* New Jersey: Prentice Hall.

Harms, S. K., Deogun, J., & Tadesse, T. (2002). Discovering sequential association rules with constraints and time lags in multiple sequences. *Lecture notes in artificial intelligence 2366: Foundations of intelligent systems.* In *Proceedings of the 13th International Symposium on Methodologies for Intelligent Systems (*pp. 432-441). Lyon, France.

Ji, L. & Peters, A. J. (2003). Assessing vegetation response to drought in the northern Great Plains using vegetation and drought indices. *Remote Sensing of Environment, 87*, 85-98.

Kottegoda, N. T., Natale, L., & Raiteri, E. (2004). Some considerations of periodicity and persistence in daily rainfalls. J*ournal of Hydrology, 296*(1-4), 23-37.

Larose, D. T. (2005). *Discovering knowledge in a data: An introduction to data mining.* New Jersey: John Wiley & Sons, Inc.

McVicar, T. R. & Bierwirth, P. N. (2001). Rapidly assessing the 1997 drought in Papua New Guinea using composite AVHRR imagery. *International Journal of Remote Sensing, 22*, 2109-2128.

Svoboda, M., LeComte, D., Hayes, M., Heim, R., Gleason, K., Angel, J., Rippey, B., Thinker, R., Palecki, M., Stooksbury, D., Miskus, D., & Stephens, S. (2002). The drought monitor. *Bulletin of the American Meteorological Society, 83*(8), 1181-1190.

Tadesse, T., Brown, J. F., & Hayes, M. J. (2005a). A new approach for predicting drought-related vegetation stress: Integrating satellite, climate, and biophysical data over the U.S. central plains. *ISPRS Journal of Photogrammetry and Remote Sensing*, 59(4), 244-253.

Tadesse, T., Wilhite, D. A., Hayes, M. J., Harms, S. K., & Goddard, S. (2005b). Discovering associations between climatic and oceanic parameters to monitor drought in Nebraska using data-mining techniques. *Journal of Climate, 18*(10), 1541-1550.

Tadesse, T., Wilhite, D. A., Harms, S. K., Hayes, M. J., & Goddard, S. (2004). Drought monitoring using data mining techniques: A case study for Nebraska, USA. *Natural Hazards, 33*(1), 137-159.

White, A. B., Kumar, P., & Tcheng, D. (2005). A data mining approach for understanding topographic control on climate-induced inter-annual vegetation variability over the United States. *Remote Sensing of Environment, 98*, 1-20.

Wilhite, D. A. (2000): Drought as a natural hazard: concepts and definitions. In D. A. Wilhite (Ed.), *Drought: A global assessment* (Vol. 1, pp. 3-18). London: Routledge Publishers.

Wilhelmi, O. V. & Wilhite, D. A. (2002). Assessing vulnerability to agricultural drought: a Nebraska case study. *Natural Hazards, 25*(1), 37-58.

Compilation of References

Abbass, H. A., Sarker, R. A., & Newton, C. S. (Eds.) (2002). *Data mining: A heuristic approach*. Hershey, PA: IGI Global.

Abbott, J. (2001). Data data everywhere – and not a byte of use? *Qualitative Market Research: An International Journal, 4*(3), 182-192.

Abiteboul, S. (1997). Querying semi-structured data. In *Proceedings of the International Conference on Database Theory*, Delphi, Greece.

Ackland, R. & Gibson, R. (2004). Mapping political party networks on the WWW. In *Proceedings of the Australian Electronic Governance Conference*, Melbourne, Australia.

Ackland, R. (2005). *Estimating the size of political Web graphs*. Revised paper presented to ISA Research Committee on Logic and Methodology Conference. Retrieved April 10, 2008, from http://acsr.anu.edu.au/staff/ackland/papers/political_ web_graphs. pdf

Ackland, R. (2005). *Mapping the U.S. political blogosphere: Are conservative bloggers more prominent?* Paper presented to BlogTalk Downunder, Sydney. Retrieved April 10, 2008, from http://acsr.anu.edu.au/staff/ackland/papers/polblogs.pdf

Ackoff, R. L. (1989). From data to wisdom. *Journal of Applied Systems Analysis, 16*, 3-9.

Adafre, S. F. & Rijke, M. D. (2005). Discovering missing links in Wikipedia. In *Proceedings of the 3rd International Workshop on Link Discovery* (pp. 90-97). ACM Press.

Adamo, Jean-Marc (2001). *Data mining for association rules and sequential patterns: Sequential and parallel algorithms*. Springer Verlag

AfricaGIS (2005). *Conference resolutions draft*. Retrieved April 13, 2008, from http://www.africagis2005.org.za/agp/africagispapers/AfricaGIS 2005 Resolutions draft 041105.doc

Agosta, L. (2000). *The essential guide to data warehousing*. Upper Saddle River, NJ: Prentice Hall.

Agrawal, R. & Srikant, R. (1994). Fast algorithms for mining association rules. In J. B. Bocca, M. Jarke & C. Zaniolo (Eds.), *Proceedings of the 20th International Conference on Very Large Data Bases* (pp. 487-499). San Francisco: Morgan Kaufmann Publishers.

Agrawal, R. & Srikant, R. (1995). Mining sequential patterns. In *Proceedings of the 1995 International Conference Data Engineering*, Taipei, Taiwan.

Agrawal, R., Imielinski, T., & Swami, A. (1993). Mining association rules between sets of items in large databases. In P. Buneman & S. Jajdia (Eds.), *Proceedings of the 1993 ACM SIGMOD International Conference on Management of Data* (pp. 207-216). New York: ACM Press.

Aguilar, A. M. (2002). Integrating GIS, circular statistics and KDSD for modelling spatial data: A case study. *Geographical and Environmental Modelling, 6*(1), 5-25.

Aksoy, S., Koperski, K., Tusk, C. & Marchisio, G. (2004). Interactive training of advanced classifiers for mining remote sensing image archives. In *Proceedings of the ACM International Conference on Knowledge Discovery and Data Mining* (pp. 773-782). Seattle, Washington.

Albertoni, R., Bertone, A., & De Martino, M. A. (2003). Visualization-based approach to explore geographic metadata. In *Proceedings of the 11th International Conference in Central Europe on Computer Graphics, Visualization and Computer Vision, WSCG 2003*, Plzen-Bory, Czech Republic.

Alexandre, F. (1997). *Connectionist-symbolic integration: From unified to hybrid approaches.* Mahwah, NJ: Lawrence Erlbaum Associates.

Ali, K., Manganaris, S., & Srikant, R. (1997). Partial classification using association rules. In D. Heckerman, H. Mannila, D. Pregibon & R. Uthurusamy (Eds.), *Proceedings of the Third International Conference on Knowledge Discovery and Data Mining* (pp. 115-118). Menlo Park, CA: AAAI Press.

Allenby, B. R., Compton, W. D., & Richards, D. J. (2007). *Information systems and the environment overview and perspectives*. Retrieved April 13, 2008, from http://books.nap.edu/openbook.php?record_id=6322&page=1

Allot (2005). The traffic management handbook. MN: Allot Communications Ltd.

Alon et al. (1999). Broad patterns of gene expression revealed by clustering analysis of tumor and normal colon tissues probed by oligonucleotide arrays. In *Proceedings of the National Academy of Sciences.*

Alpaydin, E. (2004). *Introduction to machine learning (adaptive computation and machine learning).* The MIT Press.

Altman, E., Haldeman, G., & Narayanan, P. (1977). Zeta analysis: A new model to identify bankruptcy risk of corporations. *Journal of Banking and Finance*, June, 29-54.

Alvarez, G. (2004). What's missing from RFID tests. *Information Week*. Retrieved November 20, 2004, from http://www.informationweek.com/story/showArticle.jhtml?articleID=52500193

Alves, D. S. (2002). Space-time dynamics of deforestation in Brazilian Amazonia. *International Journal of Remote Sensing, 23*(14).

Américo, M. C. S., Vieira, I. C. G., Veiga, J. B. & Araujo, R. (in press) Pecuária e Amazônia: Estratégias sociais e reestruturação do território nas frentes pioneiras: Rodovia PA-279 e região da Terra do Meio no Pará [Cattle ranching and Amazonia: Social strategies and territory reorganization in new frontiers – PA-279 and Terra do Meio region in Pará state]. In R. Araujo & P. Lena (Eds.), *Alternativas de desenvolvimento sustentável na Amazônia: Experiências recentes* [Alternatives of sustainable development in Amazônia: Recent experiences].

Anahory, S., & Murray, D. (1997). *Data warehousing in the real world: A practical guide for building decision support systems.* Harlow, UK: Addison-Wesley.

Anandarajan, M., Picheng, L. & Anandarajan, M. (2001). Bankruptcy prediction of financially stressed firms: An examination of the predictive accuracy of artificial neural networks. *International Journal of Intelligent Systems in Accounting, Finance and Management, 10*, 69-81.

Andrew, M. (2005). *The role of research in sustainable tourism policy-making.* Paper presented at the First Regional Sustainable Tourism Policy and Intersectoral Planning Workshop Grand Barbados Hotel, Barbados, West Indies.

Apoteker, T. & Barthelemythierry, S. (2005). Predicting financial crises in emerging markets using a composite non-parametric model. *Emerging Markets Review, 6*(4), 363-375.

Aspncs, J. (2002). Randomized protocols for asynchronous consensus, Ref.

Avgerou, C. (2002). *Information systems and global diversity.* London, UK: Oxford University Press.

Badia, A. & Kantardzik, M. (2005). Graph building as a mining activity: Finding links in the small. In *Proceedings of the 3rd International Workshop on Link Discovery* (pp. 17-24). ACM Press.

Bailey, T. C. (1994). A review of statistical spatial analysis in geographical information systems. In A. S. Fotheringham & P. A. Rogerson (Ed.), *Spatial analysis and GIS* (pp. 14-44). London, UK: Taylor and Francis.

Baldi, P., Frasconi, P., & Smyth, P. (2003). *Modeling the Internet and the Web: Probabilistic methods and algorithms.* West Sussex, UK: John Wiley.

Basheer, I. A., Reddi, L. N., & Najjar, Y. M. (1996). Site characterization by NeuroNets: An application to the landfill siting problem. *Ground Water, 34*, 610-617.

Bauer, E., & Kohavi, R. (1999). an empirical comparison of voting classification algorithms: Bagging, boosting, and variants. *Machine Learning, 36*, 105-139.

Baum, E.B., & Haussler, D. (1989). What net size gives valid generalisation? *Neural Computation, 1*(1), 151-160.

Beaver, W. (1966). Financial ratios as predictors of failure. *Journal of Accounting Research*, pp. 71-111.

Beck, U. (2000). *What is globalization?* Cambridge, UK: Polity Press.

Becker, B. (1997). *Amazonia*. São Paulo: Atica.

Bedard, Y., Gosselin, P., Rivest, S., Proulx, M., Nadeau, M., Lebel, G., & Gagnon, M., (2003). Integrating GIS components with knowledge discovery technology for environmental health decision support. *International Journal of Medical Informatics, 70*, 79-94.

Bellaachia, A., Portnoy, D., Chen, Y. & Elkahloun, A. G. (2002) E-CAST: A data mining algorithm for gene expression data. In *Proceedings of the BIOKDD02: Workshop on Data Mining in Bioinformatics (with SIGKDD02 Conference)*, Edmonton, Alberta, Canada.

Bellaachia, A., Vommina, E., & Berrada, B. (2006). Minel: A framework for mining e-learning logs. In *Proceedings of the 5th IASTED International Conference on Web-based Education* (pp. 259-263). Puerto Vallarta, Mexico.

Ben-Dor, A., Shamir, R. & Yakhini, Z. (1999). Clustering gene expression patterns. *Journal of Computational Biology, 6*(3/4), 281–297.

Ben-Or, M. (1983). Another advantage of free choice: Completely asynchronous agreement protocols (extended abstract). In *Proceedings of the Second Annual ACM SIGACT-SIGOPS Symposium on Principles of Distributed Computing* (pp. 27–30), Montreal, Quebec, Canada.

Bentley, R. (2005, August 25). Data with destiny. *Caterer & Hotelkeeper, 38*.

Berendt, B. (2002). Using site semantic to analyze, visualize and support navigation. *Data Mining and Knowledge Discovery, 6*, 37-59.

Berendt, B., Hotho, A., & Stumme, G. (2002). Towards semantic web mining. *Lecture Notes in Computer Science* (vol. 2342, pp. 264-278).

Berler, A., Pavlopoulos, S., & Koutsouris, D. (2005). Using key performance indicators as knowledge-management tools at a regional health-care authority level. *IEEE Trans Inf Technol Biomed, 9*(2), 184-192.

Berners-Lee, T., Hendler, J., & Lassila, O. (2001). The Semantic Web. *Scientific American, 284*(5), 34-43.

Berry, J. A. & Linoff, G. (2000). *Mastering data mining: The art and science of customer relationship management.* New York: John Wiley & Sons, Inc.

Berry, M. J. A. & Linoff, G. S. (1997). *Data mining techniques for marketing, sales and customer support.* John Wiley & Sons.

Berry, M. J. A. & Linoff, G. S. (1999). *Mastering data mining: The art and science of customer relationship management.* John Wiley & Sons.

Berry, M. J. A. & Linoff, G. S. (2000). *Mastering data mining.* John Wiley & Sons.

Berry, M. J. A. & Linoff, G. S. (2002). *Mining the Web: Transforming customer data.* John Wiley & Sons.

Berry, M. J. A. & Linoff, G. S. (2004). *Data mining techniques: For marketing, sales, and customer relationship management.* Wiley Computer Publishing.

Berson A., Smith, S. J., & Thearling, K. (1999). *Building data mining applications for CRM.* McGraw Hill.

Berson, A. & Smith, S. J. (1997). *Data warehousing, data mining, and OLAP.* McGraw Hill.

Berthold, M. & Hand, D. J. (1999). *Intelligent data analysis: An introduction*. Springer Verlag.

BESR (2004). Board on Earth Sciences and Resources (BESR), *Future challenges for the U.S. Geological survey's mineral resources program (2004)*. Washington, D.C.: The National Academies Press.

Bhat, N., & McAvoy, T. J. (1990). Use of neural nets for dynamic modelling and control of chemical process systems. *Computer Chemical Engineering, 14*(4/5), 573-583.

Bhattacharya, I. & Getoor, L. (2004). Deduplication and group detection using links. In *Proceedings of the SIGKDD Workshop on LinkAnalysis and Group Detection,* Seattle, WA.

Bins, L., Fonseca, L. & Erthal, G. (1996). Satellite imagery segmentation: A region growing approach. In *Proceedings of the 8th Brazilian Symposium on Remote Sensing* (pp.1-4).

BIS, The Bank for International Settlements. (2006). *Basel II: Revised international capital framework*. Retrieved April 13, 2008, from http://www.bis.org/publ/bcbsca.htm.

Bishop, C. M. (1995). *Neural networks for pattern recognition*. Oxford, UK: Oxford University.

Bishop, C. M. (2003). *Neural networks for pattern recognition*. Oxford University Press.

Bitzenis, A. & Nito, E. (2005). Obstacles to entrepreneurship in a transition business environment: The case of Albania. *Journal of Small Business and Enterprise Development, 12*(4), 564-578.

Blackman, R. B. and Turkey, J. W. (1958). *The measurements of power spectra*. New York: Dover Publications, Inc.

Blaschke, A. (2001). Environmental monitoring and management of protected areas through integrated ecological information systems- An EU perspective. In C. Rautenstrauch & S. Patig (Ed.), *Environmental information systems in industry and public administration* (pp. 75-100). Hershey, PA: Idea Group Publishing.

Blazewicz, J., & Kasprzak, M. (2003). Determining genome sequences from experimental data using evolutionary computation. In G. G. Fogel & D. W. Corne (Eds.), *Evolutionary computation in bioinformatics* (pp. 41-58). San Francisco: Morgan Kaufmann.

Bodie, Z., Kane, A., Marcus, A. J., & Ryan, P. J. (2003). *Investments (Fourth Canadian Edition)*. Toronto, ON (Canada): McGraw-Hill Ryerson Limited.

Boritz, E. J., & Kennery, D. (1995). Effectiveness of neural network types for predicition of business failure. *Expert Systems with Applications, 9*,503-512.

Botschen, G., Thelen, E. M. & Pieters, R. (1999).Using means-end structures for benefit segmentation: An application to services. *European Journal of Marketing, 33* (1/2).

Boulicaut, Jean-Francois, Esposito, F., Giannotti, F. & Pedreschi, D. (Eds.) (2004). Knowledge discovery in databases. In *Proceedings of the PKDD 2004: 8th European Conference on Principles and Practice of Knowledge Discovery in Databases*, Pisa, Italy.

Bozdogan, H. (Ed.) (2004). *Statistical data mining and knowledge discovery*. CRC Press.

Bracha, G. & Rachman, O. (1992). Randomized consensus in expected O(n2log n) operations. In S. Toueg, P. G. Spirakis & L. M. Kirousis (Eds.), *Lecture notes in computer science* (Vol. 579, pp. 143–150). Delphi, Greece: Springer. Retrieved April 12, 2008, from http://www.cs.yale.edu/homes/aspnes/randomized-consensus-survey.pdf

Brachman, R. J. & Anand, T. (1996). The process of knowledge discovery in databases. In U. Fayyad, G. Piatetsky-Shapiro, P. Smyth & R. Uthurusamy (Ed.), *Advances in knowledge discovery and data mining* (pp. 37-57). Cambridge, MA: AAAI/MIT Press.

Bradbury, D. (2005, August 31). Technology Jargon Buster. *Caterer & Hotelkeeper,*

Bradley, P.S., Fayyad, U.M., & Mangasarian, O.L. (1999). Mathematical programming for data mining: Formulations and challenges. *INFORMS Journal on Computing, 11*, 217-238.

Braha, D. (Ed.) (2001). *Data mining for design and manufacturing: Methods and applications*. Kluwer Publishers.

Bramer, M. (2007). *Principles of data mining: Undergraduate topics in computer science*. London, UK: Springer-Verlag.

Bramer, M. A. (Ed.) (1999). *Knowledge discovery and data mining: Theory and practice*. IEE Books.

Briassoulis, H. (2004). *Analysis of land use change: Theoretical and modeling approaches*. Retrieved April 8, 2008, from http://www.rri.wvu.edu/WebBook/Briassoulis

Brin, S. & Page, L. (1998). The anatomy of a large-scale hypertextual Web search engine. In *Proceedings of the 7th International World Wide Web Conference, Elsevier Science* (pp. 107-117), New York.

Brin, S., Motwani, R., Ullman, J. D., & Tsur, S. (1997). Dynamic itemset counting and implication rules for market basket data. In J. Peckham (Ed.), *Proceedings of the 1997 ACM SIGMOD International Conference on Management of Data* (pp. 255-264). New York: ACM Press.

Brown, J. F., Tadesse, T., & Reed, B. C. (2002). Integrating satellite data and climate data for US drought mapping and monitoring. In *Proceedings of the 15th Conference on Biometeorology and Aerobiology joint with 16th International Congress on Biometeorology*, (pp. 147-150). Kansas City, Missouri.

Buchner, A. G., Mulvenna, M. D., Anand, S. S. & Hughes, J. G. (1999). Navigation pattern discovery from Internet data. In *Proceedings of the Web Usage Analysis and User Profiling Workshop* (pp. 25-30), San Diego, CA.

Bukvic, V. & Bartlett, W. (2003). Financial barriers to SME growth in Slovenia. *Economic and Business Review, 5*(3), 161-181.

Burdick, D., Calimlim, M., & Gehrke, J. (2001). MAFIA: A maximal frequent itemset algorithm for transactional databases. In *Proceedings of the 17th International Conference on Data Engineering* (pp. 443-452). Los Alamitos, CA: IEEE Computer Society Publications.

Burn, J. M. & Loch, K. D. (2001). The societal impact of the World Wide Web—Key challenges for the 21st century. *Information Resources Management Journal, 14*(4), 4-14.

Business intelligence: Aspectos e tendências do uso de ferramentas de análise corporativa. Retrieved March 12, 2003, from www.idcbrasil.com.br.

Buytendijk, F. (2001). *Strategic BI: Its definition and effect on infrastructure*. Gartner Group.

C 5.0. (2004). Retrieved from http://www.rulequest.com/see5-info.html

Cabena, P. H., Stadler, R., Verhees, J., & Zanasi, A. (1998). *Discovering data mining: From concept to implementation*. New Jersey: IBM.

Cabena, P., Hadjinian, P., Stadler, R., Verhees, J., & Zanasi, A. (1997). *Discovering data mining: from concept to implementation*. Upper Saddle River, NJ: Prentice Hall PTR.

Cai, C. H., Fu, A. W. C., Cheng, C. H., & Kwong, W. W. (1998). Mining association rules with weighted items. In B. Eaglestone, B. C. Desai & J. Shao (Eds.), *Proceedings of the 1998 International Database Engineering and Application Symposium* (pp. 68-77). Los Alamitos, CA: IEEE Computer Society Publications.

Cai, D., Shao, Z., He, X., Yan, X., & Han, J. (2005). Minning hidden community in heterogeneous social networks. In *Proceedings of the 3rd International Workshop on Link Discovery* (pp. 58-65). ACM Press.

Câmara, G., Souza, R., Freitas, U. & Garrido, J. (1996). SPRING: Integrating Remote Sensing and GIS with object-oriented data modelling. *Computers and Graphics, 15*(6), 13-22.

Câmara, G., Vinhas, L., Souza, L., Paiva, L., Monteiro, A., Carvalho, M. & Raoult, B. (2001). Design patterns in GIS development: The Terralib experience. In *Proceedings of the III Brazilian Symposium in Geoinformatics, GeoInfo 2001*, Rio de Janeiro.

Canada Centre for Remote Sensing (2003). *Fundamentals of remote sensing.* Remote Sensing Tutorial (pp. 5-44). Retrieved April 8, 2008, from www.ccrs.nrcan.gc.ca/ccrs/learn/tutorials/fundam/fundam_e.html

Canbas, S., Onal, B. Y., Duzakin, H. G., & Kilic, S. B. (2006). Prediction of financial distress by multivariate statistical analysis: The case of firms taken into the surveillance market in the Istanbul Stock Exchange. *International Journal of Theoretical & Applied Finance, 9*(1), 133.

Carty, A. J. (2002). Scientific and technical data: Extending the frontiers of research. In *Proceedings of the Opening Address at CODATA 2002: Frontiers of Scientific and Technical Data,* Montréal, Canada.

Caterer & Hotelkeeper (2000, September 7). Hotel groups deny they're missing Web opportunities, 14.

Caterer & Hotelkeeper (2004, 24 June). Do the knowledge, 34.

Cerrito, P. (2006). *Introduction to data mining using SAS enterprise miner.* SAS Press.

Chainey, S. & Ratcliffe, J. (2005). *GIS and crime mapping.* Chichester, West Sussex: John Wiley and Sons.

Chakrabarti, S. (2003). *Mining the Web: Discovering knowledge from hypertext data.* San Francisco: Morgan Kaufmann Publishers.

Chakrabarti, S., (2000). Data mining for hypertext: A tutorial survey. *ACM SIGDDD Explorations, 1*(2), 1-11.

Chan, K. A., Menkveld, A. J., & Yang, Z. (2003). *Evidence on the foreign share discount puzzle in China: Liquidity or information asymmetry?* (Working Paper). Hong Kong, China: University of Science and Technology (HKUST).

Chan, N. H. & Wong, H. Y. (2007). Data mining of resilience indicators. *IIE Transactions, 39*, 617–627.

Chang, S., Chang, H., Lin, C., & Kao, S. (2003). The effect of organizational attributes on the adoption of data mining techniques in the financial service industry: An empirical study in Taiwan. *International Journal of Management, 20*, 497-503.

Charnes, A., & Cooper, W.W. (1961). *Management models and industrial applications of linear programming* (vols. 1 & 2). New York: John Wiley & Sons.

Chen, H. & Chau, M. (2004). Web mining: Machine learning for Web applications. *Annual Review of Information Science and Technology (ARIST), 38*, 289-329.

Chen, H., Chung, W., Xu, J. J., Wang, G., Qin, Y., & Chau, M. (2004). Crime data mining: A general framework and some examples. *Computer, 37*(4), 50-56.

Chen, H., Qin, J., Reid, E., Chung, W., Zhou, Y., Xi, W., Lai, G., Bonillas, A., & Sageman, M., (2004). The dark Web portal: Collecting and analyzing the presence of domestic and international terrorist groups on the Web. In *Proceedings of the 7th International Conference on Intelligent Transportation Systems (ITSC),* Washington D.C.

Chen, I. J. & Popovich, K. (2003). Understanding customer relationship management (CRM); People, process and technology. *Business Process Management Journal, 9*(5), 672-688.

Chen, Y., Wang, J. Z. & Krovetz, R. (2003). CLUE: Cluster-based retrieval of images by unsupervised learning. In K. A. Meraim, I. Bloch (Eds.), In *Proceedings of the IEEE Seventh International Symposium on Signal Processing and its Applications* (pp. 202-231).

Chenhall, R. H. (2005). Integrative strategic performance measurement systems, strategic alignment of manufacturing, learning and strategic outcomes: An exploratory study. *Accounting, Organizations and Society, 30*(5), 395-423.

Chetty, M. & Buyya, R. (2002). Weaving computational crids: How analogous are they with electrical grids? *IEEE Computing in Science and Engineering,* July/August, 61-71.

Chin-Sheng, H., Dorsey, R. E., & Boose, M.A. (1994). Life insurer financial distress prediction: A neural network model. *Journal of Insurance Regulation, 13*(2), 131-168.

CIDA (2005). *CIDA's strategy on knowledge for development through information and communication technologies (ICT).* Canadian International Development Agency. Retrieved April 13, 2008, from http://www.acdi-cida.gc.ca/ict

Cios, K. J. (Ed.) (2000). *Medical data mining and knowledge discovery.* Physica-Verlag (Springer).

Cios, K., Pedrycz, W., & Swiniarski, R. (1998). *Data mining methods for knowledge discovery.*

Cirasa, A., Pilato, G., Sorbello, F., & Vassallo, G. (2000). EαNet: A neural solution for Web pages classification. In *Proceedings of the 4th World Multiconference on Systemics, Cybernetics, and Informatics SCI2000*, Orlando, Florida.

Clemons, E. & Row, M. (2000, November 13).Behaviour is key to web retailing strategy. *Financial Times.*

Coats, P. K., & Frant, F. L. (1992). A neural network approach to forecasting financial distress. *The Journal of Business Forecasting Methods & Systems, 10*, 9-12.

Coats, P. K., & Frant, F. L. (1993). Recognizing financial distress patterns using a neural network tool. *Financial Management, 22*(3), 142-155.

Codata (2002). Committee on data for science and technology (CODATA). In *Proceedings of the Workshop Synthesis on Archiving Scientific and Technical Data*, Pretoria, South Africa. Retrieved April 13, 2008, from http://www.tgdc-codata.org.cn/english/Html/SA-CT.html

Coenen, F. & Leng, P. (2001). Optimising association rule algorithms using itemset ordering. In M. Bramer, F. Coenen & A. Preece (Eds.), *Research and Development in Intelligent Systems XVIII – Proceedings of the Twenty-first SGES International Conference on Knowledge Based Systems and Applied Artificial Intelligence* (pp. 53-66). London, UK: Springer-Verlag.

Coenen, F. & Leng, P. (2002). Finding association rules with some very frequent attributes. In T. Elmaa, H. Mannila & H. Toivonen (Eds.), *Principles of Data Mining and Knowledge Discovery – Proceedings of the 6th European Conference on Principles and Practice of Knowledge Discovery in Databases* (pp. 99-111). Berlin Heidelberg, Germany: Springer-Verlag.

Coenen, F. & Leng, P. (2004). An evaluation of approaches to classification rule selection. In *Proceedings of the 4th IEEE International Conference on Data Mining* (pp. 359-362). Los Alamitos, CA: IEEE Computer Society Publications.

Coenen, F., Goulbourne, G., & Leng, P. (2001). Computing association rules using partial totals. In L. D. Raedt & A. Siebes (Eds.), *Principles of Data Mining and Knowledge Discovery – Proceedings of the 5th European Conference on Principles and Practice of Knowledge Discovery in Databases* (pp. 54-66). Berlin Heidelberg, Germany: Springer-Verlag.

Coenen, F., Leng, P., & Ahmed, S. (2004). Data structure for association rule mining: T-tree and p-tree. *IEEE Transactions on Knowledge and Data Engineering, 16*(6), 774-778.

Coenen, F., Leng, P., & Goulbourne, G. (2004). Tree structures for mining association rules. *Journal of Data Mining and Knowledge Discovery, 8*(1), 25-51.

Cohen, A. & Nachmias, R. (2006). A quantitative cost effectiveness model for Web-supported academic instruction. *The Internet and Higher Education, 9*(2), 81-90.

Cohn D. & Chang, H. (2000). Learning to probabilistically identify authoritative documents. In *Proceedings of the17th International Conference on Machine Learning (ICML2000)* (pp. 167-174), Stanford, California.

COL (2003). *Find information faster: COL's "Info-mining" tools.* Retrieved April 13, 2008, from http://www.col.org/colweb/site/pid/2927

Coleman, D. J. & McLaughlin, J. D. (1997). Information access and network usage in the emerging spatial information marketplace. *Journal of Urban and Regional Information Systems Association, 9*, 8-19.

Collard, J. M. (2002). Is your company at risk? *Strategic Finance, 84*(1), 37-39.

Connelly, R., McNeil, R., & Mosimann, R. (1998). *The multidimensional manager - 24 ways to impact your bottom line in 90 days*. Ottawa, ON: Cognos Incorporated.

Cooley, R., Mobasher, B., & Srivastava, J. (1997). Web mining: Information and pattern discovery on the World Wide Web. In *Procceding of the 9th International Conference on Tools with Artificial Intelligence(ICTAI '97)* (pp. 558-567), New Port Beach, CA: IEEE Computer Society.

CORDIS (2006). *GRID technologies and applications through CORDIS*. Community Research & Development Information Service. Retrieved April 13, 2008, from http://www.environmonument.com/projects.htm

Couldwell, C. (1998, May 21). A data day battle. *Computing*, 64-66.

Cox, E. (2004). *Fuzzy modeling and genetic algorithms for data mining and exploration*. Morgan Kaufmann.

Curotto, C. L. & Ebecken, N. F. F. (2005). *Implementing data mining algorithms in Microsoft® SQL Server™*. WIT Press.

Cuthbertson, K. & Nitzsche, D. (2001). *Investments: Spot and derivatives markets*. Chichester, West Sussex, UK: John Wiley & Sons, Ltd.

Dai, H. & Mobasher, B. (2003). A road map to more effective Web personalization; Integrating domain knowledge with Web usage mining. In *Proceedings of the International Conference on Internet Computing* (IC 2003), Las Vegas, Nevada.

Damodaran, A. (2001). *Corporate finance theory and practice* (2nd ed.). New York: John Wiley & Sons, Inc.

Davidse, R. J. & Van Raan, A. F. J. (1997). Out of particles: Impact of CERN, DESY, and SLAC research to fields other than physics. *Scientometrics, 40*(2), 171-193.

Davies, A. (2000, 29 June). Data's the way to do it, *Caterer & Hotelkeeper*, 31-32.

Davies, A. (2001, 26 July). On-line, on course. *Caterer & Hotelkeeper*, 37-39.

de Ville, Barry (2006). *Decision trees for business intelligence and data mining: Using SAS enterprise miner*. SAS Press.

de Ville, Barry. (2001). Microsoft data mining, Integrated business intelligence for e-commerce and knowledge management.

De'ath, G. & Fabricus, K. E. (2000). Classification and regression trees – A powerful yet simple technique for ecological data analysis. *Ecology, 8*(11), 3178-3192.

Deakin, E. B. (1972). A discriminant analysis of predictors of business failure. *Journal of Accounting Research, 10*(1), 167-179.

Delmater, R. & Hancock, M. (2001). *Data mining explained: A manager's guide to customer-centric business intelligence*. Digital Press.

Demertzis, N., Diamantaki, K., Gazi, A., & Sartzetakis, N. (2005). Greek political marketing on-line: An analysis of parliament members' Web sites. *Journal of Political Marketing, 4*(1), 51-74.

Derby, B. L. (2003). Data mining for improper payments. *The Journal of Government Financial Management, 52*, 10-13.

Desikan, P., Srivastava, J., Kumar, V., & Tan, P. N. (2002). *Hyperlink analysis: Techniques and applications* (Tech. Rep. TR 2002-0152). Army High Performance Computing Center.

Dhar, V. & Stein, R. (1996). *Seven methods for transforming corporate data into business intelligence*. Upper Saddle River, NJ: Prentice Hall.

Dietterich, T. (2000). *Ensemble methods in machine learning*. In Kittler & Roli (Eds.), Multiple classifier systems (pp. 1-15). Berlin: Springer-Verlag (Lecture Notes in Pattern Recognition 1857).

Dietterich, T. G., Lathrop, R. H., & Lozano-Perez, T. (1997). Solving the multiple-instance problem with axis-parallel rectangles. *Artificial Intelligence, 89*(1-2), 31-71.

Diplaris, S., Tsoumakas, G., Mitkas, P. A., & Vlahavas, I. (2005). Protein classification with multiple algorithms. In *Proceedings of 10th Panhellenic Conference in Informatics*. Volos, Greece: Springer-Verlag.

Do, T. D., Chang, K., & Hui, S. C. (2004). Web mining for cyber monitoring and filtering. In *Proceedings of the 2004 IEEE Conference on Cybernetics and Intelligent Systems Vol. 1* (pp. 399-404). Singapore.

Dong, G. & Li, J. (1999). Efficient mining of emerging patterns: Discovering trends and differences. In *Proceedings of the Fifth ACM SIGKDD International Conference on Knowledge Discovery and Data Mining* (pp. 43-52). New York: ACM Press.

Dorian, P. (1999). *Data preparation for data mining.* Morgan Kaufmann.

Drewry et al. (2002). *Current state of data mining.* Department of Computer Science, University of Virginia.

Dubé, L. & Paré, G. (2003). Rigor in information systems positivist case research: Current practices, trends and recommendations. *MIS Quarterly, 27*(4), 597-635.

Duda, R. O., Hart, P. E., & Stork, D. G. (2001). *Pattern classification.* New York: John Wiley.

Dunham, M. (2003). *Data mining introductory and advanced topics.* Prentice Hall.

Dunn, C. E., Atkins, P. J., & Townsend, J. G. (1997). GIS for development: A contradiction in terms? *Area, 29*(2), 151-159.

Dyer, N. A. (1998). What's in a relationship (other than relations)? *Insurance Brokers Monthly & Insurance Adviser, 48*(7), 16-17.

Earth Institute News (2005). Scientific community must develop cross-disciplinary standards and practices in academia. Retrieved April 13, 2008, from http://www.earthinstitute.columbia.edu/news/2005/story05-01-05c.html

Easterby-Smith, M., Araujo, L., & Burgoyne, J. (1999). *Organizational learning and the learning organization: Developments in theory and practice.* London, UK: Sage Publications.

Ebecken, N. F. F., Brebbia, C. A., & Weigend, A. (2000). *Data mining II* (1st ed.). Computational Mechanics, Inc.

Edlington, S. (2003, January 20). Future perfect? *Caterer & Hotelkeeper*, 26.

Eirinaki, M. & Vazirgiannis, M. (2003). Web mining for Web personalization. *ACM Transactions on Internet Technology, 3*(1), 1-27.

Eisenhardt, K. M. & Sull, D. N. (2001). Strategy as simple rules. *Harvard Business Review, 79*(1), 106-117.

Eklund, T., Back, B., Vanharanta, H., & Visa, A. (2003). Using the self- organizing map as a visualization tool in financial benchmarking. *Information Visualization, 2*, 171-181.

El-Hajj, M. & Zaiane, O. R. (2003). Inverted matrix: Efficient discovery of frequent items in large datasets in the context of interactive mining. In L. Getoor, T. E. Senator, P. Domingos & C. Faloutsos (Eds.), *Proceedings of the Ninth ACM SIGKDD International Conference on Knowledge Discovery and Data Mining* (pp. 109-118). New York: ACM Press.

Enke, D. & Thawornwong, S. (2005). The use of data mining and neural networks for forecasting stock market returns. *Expert Systems with Applications, 29*(2005), 927-940.

Escada, M. I. S., Monteiro, A. M., Aguiar, A. P., Carneiro, T. & Câmara, G. (2005). Análise de padrões e processos de ocupação para a construção de modelos na Amazônia [Analysis of land use patterns and processes for the construction of models in Amazonia]. In *Proceedings of the XII Brazilian Symposium on Remote Sensing* (pp. 2973-2983), Goiania, Brazil.

Escada, M. I. S., Vieira, I. C. G., Amaral, S., Araújo, R., Veiga, J. B. D., Aguiar, A. P. D., Veiga, I., Oliveira, M., Pereira, J. L. G., Filho, A. C., Fearnside, P. M., Venturieri, A., Carriello, F., Thales, M., Carneiro, T. S., Monteiro, A. M. V., & Câmara, G. (2005). Padrões e processos de ocupação nas novas fronteiras da Amazônia: O interflúvio do Xingu/Iriri [Land use patterns and processes in Amazonian new frontiers: The Xingu/Iriri region]. *Estudos Avançados* [Advanced Studies], *19*, 9-23.

Esposito, F., Malerba, D., Di Pace, L., & Leo, P. (1999). A learning intermediary for automated classification of Web pages. In *Proceedings of the 16th International Workshop on Machine Learning in Text Data Analysis (ICML1999)* (pp. 37-46).

ESRI (2004). *ArcGIS 9: What is ArcGIS?* A White Paper. Redlands, CA: Environmental Systems Research Institute.

ESRI (2004). *ArcSDE: Advanced spatial data server.* White Paper. Retrieved May 8, 2008 from http://esri.com/library/whitepapers/pdfs/arcgis_spatial_analyst.pdf

European Commission (2003). *2003 observatory of European SMEs: SMEs in Europe* (Tech. Pep. No.7). European Commission.

European Commission (2005). *Specific programme for research technological development and demonstration: Integrating and strengthening the European research area, 2005 Work Programme (SP1-10).*

Faca, F. M. & Lanzi, P. L. (2005). Mining interesting knowledge from Weblogs: A survey. *Data Knowledge Engineering, 53*(3), 225-241.

Fang, X., Sheng, O. R. L. (2005). Designing a better Web portal for digital government: A Web-mining based approach. In *Proceedings of the 2005 National Conference on Digital Government Research* (pp. 277-278), Atlanta, Greorgia.

Farrell, M. (2006). Create a diversified portfolio. *©2006 Path to Investing □ Leading the way to financial knowledge®.* New York: Lightbulb Press, Inc.

Fayyad, U., G. Piatetsky-Shapiro, & P. Smyth. (1996). From data mining to knowledge discovery in databases (a survey). *AI Magazine, 17*(3), 37-54.

Fayyad, U., Grinstein, G. & Wierse, A. (2001). *Information visualization in data mining and knowledge discovery.* Morgan Kaufmann.

Fayyad, U., Piatetsky-Shapiro, G., & Smyth, P. (1996). *From data mining to knowledge discovery in databases. AI Magazine, 17*, 37-54.

Fayyad, U., Piatetsky-Shapiro, G., Smyth, P., & Uthurusamy, R. (Eds) (1996). *Advances in knowledge discovery and data mining.* AAAI/MIT Press.

Fayyad, U.M., Piatetsky-Shapiro, G., & Uthurusamy, R. (2003). Summary from the KDD-03 Panel: Data mining: The next 10 years. *ACM SIGKDD Explorations Newsletter, 5*(2), 191-196.

Feeney, M. F. (2003). SDIs and decision support. In I. Williamson, A. Rajabifard, & M. F. Feeney (Ed.), *Developing spatial data infrastructures: From concept to reality* (pp. 195-210). London, UK: Taylor & Francis.

*Ferramentas de business intelligence no Brasil (*2003). Retrieved March 12, 2003, from www.idcbrasil.com.br.

Ficenec, C. (2003, June). Explorations of participatory GIS in three Andean watersheds. *Paper presented at the University Consortium of Geographic Information Science (UCGIS) Summer Assembly 2003*, Pacific Grove, CA.

Fisher, D. (1987). Improving inference through conceptual clustering. In *Proceedings of the 1987 AAAI Conference* (pp. 461-465). Seattle, Washington.

Fisher, R.A. (1936). The use of multiple measurements in taxonomic problems. *Annals of Eugenics, 7*, 179-188.

Foot, K., Schneider, S., Dougherty, M., Xenos, M., & Larsen, E. (2003). Analyzing linking practices: Candidate sites in the 2002 U.S. electoral Web sphere. *Journal of Mediated Communication, 8*(4).

Fraser, J., Fraser, N., & McDonald, F. (2000). The strategic challenge of electronic commerce. *Supply Chain Management: An International Journal, 5*(1), 7-14

Freed, N., & Glover, F. (1981). Simple but powerful goal programming models for discriminant problems. *European Journal of Operational Research, 7*, 44-60.

Freed, N., & Glover, F. (1986). Evaluating alternative linear programming models to solve the two-group discriminant problem. *Decision Science, 17*, 151-162.

Freitas, A. A. (2002). *Data mining and knowledge discovery with evolutionary algorithms.* Springer-Verlag.

Frigo, M. L. (2002). Strategy-focused performance measures. *Strategic Finance, 84*(3), 10-13.

Fukuda, H., Passos, E., Pacheco, A. M., Neto, L. B., Valerio, J., Roberto, V. J. D., Antonio, E. R., & Chigener, L. (2000) Web text mining using a hybrid system. In *Proceedings of the 6th Brazilian Symposium on Neural Networks* (pp.131–136).

Galbreath, J. & Rogers, T. (1999). Customer relationship leadership: A leadership and motivation model for the twenty-first century business. *The TQM Magazine, 11*(3), 161-171.

Garfield, E. (1985). History of citation indexes for chemistry - A brief review. *JCICS, 25*(3), 170-174.

Garofalakis, J., Kappos, P., & Mourloukos, D. (1999). Website optimization using page popularity. *IEEE Internet Computing, 3*(4), 22-29.

Gatrell, A. & Rowlingson, B. (1994). Spatial point process modeling in a GIS environment. In A.S. Fotheringham & P.A. Rogerson (Ed.), *Spatial analysis and GIS* (pp. 148-163). London, UK: Taylor and Francis.

Gaytan, A. & Johnson, A. J. (2002). *A review of the literature on early warning systems for banking crises* (Working papers No: 183). Central Bank of Chile.

Getoor, L. & Diehl, C. P. (2005). Link mining: A survey. *ACM SIGKDD Explorations Newsletter, 7*(2), 3-12.

Getoor, L., Segal, E., Tasker, B., & Koller, D. (2001). Probabilistic models of text and link structure for hypertext classification. In *Proceedings of the IJCAI Workshop on Text Learning: Beyond Supervision*, Seattle, Washington.

Gibson, R. K. & Ward, S. J. (2000). A proposed methodology for studying the functions and effectiveness of party and candidate Web-sites. *Social Science Computer Review, 18*(3), 301-319.

Gilad, B. & Gilad, T. (1988). *The business intelligence system: A new tool for competitive advantage*. New York: Amacom.

Giles, C. L., Bollacker, K., & Lawrence, S. (1998). CiteSeer: An automatic citation indexing system. In *Proceedings of the 3rd ACM Conference on Digital Libraries*, 89-98.

Gillani, B. (1998). The Web as a delivery mechanism to enhance instruction. *Educational Media International, 35*(3), 197-202.

Giovinazzo, W. A., (2002). *Internet-enabled business intelligence*. Upper Saddle River, NJ: Prentice Hall.

Giuffrida, G., Cooper, L. G., & Chu, W. W. (1998). A scalable bottom-up data mining algorithm for relational databases. In *Proceedings of the Tenth International Conference on Scientific and Statistical Database Management* (pp. 206-209)

Gledhill, B. (2002, February 28). Learning from history. *Caterer & Hotelkeeper*, 33.

Gleim, R., Mehler, A., & Dehmer, M. (2006). Web corpus mining by instance of Wikipedia. In *Proceedings of the EACL 2006 Workshop on Web as Corpus*, Trento, Italy.

Goddard, S., Harms, S. K., Reichenbach, S. E., Tadesse, T., & Waltman, W. J. (2003). Geospatial decision support for drought risk management. *Communication of the ACM, 46*(1), 35-37.

Gökmen, A. et al. (2004). *Balaban Valley Project: Improving the quality of life in rural area in Turkey, 7*(Dec 2004). Retrieved April 13, 2008, http://www.geocities.com/doriendetombe/detombevol7menmbalabanabstract.html

Goldenberg, A. & Moore, A. W. (2005). Bayes net graphs to understand co-authorship networks? In *Proceedings of the 3rd International Workshop on Link Discovery* (pp. 1-8). ACM Press.

Goldman, J. A., Chu, W. W., Parker, D. S., & Goldman, R. M. (1999). Term domain distribution analysis: A data mining tool for text databases. *Methods of Information in Medicine, 38*, 96-101.

Goodchild, M. F., Haining, R., & Wise, S. M. (1991). Integrating GIS and spatial data analysis: Problems and possibilities. *International Journal of Geographic Information Systems, 6*, 407-423.

Gordon, M. D. & Dumais, S. (1998). Using latent semantic indexing for literature based discovery. *Journal of the American Society for Information Science, 49*(8), 674-685.

Gouda, K. & Zaki, M. J. (2001). Efficiently mining maximal frequent itemsets. In N. Cercone, T. Y. Lin & X. Wu (Eds.), *Proceedings of the 2001 IEEE International Conference on Data Mining* (pp. 163-170). Los Alamitos, CA: IEEE Computer Society Publications.

Goymour, A. (2001, 26 July). Host in the machine. *Caterer & Hotelkeeper*, 43-45.

Greenburg, E. F. (2004). Who turns on the RFID faucet, and does it matter? *Packaging Digest*, 22. Retrieved January 24, 2005, from http://www.packagingdigest.com/articles/200408/22.php

Greengrass, E. (1997). Information retrieval: An overview. *National Security Agency*. TR-R52-02-96.

Grönroos, C. (1994). From scientific management to service management: A management perspective for the age of service competition. *International Journal of Service Management, 5*(1), 5-20.

Groot, R. & McLaughlin, J. (2000). Introduction. In R. Groot & J. McLaughlin (Eds.), *Geospatial data infrastructure: Concepts, cases and good practice* (pp. 1-12). Oxford, UK: Oxford University Press.

Grossman, R. L., Kamath, C., Kegelmeyer, P., Kumar, V., & Namburu, R. (Eds.) (2006). *Data mining for scientific and engineering applications (Massive computing)* (1st ed.). Springer.

Groth, R. (1998). *Data mining: A hands-on approach for business professionals*. New Jersey: Prentice Hall.

Gulati, R. & Garino, J. (2000, May-June). Get the right mix of bricks and mortar. *Harvard Business Review*, 107-114.

Gunther, J. W. & Moore, R. R. (2003). Early warning models in real time. *Journal of Banking, 27*(10), 1979-2001.

Hackathorn, R. D. (1998). *Web farming for the data warehouse: Exploiting business intelligence and knowledge management*. San Francisco: Morgan Kaufmann Publishers.

Haining, R. (1994). Designing spatial data analysis modules for GIS. In A.S. Fotheringham & P.A. Rogerson (Eds.), *Spatial analysis and GIS* (pp. 46-63). London, UK: Taylor and Francis.

Hale, J., Threet, J., & Shenoi, S. (1994). A practical formalism for imprecise inference control. *Ifip Trans. A-Computer Science And Technology, 60*, 139-156.

Halpern, J. Y. (2003). *Reasoning about uncertainty*. MIT Press.

Hamer, M. (1983). Failure prediction: Sensitivity of classification accuracy to alternative statistical method and variable sets. *Journal of Accounting and Public Policy, 2*, 289-307.

Han, E. H., Karypis, G., & Kumar, V. (1997). Scallable parallel data mining for association rules. In *Proceedings of the ACM SIGMOD Conference Management of Data.*

Han, J. & Kamber, M. (2001). *Data mining concepts and techniques*. San Francisco: Morgan Kaufmann Publishers.

Han, J. & Kamber, M. (2006). *Data mining concepts and techniques* (2nd ed.). San Francisco: Morgan Kaufmann Publishers.

Han, J. (1999). Data mining. In J. Urban & P. Dasgupta (Eds.), *Encyclopedia of distributed computing*. Kluwer Academic Publishers.

Han, J., Cai, Y., & Cercone, N., (1993). Data-driven discovery of quantitative rules in relational databases. *IEEE Trans. Knowledge and Data Engineering, 5*, 29-40.

Han, J., Kamber, M. & Chiang, J. (1997). Metarule-guided mining of multi-dimensional association rules using data cubes. In *Proceedings of international conference on knowledge discovering and data mining (KDD'97)*, pp. 207-210.

Han, J., Koperski, K. & Stefanovic, N. (1997). GeoMiner: A system prototype for spatial data mining. In *Proceedings of the ACM SIGMOD International Conference on Management of Data* (pp. 553-556).

Han, J., Pei, J., & Yin, Y. (2000). Mining frequent patterns without candidate generation. In W. Chen, J. F. Naughton & P. A. Bernstein (Eds.), *Proceedings of the 2000 ACM SIGMOD International Conference on Management of Data* (pp. 1-12). New York: ACM Press.

Han, J.W., & Kamber, M. (2000). *Data mining: Concepts and techniques*. San Diego: Academic Press.

Hand, D. J., Mannila, H., & Smyth, P. (2000). *Principles of data mining*. MIT Press.

Hand, D., Mannila, H., & Smyth, P. (2001). *Principles of data mining*. Cambridge: MIT Press.

Hannula, M. & Pirttimaki, V. (2003). Business intelligence empirical study on the top 50 Finnish companies. *Journal of American Academy of Business, 2*(2), 593-599.

Hardfield, R. (2004). The RFID power play. *Supply Chain Resource Consortium*. Retrieved October 23, 2004, from http://scrc.ncsu.edu/public/APICS/APICSjan04.html

Harms, S. K., Deogun, J., & Tadesse, T. (2002). Discovering sequential association rules with constraints and time lags in multiple sequences. *Lecture notes in artificial intelligence 2366: Foundations of intelligent systems*. In *Proceedings of the 13th International Symposium on Methodologies for Intelligent Systems* (pp. 432-441). Lyon, France.

Harvey, C. D. (1988). Telephone survey techniques. *Canadian Home Economics Journal, 38*(1), 30-35

Hastie, T.J., & Tibshirani, R.J. (1990). *Generalized additive models*. New York: Chapman and Hall.

Hayes, C., Avesani, P., & Veeramachaneni, S. (2006). An analysis of bloggers and topics for a blog recommender system. In *Proceedings of the Workshop on Web Mining, 7th European Conference on Machine Learning and the 10th European Conference on Principles and Practice of Knowledge Discovery in Databases (ECML/PKDD)*, Berlin, Germany.

Haykin, S. (1994). *Neural networks, a comprehensive foundation*. New York: Macmillan.

He, J., Liu, X., Shi, Y., Xu, W., & Yan, N. (2004). Classifications of credit cardholder behavior by using fuzzy linear programming. *International Journal of Information Technology and Decision Making, 3*, 633-650.

Hearst, M. A. (1999). Untangling text data mining. In *Proceedings of ACL 99, the 37th Annual Meeting of the Association for Computational Linguistics*, University of Maryland.

Hecht-Nielsen, R. (1990). *Neurocomputing*. Reading, MA: Addison Wesley.

Heeks, R. (2002). Information systems and developing countries: Failure, success and local improvisation. *Information Society, 18*(2), 101-112.

Hernández, V., Göhring, W., & Hopmann, C. (2004). Sustainable decision support for environmental problems in developing countries: Applying multi-criteria spatial analysis on the Nicaragua Development Gateway niDG. *Research on computing science* (Vol. 11, pp.136-150). México: Instituto Politécnico Nacional.

Herrera-Viedma, E. & Pasi, G. (2006). Soft approaches to information retrieval and information access on the Web: An introduction to the special topic section. *Journal of the American Society for Information Science and Technology, 57*(4), 511-514.

Hertz, J., Krogh, A., & Palmer, R. G. (1991). *Introduction to the theory of neural computation*. Redwood City, CA: Addison Wesley

Hess, A. & Kushmerick, N. (2004). Machine learning for annotating semantic Web services. In *Proceedings of the AAAI Spring Symposium on Semantic Web Services*, Palo Alto, California.

Hesselgesser, J., Taub, D., Baskar, P., Greenberg, M., Hoxie, J., Kolson, D.L., & Horuk, R. (1998). Neuronal apoptosis induced by HIV-1 gp120 and the Chemokine SDF-1alpha mediated by the Chemokine receptor CXCR4. *Curr Biol, 8*, 595-598.

Hidber, C. (1999). Online association rule mining. In A. Delis, C. Faloutsos & S. Ghandeharizadeh (Eds.), *Proceedings of the 1999 ACM SIGMOD International*

Conference on Management of Data (pp. 145-156). New York: ACM Press.

Ho, K. & Robinson, C. (2001). *Personal financial planning* (3rd ed.). North York, ON (Canada): Captus Press Inc.

Holsheimer, M., Kersten, M. L., Mannila, H., & Toivonen, H. (1995). A perspective on databases and data mining. In U. M. Fayyad & R. Uthurusamy (Eds.), *Proceedings of the First International Conference on Knowledge Discovery and Data Mining* (pp. 150-155). Menlo Park, CA: AAAI Press.

Hong, G. H. & Lee, J. H. (2005). Designing an intelligent Web information system of government based on Web mining. *Lecture notes in computer science* (Vol. 3614, pp. 1071-1078).

Hopkins, L. D. (1984). Evaluation of methods for exploring Ill-defined problems. *Environmental Planning B: Planning and Design, 11*, 339-348.

Hoppszallern, S. (2003). Healthcare benchmarking. *Hospitals & Health Networks, 77*, 37-44.

Hoss, D. (2000). The e-business explosion: Strategic data solutions for e-business success. *DM Review, 10*(8), 24-28.

Houtsma, M. & Swami, A. (1995). Set-oriented mining of association rules in relational databases. In P. S. Yu & A. L. Chen (Eds.), *Proceedings of the Eleventh International Conference on Data Engineering* (pp. 25-33). Los Alamitos, CA: IEEE Computer Society Publications.

Hsu, K-L., Gupta, H. V., & Soroosian, S. (1995). Artificial neural network modeling of the rainfall-runoff process. *Water Resour. Res., 31*, 2517-2530.

Hu, W. & Meng, B. (2005). Design and implementation of Web mining system based on multi-agent. *Lecture notes on artificial intelligence* (Vol. 3584, pp.491-498).

Hung, S.-Y., Liang, T.-P., & Liu, V. W.-C. (1996). Integrating arbitrage pricing theory and artificial neural networks to support portfolio management. *Decision Support Systems, 18*(1996), 301-316.

ICDM (2003). ICDM 2003 tutorial. In *Proceedings of the Third IEEE International Conference on Data Mining, Sponsored by the IEEE Computer Society*, Melbourne, Florida. Retrieved April 13, 2008, from http://www.cs.sfu.ca/~ester/ICDM2003/Lazarevic.abstract.htm

Inegbenebor, A. U. (2006). Financing small and medium industries in Nigeria-case study of the small and medium industries equity investment scheme: Emprical research finding. *Journal of Financial Management & Analysis, 19*(1), 71-80.

Inmon, W. H., & Inmon, W. H. (2002). *Building the data warehouse* (3rd ed.). New York: John Wiley & Sons.

INPE, National Institute for Space Research (2005). *PRODES project - Monitoring the Brazilian Amazon forest using satellites.* National Institute for Space Research. Retrieved April 8, 2008, from http://www.obt.inpe.br/prodes

Intransa (2005). *Managing storage growth with an affordable and flexible IP SAN: A highly cost-effective storage solution that leverages existing IT resources.* CA: Intransa, Inc.

Irani, Z., Al-Sebie, M., & Elliman, T. (2006). Transaction stage of e-government systems: Identification of its location & importance. In *Proceedings of the 39th Hawaii International Conference on System Sciences,* Hawai.

IUPAC (2005). *Chemistry and human health council report: 2003-2005.* International Union of Pure and Applied Chemistry, IUPAC Division VII. Retrieved April 13, 2008, from http://www.iupac.org/news/archives/2005/43rd_council/Item_09_Div_VII.pdf

Jacobs, L. J. & Kuper, G. H. (2004). Indicators of financial crises do work! An early-warning system for six Asian countries. *International Finance, 0409001*, 39.

Jazayeri-Rad, H. (2004). The nonlinear model-predictive control of a chemical plant using multiple neural networks. *Neural Computing and Applications, 13*(1), 2-15.

Jeffery, K. G. (2000). *The grid for e-science: E-commerce benefits, information technology department.* CLRC, ITD.

Jepson, B., Collins, A., & Evans, A. (1993). Post-neural network procedure to determine expected prediction values and their confidence limits. *Neural Computing and Applications, 1*(3), 224-228.

Ji, L. & Peters, A. J. (2003). Assessing vegetation response to drought in the northern Great Plains using vegetation and drought indices. *Remote Sensing of Environment, 87*, 85-98.

Jiang, Z., Piggee, C., Heyes, M.P., Murphy, C., Quearry, B., Bauer, M., Zheng, J., Gendelman, H.E., & Markey, S.P. (2001). Glutamate is a mediator of neurotoxicity in secretions of activated HIV-1-infected macrophages. *Journal of Neuroimmunology, 117*, 97-107.

Jobber, D. (1998). *Principles of marketing* (2nd ed.). McGraw-Hill

John, G. H., Miller, P., & Kerber, R. (1996). Stock selection using rule induction. *IEEE Expert, 11*(5), 52-58.

Jones, F. (1987). Current techniques in bankruptcy prediction. *Journal of Accounting Literature, 6*, 131-164.

Jong Soo Park, Ming-Syan Chen, & Philip S. Yu. (1997). Using a hash-based method with transaction trimming for mining association rules. *IEEE Transactions on Knowledge and Data Engineering, 9*(5), 813-825.

Joplin, B. (2001, March/April). Are we in danger of becoming CRM lemmings? *Customer Management*, 81- 85

Juma, C. (2006, April). *Reinventing African economies: Technological innovation and the sustainability tansition.* Paper presented at The John Pesek Colloquium on Sustainable Agriculture, Ames, Iowa

Kalakota, R. & Robinson, M. (2001). *E-business 2.0—Roadmap for success.* New York: Addison-Wesley.

Kalakota, R., & Robinson, M. (2003). *From e-business to services: Why and why now?* Addison-Wesley. Retrieved January 24, 2005, from http://www.awprofessional.com/articles/article.asp?p=99978&seqNum=5

Kamin, S. B., Schindler, J., & Samuel, S. (2007). The contribution of domestic and external factors to emerging market currency crises: An early warning system. *International Journal of Finance and Economics, 12*(3), 317-322.

Kandampully, J. & Duddy, R. (1999). Relationship marketing: a concept beyond primary relationship. *Marketing Intelligence & Planning, 17*(7), 315-323.

Kang, K. (2006) *Outlook and reforms for the Korean economy in 2006.* Retrieved April 13, 2008, from http://www.keia.org/2-Publications/2-2-Economy/Economy2006/01cover.pdf

Kantardzic, M. (2002). *Data mining: Concepts, models, methods, and algorithms.* Wiley-IEEE Press.

Kao, J-J. (1996). Neural net for determining DEM-based model drainage pattern. *Journal of Irrigation and Drainage Engineering, 122*, 112-121.

Kaplan, R. & Norton, D. (1992). The balanced scorecard—Measures that drive performance. *Harvard Business Review, 70*(1), 71-79.

Kaplan, R. (1996). *The balanced scorecard: Translating strategy into action.* Boston: Harvard Business School Press.

Kargupta, H. & Chan, P. (Eds.) (2001). *Advances in distributed and parallel knowledge discovery.* MIT/AAAI Press.

Kargupta, H., Joshi, A., Sivakumar, K., & Yesh, Y. (Eds) (2004). *Data mining: Next generation challenges and future directions.* AAAI Press.

Karypis, G. (2006). *CLUTO—A clustering toolkit.* Retrieved April 13, 2008, from http://www.cs.umn.edu/~cluto.

Kaul, M., Garden, G.A., & Lipton, S.A. (2001). Pathways to neuronal injury and apoptosis in HIV-associated dementia. *Nature, 410*, 988-994.

Kautz, H., Selman, B., & Shah, M. (1997). Referral Web: Combining social networks and collaborating filtering. *Communications of the ACM, 40*(3), 63-65.

Keim, D. A. (2002). Information visualization and visual data mining. *IEEE Transactions on Visualization and Computer Graphics, 7*, 100-107.

Key Note (2002), Customer Relationship Management

Key Note (2002), Hotels

Key Note (2003), Hotels

Khalil, O. E. M. & Harcar, T. D. (1999). Relationship marketing and data quality management. *SAM Advanced Management Journal, 64* (2).

Khatri, V., Ram, S., & Snodgrass, R. T. (2004). Augmenting a conceptual model with geospatiotemporal annotations. *IEEE Transactions on Knowledge and Data Engineering, 16*, 1324-1338.

Kim, K., & Lee, W. B. (2004). Stock market prediction using artificial neural networks with optimal feature transformation. *Neural Computing and Applications, 13*(3), 255-260.

Kimball, R., & Ross, M. (2002). *The data warehouse toolkit: The complete guide to dimensional modeling* (2nd ed.). New York: John Wiley & Sons.

Klersey, G. F. & Dugan, M.T. (1995). Substantial doubts: Using artificial neural networks to evaluate going concern. In *Advanced in Accounting Information Systems*. Greenwich: JAI Press.

Kloesgen, W. & Zytkow, J. (Eds.) (2002). *Handbook of data mining and knowledge discovery*. Oxford University Press.

Kloptchenko, A., Eklund, T., Karlsson. J., Back, B., Vanhatanta, H., & Visa, A. (2004). Combinig data and text mining techniques for analyzing financial reports. *Intelligent Systems in Accounting Finance and Management, 12*, 29-41.

Ko, P. C. & Lin, P. C. (2005). An evolutionary modularized data mining mechanism for financial distress forecasts. In A. Ghosh, & L.C. Jain (Eds.), *Evolutionary Computation in Data Mining* (pp. 249-263). Berlin Heidelberg, Germany: Springer-Verlag.

Kohara, K., Ishikawa, T., Fukuhara, Y., & Nakamura, Y. (1997). Stock price prediction using prior knowledge and neural networks. *International Journal of Intelligent Systems in Accounting, Finance and Management, 6*(1), 11-22.

Kohonen, T. (1990). The self-organizing maps. *Proceedings of the IEEE*, 78, 1464-1480.

Kommers, P., Kinelev, V., & Kotsik, B. (2003). ICT in secondary education for the knowledge society. In T. Varis, T. Utsumi & W. R. Klemm (Eds), *Global peace through the global university system*. The Finnish National Commission for UNESCO, University of Tampere, Hämeenlinna, Finland.

Kosala, R. & Blockeel, H. (2000). Web mining research: A survey. *ACM, 2*(1), 1-15.

Kostoff, R. N,. Koytcheff, R., & Lau, C. G. Y. (2007). *Structure of the global nanoscience and nanotechnology research literature* (DTIC Tech. Rep. No. ADA461930), Defense Technical Information Center, Fort Belvoir, VA. Retrieved April 13, 2008, from http://www.dtic.mil/

Kostoff, R. N. (1997). Accelerating the conversion of science to technology: Introduction and overview. *Journal of Technology Transfer* [Special Issue on Accelerating the Conversion of Science to Technology], *22*(3) .

Kostoff, R. N. (2003). Text mining for global technology watch. In M. Drake (Ed.), *Encyclopedia of library and information science* (2nd ed) (Vol. 4, pp. 2789-2799). New York: Marcel Dekker, Inc.

Kostoff, R. N. (2003). Stimulating innovation. In L. V. Shavinina (Ed.), *International handbook of innovation* (pp. 388-400). Oxford, UK: Elsevier Social and Behavioral Sciences.

Kostoff, R. N. (2003). Bilateral asymmetry prediction. *Medical Hypotheses, 61*(2), 265-266.

Kostoff, R. N. (2006). Systematic acceleration of radical discovery and innovation in science and technology. *Technological Forecasting and Social Change, 73*(8), 923-936.

Kostoff, R. N., Del Rio, J. A., García, E. O., Ramírez, A. M., & Humenik, J. A. (2001). Citation mining: Integrating text mining and bibliometrics for research user profiling. *Journal of the American Society for Information Science and Technology, 52*(13), 1148-1156.

Kostoff, R. N., Eberhart, H. J., & Toothman, D. R. (1997). Database tomography for information retrieval. *Journal of Information Science, 23*(4), 301-311.

Kostoff, R. N., Green, K. A., Toothman, D. R., & Humenik, J. A. (2000). Database tomography applied to an aircraft science and technology investment strategy. *Journal of Aircraft, 37*(4), 727-730.

Kostoff, R. N., Johnson, D., Bowles, C. A., & Dodbele, S. (2006). *Assessment of India's research literature* (DTIC Tech. Rep. No. ADA444625), Defense Technical Information Center, Fort Belvoir, VA. Retrieved April 13, 2008, from http://www.dtic.mil/

Kostoff, R. N., Murday, J., Lau, C., & Tolles, W. (2005). *The seminal literature of global nanotechnology research* (DTIC Tech. Rep. No. ADA435986), Defense Technical Information Center, Fort Belvoir, VA. Retrieved April 13, 2008, from http://www.dtic.mil/

Kostoff, R. N., Murday, J., Lau, C., & Tolles, W. (2006). The seminal literature of global nanotechnology research. *Journal of Nanoparticle Research, 8*(2), 193-213.

Kostoff, R. N., Shlesinger, M., & Malpohl, G. (2004). Fractals roadmaps using bibliometrics and database tomography. *Fractals, 12*(1), 1-16.

Kostoff, R. N., Shlesinger, M., & Tshiteya, R. (2004). Nonlinear dynamics roadmaps using bibliometrics and database tomography. *International Journal of Bifurcation and Chaos, 14*(1), 61-92.

Kostoff, R. N., Stump, J.A., Johnson, D., Murday, J., Lau, C., & Tolles, W. (2006). The structure and infrastructure of the global nanotechnology literature. *Journal of Nanoparticle Research, 8*(3-4), 301-321.

Kostoff, R. N., Tshiteya, R., Pfeil, K. M., & Humenik, J. A. (2002). *Power source text mining using bibliometrics and database tomography.*

Kottegoda, N. T., Natale, L., & Raiteri, E. (2004). Some considerations of periodicity and persistence in daily rainfalls. *Journal of Hydrology, 296*(1-4), 23-37.

Kou, G., & Shi, Y. (2002). *Linux-based Multiple Linear Programming Classification Program:* (*Version 1.0.*) College of Information Science and Technology, University of Nebraska-Omaha, USA.

Kou, G., Liu, X., Peng, Y., Shi, Y., Wise, M., & Xu, W. (2003). Multiple criteria linear programming approach to data mining: Models, algorithm designs and software development. *Optimization Methods and Software, 18*, 453-473.

Kou, G., Peng, Y., Chen, Z., Shi, Y., & Chen, X. (2004, July 12-14). A multiple-criteria quadratic programming approach to network intrusion detection. In *Proceedings of the Chinese Academy of Sciences Symposium on Data Mining and Knowledge Management*, Beijing, China.

Kou, G., Peng, Y., Shi, Y., & Chen, Z. (2006). *A new multi-criteria convex quadratic programming model for credit data analysis.* Working Paper, University of Nebraska at Omaha, USA.

Kou, G., Peng, Y., Yan, N., Shi, Y., Chen, Z., Zhu, Q., Huff, J., & McCartney, S. (2004, July 19-21). Network intrusion detection by using multiple-criteria linear programming. In *Proceedings of the International Conference on Service Systems and Service Management*, Beijing, China.

Koua, E. L. & Kraak, M. J. (2004). Geovisualization to support the exploration of large health and demographic survey data. *International Journal of Health Geographics, 3*,12.

Kovalerchuk, B. & Vityaev, E. (2000). *Data mining in finance: Advances in relational and hybrid methods.* New York: Kluwer Academic Publisher.

Koyuncugil, A. S. & Ozgulbas, N. (2006). Financial profiling of SMEs: An application by data mining. *The European Applied Business Research (EABR) Conference*, Clute Institute for Academic Research.

Koyuncugil, A. S. & Ozgulbas, N. (2006). Is there a specific measure for financial performance of SMEs? *The Business Review, 5*(2), 314-319.

Koyuncugil, A. S. & Ozgulbas, N. (2006). Determination of factors affected financial distress of SMEs listed in ISE by data mining. In *Proceedings of the 3rd Congress of SMEs and Productivity*, KOSGEB and Istanbul Kultur University, Istanbul.

Koyuncugil, A. S. (2006). *Fuzzy data mining and its application to capital markets*. Unpublished doctoral dissertation, Ankara University, Ankara.

Kraak, M.-J. (2000). Access to GDI and the function of visualization tools. In R. Groot & J. McLaughlin (Eds.), *Geospatial data infrastructure: Concepts, cases and good practice* (pp. 217-321). Oxford, UK: Oxford University Press.

Krol, C. (1999, May). A new age: It's all about relationships. *Advertising Age, 70*(21), S1-S4.

Kudyba, S. & Hoptroff, R. (2001). *Data mining and business intelligence: A guide to productivity*. Hershey, PA: Idea Group Publishing.

Kufoniyi, O., Huurneman, G., & Horn, J. (2005, April). *Human and institutional capacity building in geoinformatics through educational networking*. Paper presented at the International Federation of Surveyors Working Week 2005, Cairo, Egypt.

Kuncheva, L.I. (2000). Clustering-and-selection model for classifier combination. In *Proceedings of the 4th International Conference on Knowledge-Based Intelligent Engineering Systems and Allied Technologies* (KES'2000).

Kuo, R. J., Liao, J. L., & Tu, C. (2005). Integration of ART2 neural network and genetic k-means algorithm for analyzing Web browsing paths in electronic commerce. *Decision Support Systems, 40*, 355-374.

Kutman, O. (2001). Researching the early warning signals for the enterprises in Turkey. *Journal of Dogus University, 4*, 59-70.

Kwak, W., Shi, Y., Eldridge, S., & Kou, G. (2006). Bankruptcy prediction for Japanese firms: Using multiple criteria linear programming data mining approach. In *Proceedings of the International Journal of Data Mining and Business Intelligence*.

Lam, L. (2000). *Classifier combinations: Implementations and theoretical issues*. In Kittler & Roli (Eds.), Multiple classifier systems (pp. 78-86). Berlin: Springer-Verlag (Lecture Notes in Pattern Recognition 1857).

Lambin, E. (1999). *Land-use and land-cover change implementation strategy*. Retrieved April 8, 2008, from http://www.geo.ucl.ac.be/LUCC/lucc.html

Lambin, E. F., Geist, H. J. & Lepers, E. (2003). Dynamics of land use and land cover change in Tropical Regions. *Annual Review of Environment and Resources, 28*(1) 205-241.

Lansiluoto, A., Eklund, T., Barbro, B., Vanharanta, H., & Visa, A. (2004). Industry-specific cycles and companies' financial performance comparison using self-organising maps. *Benchmarking, 11*, 267-286.

Lappas, G. & Yannas, P. (2006). A framework to evaluate political party Websites. In *Proceedings of the 4th International Conference on Politics and Information Systems: Technologies and Applications Vol. II* (pp. 226-231), Orlando, Florida.

Larose, D. T. (2004). *Discovering knowledge in data: An introduction to data mining*. Wiley-Interscience.

Law, R. (1999). Demand for hotel spending by visitors to Hong Kong: A study of various forecasting techniques. *Journal of Hospitality and Leisure Marketing, 6*(4), 17-29.

Lawlor, L. R. (1980). Structure and stability in natural and randomly constructed competitive communities. *The American Naturalist, 116*(3), 394-408.

Lawrence, S. & Giles, C. L. (1999). Accessibility of information on the Web. *Nature, 400*, 107-09.

Lazo, J. G., Maria, M., Vellasco, R., Aurelio, M., & Pacheco, C. (2000). A hybrid genetic-neural system for portfolio selection and management. In *Proceedings of the 7th International Conference on Engineering Applications of Neural Networks*. Kingston Upon Thames, UK: Kingston University.

Lee, H. Y., Ong, H. L., & Quek, L. H. (1995). *Exploiting visualization in knowledge discovery*. In *Proceedings of the 1st International Conference on Knowledge Discovery and Data Mining* (pp. 198 – 201), Montreal, Canada.

Lee, H. Y., Ong, H. L., Toh, E. W., & Chan, S. K. (1996). A multi-dimensional data visualization tool for knowledge discovery in databases. In *Proceedings of IEEE Conference on Visualization*, pp. 26–31.

Lee, S.M. (1972). *Goal programming for decision analysis*. Auerbach.

Lempel, R. & Moran, S. (2001). SALSA: The stochastic approach for link-structure analysis. *ACM Transactions on Information Systems*, 19(2), 131-160.

Lester, T. (2004, March 31). Pitfalls of precision bombing. *FT Management*, 4.

Levene, M. & Loizou, G. (1999). *Computing the entropy of user navigation in the Web* (Tech. Rep. No. RN/99/42), University College London.

Li, X. (1998). Web page design and graphic use of three U.S. newspapers. *Journalism and Mass Communication Quarterly*, *75*(2), 353-365.

Liautaud, B. (2000). *E-business intelligence: turning information into knowledge into profit*. New York: McGraw-Hill.

Lin, D.-I., & Kedem, Z. M. (1998). Pincer search: A new algorithm for discovering the maximum frequent set. In H.-J. Schek, F. Saltor, I. Ramos & G. Alonso (Eds.), *Advances in Database Technology – Proceedings of the 6th International Conference on Extending Database Technology* (pp. 105-119). Berlin Heidelberg, Germany: Springer-Verlag.

Lindgreen, A. & Crawford, I. (1999). Implementing, monitoring and measuring a programme of relationship marketing. *Marketing Intelligence & Planning*, *17*(5), 231-239.

LINDO Systems Inc. (2003). *An overview of LINGO 8.0*. Retrieved from http://www.lindo.com/cgi/frameset.cgi?leftlingo.html;lingof.html

Lindsay, P.H., & Norman, D.A. (1972). *Human information processing: An introduction to psychology*. New York: Academic Press.

Liu, H. & Motoda, H. (1998). *Feature selection for knowledge discovery and data mining*. Kluwer.

Liu, H. & Motoda, H. (1998). *Feature extraction, construction and selection: A data mining perpective*. Kluwer

Liu, J. W., Yu, S. J., & Le, J. J. (2005). Online mining dynamic Web news patterns using machine learn methods. *Lecture notes on artificial intelligence* (Vo. 3614, pp. 462-465).

Liu, J., Pan, Y., Wang, K., & Han, J. (2002). Mining frequent item sets by opportunistic projection. In *Proceedings of the Eighth ACM SIGKDD International Conference on Knowledge Discovery and Data Mining* (pp. 229-238). New York: ACM Press.

Liu, S. & Lindholm, C. K. (2006), Assessing early warning signals of currency crises: A fuzzy clustering approach. *Intelligent Systems in Accounting, Finance and Management, 14*(4), 179-184.

Long, G., Hogg, M. K., Hartley, M. & Angold, S. J. (1999). Relationship marketing and privacy: Exploring the thresholds. *Journal of Marketing Practice: Applied Marketing Science, 5*(1), 4-20.

Long, M. M. & Schiffman, L. G. (2000). Consumption values and relationships: Segmenting the market for frequency programs. *Journal of Consumer Marketing, 17*(3).

Longley, P. A., Goodchild, M. F., Maguire, D.J., & Rhind, D. W. (2001). *Geographic information systems and science*. West Sussex, England: John Wiley and Son, Ltd.

Lopez, A., Bauer, M.A., Erichsen, D.A., Peng, H., Gendelman, L., Shibata, A., Gendelman, H.E., & Zheng, J. (2001). The regulation of neurotrophic factor activities following HIV-1 infection and immune activation of mononuclear phagocytes. In *Proceedings of Soc. Neurosci. Abs*., San Diego, CA.

Losiewicz, P., Oard, D., & Kostoff, R. N. (2000). Textual data mining to support science and technology management. *Journal of Intelligent Information Systems, 15*, 99-119.

Lu, Q. & Getoor, L. (2003). Link-based text classification. In *Proceedings of the 3rd International Workshop on Link Discovery* (pp. 1-8). ACM Press.

Lu, S., Hu, H., & Li, F. (2001). Mining weighted association rules. *Intelligent Data Analysis*, *5*(2001), 211-255.

Luck, D. & Lancaster, G. (2003). E-CRM: Customer relationship marketing in the hotel industry. *Managerial Auditing Journal – Accountability and the Internet, 18*(3), 213-232.

MacDonald, J. (2002). The Earth observation business and the forces that impact it. In D. Couts (Ed.), *Earth observation business network 2002*. Vancouver, CA: MacDonald Dettwiler.

MacEachren, A. M. & Kraak, M.-J. (1997). Exploratory carthographic visualization: Advancing the agenda. *Computer and Geosciences, 23*, 335-343.

Magnusson, C., Arppe, A., Eklund, T., & Back, B. (2005). The language of quarterly reports as an indicator of change in the company's financial staus. *Information & Management, 42*, 561-570.

Maier, H. R. & Dandy, G. C. (1996). The use of artificial neural networks for the prediction of water quality parameters. *Water Resour. Res., 32*, 1013-1022.

Maimon, O. & Last, M. (2000). *Knowledge discovery and data mining - The Info-Fuzzy Network (IFN) Methodology*. Kluwer Publishers, Massive Computing.

Mangasarian, O.L. (2000). Generalized support vector machines. In A. Smola, P. Bartlett, B. Scholkopf, & D. Schuurmans (Eds.), Advances in large margin classifiers (pp. 135-146). Cambridge, MA: MIT Press.

Mannila, H. & Raiha, K. J. (1987). Dependency inference. In *Proceedings of the 1987 International Conference Very Large Data Bases,* (pp. 155-158). Brighton, England.

Mannila, H., Toivonen, H., & Verkamo, A. I. (1994). Efficient algorithms for discovering association rules. In U. M. Fayyad & R. Uthurusamy (Eds.), *Knowledge Discovery in Databases: Papers from the 1994 AAAI Workshop* (pp. 181-192). Menlo Park, CA: AAAI Press.

Mannila, H., Toivonen, H., & Verkamo, I. (1994). Efficient algorithms for discovering association rules. In *Proceedings of the AAAI Workshop, Knowledge Discovery in Databases.*

Margolis, M., Resnick, D., & Tu, C.-C. (1997). Campaigning on the Internet: Parties and candidates on the World Wide Web in the 1996 primary season. *Harvard International Journal of Press/Politics*, *2*(1), 59-78.

Markus, B. (2005). *Building spatial knowledge infrastructure*. Paper presented at the ISPRS Workshop on Service and Application of Spatial Data Infrastructure, XXXVI, Hangzhou, China. Retrieved April 13, 2008, from http://www.commission4.isprs.org/workshop_hangzhou/papers/65-70%20Bela%20markus-A103.pdf

Martin-Guerrero, J. D., Palomares, A., Balaguer-Ballester, E., Soria-Olivas, E., Gomez-Sanchis, J., & Soriano-Asensi, A. (2006). Studying the feasibility of a recommender in a citizen Web portal based on user modeling and clustering algorithms. *Expert Systems with Applications, 30*, 299-312.

Matsuo, Y., Ohsawa, Y., & Ishizuka, M. (2001). Average-clicks: A new measure of distance on the WWW. In *Proceedings of First Asia-Pacific Conference, Web Intelligence*, Japan.

Mattison, R. M. (1997). *Data warehousing and data mining, for telecommunications*. Artech House.

Maule, R. W. (1998). Content design frameworks for Internet studies curricula and research. *Internet Research: Electronic Networking Applications and Policy, 8*(2), 174-184.

May, M. & Savinov, A. (2002). An integrated platform for spatial data mining and interactive visual analysis. In *Proceedings of the International Conference on Data Mining Methods and Databases for Engineering* (pp. 90-101).

Mayer, M. A., Karkaletsis, V., Stamatakis, K., Leis, A., Villarroel, D., Thomeczek, C., Labsky, M., Lopez-Ostenero, F., & Honkela, T. (2006). MedIQ-Quality labelling of medical Web content using multilingual information extraction. *Studies in Health Technology and Informatics, 121*, 133-190.

McDonald, W. J. (1998). *Direct marketing: An integrated approach*. McGraw-Hill International Editions.

McGarigal, K. & Marks, B. (1995). *FRAGSTATS: Spatial pattern analysis program for quantifying landscape structure*. USDA Forestry Service Technical Report PNW-351. Washington, DC.

McGarigal, K. (2002). Landscape pattern metrics. In A.H. El-Shaarawi & W.W. Piegorsch (Eds.), *Encyclopedia of environmentrics* (pp. 1135-1142). Sussex, England: John Wiley & Sons.

McGonagle, J. J. & Vella, C. M. (1990). *Outsmarting the competition*. Naperville, IL: Sourcebooks.

McVicar, T. R. & Bierwirth, P. N. (2001). Rapidly assessing the 1997 drought in Papua New Guinea using composite AVHRR imagery. *International Journal of Remote Sensing, 22*, 2109-2128.

Meier, R. L. (2000). Late-blooming societies can be stimulated by information technology. *Futures, 32*(2), 163.

Meinel, G. & Neubert, M. (2004). A comparison of segmentation programs for high resolution remote sensing data. *International Archives of Photogrammetry and Remote Sensing, 35*(1), 1097-1105.

Melville, P., Mooney, R. J., & Nagarajan, R. (2002). Content-boosted collaborative filtering for improved recommendations. In *Proceedings of the 18th National Conference on Artificial Intelligence* (pp. 187-192).

Mena, J. (2003). *Investigative data mining for security and crimianal detection*. USA: Elsevier Science.

Meyer, P. A., & Pifer, W. H. (1970). Prediction of bank failures. *The Journal of Finance, 25*(4),853-868.

Michalski, R. S. & Tecuci, G. (1994). *Machine learning: A multistrategy approach* (Vol. IV). Morgan Kaufmann

Miles, M. B. & Huberman, A. M. (1990). *Qualitative data analysis*. London: Sage Publications.

Miller, C. (1995). In-depth interviewing by telephone: Some practical considerations. *Evaluation and Research in Education, 9*(1), 29-38.

Miller, H. J. & Han, J. (2001). *Geographic data mining and knowledge discovery*. London: Taylor and Francis.

Miller, J. (2002). *O milênio da inteligência competitiva*, Brazil: Bookman.

Mirkin, B. & Mirkin, B. G. (2005). *Clustering for data mining: A data recovery approach*. Virginia Beach, VA: Chapman & Hall / CRC.

Mishra, A., Ray, C., & Kolpin, D. W. (in press). Use of qualitative and quantitative information in neural networks for assessing agricultural chemical contamination of domestic wells. *Journal of Hydrological Engineering*.

Mitchell, T. (1997). *Machine learning*. McGraw Hill.

Mitra, S. & Acharya, T. (2003). *Data mining: Multimedia, soft computing and bioinformatics*. Hoboken, NJ: John Wiley and Sons, Inc.

MITRE (2001). *Stopping traffic: Anti drug network (ADNET)*. MITRE Digest Archives. Retrieved April 13, 2008, from http://www.mitre.org/news/digest/archives/2001/adnet.html

Mladenic, D. & Grobelnik, M. (1999). Predicting content from hyperlinks. In *Proceedings of the 16th International ICML99 Workshop on Machine Learning in Text Data Analysis* (pp. 109-113).

Mobasher, B., Cooley, R., & Srivastava, J. (1999). Creating adaptive Web sites through usage based clustering of URLs. In *Proceedings of the IEEE Knowledge and Data Engineering Exchange Workshop* (KDEX99), Chicago, Illinois.

Mobasher, B., Dai, H., Luo, T., Sung, Y., & Zhu, J. (2000). Integrating Web usage and content mining for more effective Web personalization. In *Proceedings of the International Conference on E-Commerce and Web Technologies (ECWeb 2000)* (pp. 165-176). Greenwich, UK.

Mobasher, B., Jain, N., Han, E., & Srivastava, J. (1996). *Web Minning: Pattern discovery from WWW transaction* (Tech. Rep. TR-96050). Department of Computer Science, University of Minnesota, Minneapolis. Retrieved April 12, 2008, from http://citeseer.ist.psu.edu/mobasher96web.html

Mokyr, J. (2002). *The gifts of Athena: Historical origins of the knowledge economy.* New Haven: Princeton University Press.

Moncrief, W. C. & Cravens, D. (1999). Technology and the changing marketing world. *Marketing Intelligence and Planning, 17*(7), 329-332.

Mooney, R. J. & Roy, L. (2000). Content-based book recommending using learning for text categorization. In *Proceedings of the 5th ACM Conference on Digital Libraries* (pp. 195-204). ACM Press.

Moore, G. (1999). *Crossing the chasm: Marketing and selling high-tech products to mainstream customers.* Oxford, UK: Capstone.

Mukherjee, D. (2006). Promote scientific research. *Central Chronicle.*

Murphy, J. M. (2001, March-April). Customer excellence: From the top down. *Customer Management*, 36-41.

Mursu, A., Soriyan, H. A., Olufokunbi, K., & Korpela, M. (2000). Information systems development in a developing country: Theoretical analysis of special requirements in Nigeria and Africa. In *Proceedings of the 33rd Hawaii International Conference on System Sciences.* Maui, Hawaii: IEEE.

Myatt, G. J. (2006). *Making sense of data: A practical guide to exploratory data analysis and data mining.* John Wiley.

Nagao, M. & Matsuyama, T. (1980*). A structural analysis of complex aerial photographs.* New York: Plenum Press.

Nanopoulos, A., Katsaros, D., & Manolopoulos, Y. (2003). A data mining algorithm for generated Web prefetching. *IEEE Transactions on Knowledge and Data Engineering, 15*(5), 1155-1169.

Narin, F. (1976). *Evaluative bibliometrics: the use of publication and citation analysis in the evaluation of scientific activity* (monograph). *NSF C-637.* National Science Foundation. Contract NSF C-627. NTIS Accession No. PB252339/AS.

Narin, F., Olivastro, D., & Stevens, K. A. (1994). Bibliometrics theory, practice and problems. *Evaluation Review, 18*(1), 65-76.

Nasraoui, O. & Pavuluri, M. (2004). Complete this puzzle : A connectionist approach to accurate Web recommendations based on a committee of predictors. In *Proceedings of the 6th WEBKDD Workshop*, Seattle, Washington.

National Academy of Sciences (NAS) (2003). *IT roadmap to a geospatial future.*, Washington, D.C.: The National Academies Press.

Neftci, S. N. (2004). *Principles of financial engineering.* Burlington, MA: Elsevier Academic Press.

Nemati, H. R. & Barko, C. D. (2003). Key factors for achieving organizational data-mining success. *Industrial Management & Data Systems, 103*(4), 282-292.

Ngu, D. S. W. & Wu, X. (1997). Sitehelper: A localized agent that helps incremental exploration of the World Wide Web. *Computer Networks, 29*(8-13), 1249-1255.

Nitsche, M. (2002, January-March). Developing a truly customer-centric CRM system: Part One – Strategic and artchitectural implementation. *Interactive Marketing, 3*(3), 207-217.

Nittel, S. & Stefanidis, A. (2005). GeoSensor networks and virtual GeoReality. In S. Nittel & A. Stefanidis (Eds.), *GeoSensor networks* (pp. 1-9). Boca Raton, FL: CRC Press.

Niven, P. R. (2002). *Balanced scorecard step-by-step: Maximizing performance and maintaining results.* New York: J. Wiley & Sons.

NSI Software (2004). *Six tips small and midsize businesses can use to protect their critical data.* NJ: NSI Software.

Nwabueze, K. (November 30, 2003). A case study: Role of technology venture capitalist market in developing countries, data mining, integration, and analysis. *Timbuktu Chronicles.* Retrieved April 13, 2008, http://timbuktu-chronicles.blogspot.com/2003_11_01_archive.html

O'Bada, A. (2002). Local adaptations to global trends: A study of an IT-based organizational change program in a Nigerian bank. *Information Society, 18*(2), 77.

O'Kelly, M. E. (1994). *Spatial analysis and GIS.* In A.S. Fotheringham & P.A. Rogerson (Eds.), *Spatial analysis and GIS* (pp. 66-79). London, UK: Taylor and Francis.

Oberle, D., Berendt, B., Hotho, A., & Gonzalez, J. (2003). Conceptual user tracking. *Lecture notes on artificial intelligence* (Vol. 2663, pp. 155-164).

OECD (2000). *Policy briefs small and medium-sized enterprises: Local strength, global reach.* Retrieved May 9, 2008, from www.oecd.org/dataoecd/3/30/1918307.pdf

Oksay, S. (2006). *Publication of insurance research and analysis.* Turkey: TSRSB.

Olson, D., & Shi, Y. (2005). *Introduction to business data mining.* New York: McGraw-Hill/Irwin.

Onwu, I. (2005). *Knowledge discovery interface for environmental applications.* Unpublished master's thesis, Iowa State University, Ames.

Opitz, D., & Maclin, R. (1999). Popular ensemble methods: An empirical study. *Journal of Artificial Intelligence Research, 11*, 169-198.

Orlikowski, W. J. & Iacono, C. S. (2001). Research commentary: Desperately seeking "IT" in IT research—A call to theorizing the IT artifact. *Information Systems Research, 12*(2) 121-156.

Overell, S. (2004, March 31). Customers are not there to be hunted. *FT Management*, 2.

Ozgulbas, N. & Koyuncugil, A. S. (2006). Profiling and determining the strengths and weaknesses of SMEs listed in ISE by the data mining decision trees algorithm CHAID. In *Proceedings of the 10th National Finance Symposium,* Izmir.

Ozgulbas, N., Koyuncugil, A. S., & Yılmaz, F. (2006). Identifying the effect of firm size on financial performance of SMEs. *The Business Review, 5*(2), 162-167.

Pal, S. K. & Mitra, P. (2004). *Pattern recognition algorithms for data mining.* Chapman & Hall/CRC.

Pal, S., Talwar, V., & Mitra, P. (2002). Web mining in soft computing framework: Relevance, state of the art and future directions. *IEEE Transactions on Neural Networks, 13*(5), 1163-1177.

Palmer, A. (1996). Relationship marketing: A universal paradigm or management fad? *The Learning Organisation, 3*(3), 18-25.

Palmer, A., McMahon-Beattie, U. & Beggs, R. (2000). A structural analysis of hotel sector loyalty programmes. *International Journal of Contemporary Hospitality Management, 12*(1), 54-60.

Pantalone, C., & Platt, M. (1987). Predicting failures of savings and loan associations. *AREUEA Journal, 15*, 46-64.

Parhami, B. (1994). *Voting algorithms. IEEE Transactions on Reliability, 43*, 617-629.

Park, H. W. (2003). Hyperlink network analysis: A new method for the study of social structure on the Web. *Connections, 25*(1), 49-61.

Park, J. S., Chen, M.-S., & Yu, P. S. (1995). An effective hash based algorithm for mining association rules. In M. J. Carey & D. A. Schneider (Eds.), *Proceedings of the 1995 ACM SIGMOD International Conference on Management of Data* (pp. 175-186). New York: ACM Press.

Parthasarathy, S., Zaki, M. J., & Li, W. (1997). *Application driven memory placement for dynamic data structures* (Tech. Rep. URCS TR 653). University of Rochester.

Pawlak, Z. (1982). Rough sets. *Journal of Computer and Information Science, 11*(5), 341-356, 1982.

Pazzani, M. & Billsus, D. (1997). Learning and revising user profiles: The identification of interesting Web sites. *Machine Learning, 27*(3), 313-331.

Pei, J., Han, J., & Lakshmanan, L. V. S. (2001). *Mining frequent itemsets with convertible constraints.* Paper presented at the Proceedings of the 17th International Conference on Data Engineering (pp. 433–332), Heidelberg, Germany.

Pei, J., Han, J., & Mao, R. (2000, May). CLOSET: An efficient algorithm for mining frequent closed itemsets. In D. Gunopulos & R. Rastogi (Eds.), *Proceedings of the 2000 ACM SIGMOD Workshop on Research Issues in Data Mining and Knowledge Discovery* (pp. 21-30), Dallas, TX.

Peng, Y., Kou, G., Chen, Z., & Shi, Y. (2004). Cross-validation and ensemble analyses on multiple-criteria linear programming classification for credit cardholder behavior. In *Proceedings of ICCS 2004* (pp. 931-939). Berlin: Springer-Verlage (LNCS 2416).

Perner, P. & Petrou, M. (Eds.). *Machine learning and data mining in pattern recognition.* Springer Verlag.

Piatetsky-Shapiro, G., Djeraba, C., Getoor, L., Grossman, R., Feldman, R. & Zaki, M. (2006). What are the grand challenges for data mining? - KDD-2006 Panel Report. *SIGKDD Explorations, 8*(2), 70-77.

Pierrakos, D., Paliouras, G., Papatheodorou, C., Karkaletsis, V., & Dikaiakos, M. (2003). Web community directories: A new approach to Web personalization. *Lecture notes on artificial intelligence* (Vol. 3209, pp. 113-129).

Pilato, G., Vitabile, S., Vassallo, G., Conti, V., & Sorbello, F. (2003). A concurrent neural classifier for HTML documents retrieval. *Lecture notes in computer science* (Vol. 2859, pp. 210-217).

Potgieter, J. (2003). *OLAP data scalability: Ignore OLAP data explosion at great cost.* NSW Australia: SPF Pty Ltd.

Pozzebon, M. (2003). *The implementation of configurable technologies: Negotiations between global principles and local contexts.* Unpublished doctoral dissertation, McGill University, Montreal, Canada.

Prabhaker, P. (2001). Integrated marketing-manufacturing strategies. *Journal of Business & Industrial Marketing, 16*(2), 113-128.

Prashanth, K. (2004). *Wal-Mart's supply chain management practices (B): Using IT/Internet to manage the supply chain.* Hyderabad, India: ICFAI Center for Management Research.

Pyle, D. (1999). *Data preparation for data mining.* Morgan Kaufmann.

Quah, T. S. & Srinivasan, B. (1999). Improving returns on stock investment through neural network selection. *Expert Systems with Applications, 17*(4), 295-301.

Quéau, P. (2001). *The information society and the global good.* Retrieved April 13, 2008, from http://goanna.cs.rmit.edu.au/~aym/rinseap/bali/QueauTalk.html

Quinlan, J. R. (1993). *C4.5: Programs for machine learning.* San Francisco: Morgan Kaufmann Publishers.

Quinlan, R. (1993). *Programs for machine learning.* San Francisco: Morgan Kaufmann.

Raghavan, S. N. R. (2005). Data mining in e-commerce: A survey. *Sadhana, 30*(2&3), 275-289.

Rahman, H. (2004). Information dynamics in developing countries. In *Proceedings of the 5th International Conference on IT in Regional Areas*, Caloundra, Queensland, Australia.

Rahman, H. (2006). Role of ICTs in socio-economic development and poverty reduction. In H. Rahman (Ed.), *Information and communication technologies for economic and regional developments*

Rao, M. (2002). *Systems design of a national spatial data.* Bangalore: Indian Space Research Organisation Headquarters.

Rasmussen, N., Goldy, P. S., & Solli, P. O. (2002). *Financial business intelligence—Trends, technology, software selection, and implementation.* New York: John Wiley and Sons.

Ratcliffe, J. (2004). *Strategic thinking in criminal intelligence.* Sydney: Federation Press.

Rautenstrauch, C. & Page, B. (2001). *Environmental informatics-methods, tools and applications in environmental information processing*. In C. Rautenstrauch & S. Patig (Eds.), *Environmental information systems in industry and public administration* (pp. 2-11). Hershey, PA: Idea Group Publishing.

Reeves, T C. & Dehoney, J. (1998). Cognitive and social functions of courseWeb sites. In H. Maurer & R.G. Olson (Eds.), *Proceedings of WebNet World Conference 98—World Conference of the WWW, Internet & Intranet*. Orlando, FL: Association for the Advancement of Computing in Education.

Reich, B. & Benbasat, I. (2000). Factors that influence the social dimension of alignment between business and information technology objectives. *MIS Quarterly, 24*(1), 81-113.

Reichheld, F. & Schefter, P. (2000, July/ August). E-loyalty. *Harvard Business Review*, 105-113.

Rennie, J. & McCallum, A. K. (1999). Using reinforcement learning to spider the Web efficiently. In *Proceedings of the 16th International ICML99 Workshop on Machine Learning in Text Data Analysis* (pp. 335-343).

Resig, J., Dawara, S., Homan, C. M., & Teredesai, A. (2004). Extracting social networks from instant messaging populations. In *Proceedings of LinkKDD'04*, Seattle, Washington.

Rich, M. K. (2000). The direction of marketing relationships. *The Journal of Business & Industrial Marketing, 15*(2/3), 170-179.

Richardson, M. & Domingos, P. (2002). The intelligent surfer: Probabilistic combination of link and content information in PageRank. *Advances in Neural Information Processing Systems, 14*.

Riedl, R. (2003). Design principles for E-government services. In *Proceedings of eGov Day 2003*, Vienna, Austria.

Roberto, J. & Bayardo, Jr. (1998). Efficiently mining long patterns from databases. In L. M. Hass, & A. Tiwary (Eds.), *Proceedings of the 1998 ACM SIGMOD International Conference on Management of Data* (pp. 85-93). New York: ACM Press.

Robey, D., Ross, J., & Boudreau, M. (2002). Learning to implement enterprise systems: An exploratory study of the dialectics of change. *Journal of Management Information Systems, 19*(1), 17.

Ross, S. (1976). The arbitrage theory of capital asset pricing. *Journal of Economic Theory, 13*, 341-360.

Rud, O. P. (2001). *Data mining cookbook: Modeling data for marketing, risk, and CRM*. Wiley.

Rui, Y., Huang, T. S. & Chang, S. F. (1999). Image retrieval: Current techniques, promising directions and open issues. *Journal of Visual Communication and Image Representation, 10*(1), 39-62.

Rushing, J., Ramachandran, R., Nair, U. J., Graves, S. J., Welch, R. & Lin, A. (2005). ADaM: A data mining toolkit for scientists and engineers. *Computers and Geosciences, 31*(5), 607-618.

Rüther, H. (2001, October). EIS education in Africa – The geomatics perspective. *Paper presented at the International Conference on Spatial Information for Sustainable Development*, Nairobi, Kenya

Rymon, R. (1992). Search through systematic set enumeration. In B. Nebel, C. Rich & W. R. Swartout (Eds.), *Proceedings of the 3rd International Conference on Principles of Knowledge Representation and Reasoning* (pp. 539-550). San Francisco: Morgan Kaufmann Publishers.

Sahay, S. & Avgerou, C. (2002). Information and communication technologies in developing countries. *Information Society, 18*(2), 1-5.

Sammon, W. L., Kurland, M. A., & Spitalnic, R. (1984). *Business competitor intelligence: Methods for collecting, organizing, and using information*. New York: John Wiley & Sons.

Sanchez, A. & Marin, G. S. (2005). Strategic orientation, management characteristics, and performance: A study of Spanish SMEs. *Journal of Small Business Management, 43*(3), 287-309.

Sarwar, B., Karypis, G., Konstan, J., & Riedl, J. (2000). Analysis of recommendation algorithms for e-commerce. In *Proceedings of the ACM Conference on Electronic Commerce* (pp. 158-162).

Savasere, A., Omiecinski, E., & Navathe S. (1995). An efficient algorithm for mining association rules in large databases. In U. Dayal, P. M. D. Gray & S. Nishio (Eds.), *Proceedings of the 21st International Conference on Very Large Data Bases* (pp. 432-444). San Francisco: Morgan Kaufmann Publishers.

Schaap, B. D. & Linhart, S.M. (1998). *Quality of ground water used for selected municipal water supplies in Iowa, 1982-96 water years* (p. 67). Iowa City, IA: U.S. Geological Survey Open File Report 98-3.

Schaap, M. G. & Bouten, W. (1996). Modeling water retention curves of sandy soils using neural networks. *Water Resour. Res., 32*, 3033-3040.

Scheffer, T. (2004). Email answering assistance by semi-supervised text classification. *Intelligent Data Analysis, 8*(5), 2004.

Schneider, S. & Foot, K. (2004). The Web as an object of study. *New Media & Society, 6*(1), 114-122.

Schonberg, E., Cofino, T., Hoch, R., Podlaseck, M., & Spraragen, S. (2000). Measuring success. *Communications of the ACM, 43*(8), 53-57.

Schröder, M., Rehrauer, H., Seidel, K. & Datcu, M. (2000). Interactive learning and probabilistic retrieval in remote sensing image archives. *IEEE Transactions on Geoscience and Remote Sensing, 23*(1), 2288-2298.

Schubert, A., Glanzel, W., & Braun, T. (1987). Subject field characteristic citation scores and scales for assessing research performance. *Scientometrics, 12*(5-6), 267-291.

SCI (2006). Certain data included herein are derived from the *Science Citation Index/Social Science Citation Index* prepared by the THOMSON SCIENTIFIC ®, Inc. (Thomson®), Philadelphia, Pennsylvania, USA: © Copyright THOMSON SCIENTIFIC ® 2006. All rights reserved.

Science Blog (2002). Partnerships, finance, sustainable production and consumption patterns. Press Release: United Nations. Retrieved April 12, 2008, from http://www.scienceblog.com/community/older/archives/L/2002/A/un020319.html

Scime, A. (2005). *Web mining: Application and techniques.* Hershey, PA: Idea Group Inc.

SCN Education B. V. (2001). *Data warehousing — The ultimate guide to building corporate business intelligence* (1st ed.). Vieweg & Sohn Verlagsgesellschaft mBH.

Scoggins, J. (1999). A practitioner's view of techniques used in data warehousing for sifting through data to provide information. In *Proceedings of The Eight International Conference on Information and Knowledge Management*, Kansas City, MI.

SEARCH (2006). *TechOasis.* Norcross, GA: Search Technology Inc.

Sebastiani, F. (2002). Machine learning in automated text categorization. *ACM Computing Surveys, 34*(1), 1–47.

Semeraro, G., Basile, P., Degemmis, M., & Lops, P. (2006). Discovering user profiles from papers by using word sense disambiguation. In *Proceedings of the ECML/PKDD Workshop on Web Mining* (pp. 69-79), Berlin, Germany.

Senthil Kumar, A. V. & Wahidabanu, R. S. D. (2006). Directed graph approach for association rule mining. In *Proceedings of the 2nd International Conference ICTS,* Indonesia.

Servaes, J. E. J. (2004). Knowledge is power (revisited): Internet and democracy. In P. Lee (Ed.), *Proceedings of the International Conference on Internet Communication in Intelligent Societies* (pp. 1 – 16). Chinese University of Hong Kong, Hong Kong. Retrieved April 13, 2008, http://www.com.cuhk.edu.hk/conference/2004/)

Shamseldin, A. Y. (1997). Application of a neural network technique to rainfall-runoff modeling. *Journal of Hydrology, 199*, 272-294.

Sharma, A. & Woodward, R. (2001). Political economy Websites: A researcher's guide. *New Political Economy, 6*(1), 119-130.

Shi, Y, Peng, Y., Kou, G., & Chen, Z. (2005). Classifying credit card accounts for business intelligence and decision making: A multiple-criteria quadratic programming approach. *International Journal of Information Technology and Decision Making, 4*, 581-600.

Shi, Y, Peng, Y., Xu, W., & Tang, X. (2002). Data mining via multiple criteria linear programming: Applications in credit card portfolio management. *International Journal of Information Technology and Decision Making, 1*, 131-151.

Shi, Y. (2001). *Multiple criteria and multiple constraint levels linear programming: Concepts, techniques and applications.* NJ: World Scientific.

Shi, Y., & Yu, P.L. (1989). *Goal setting and compromise solutions.* In B. Karpak & S. Zionts (Eds.), Multiple criteria decision making and risk analysis using microcomputers (pp. 165-204). Berlin: Springer-Verlag.

Shi, Y., Wise, W., Luo, M., & Lin, Y. (2001). *Multiple criteria decision making in credit card portfolio management.* In M. Koksalan & S. Zionts (Eds.), Multiple criteria decision making in new millennium (pp. 427-436). Berlin: Springer-Verlag.

Shibata, A., Zelivyanskaya, M., Limoges, J., Carlson, K.A., Gorantla, S., Branecki, C., Bishu, S., Xiong, H., & Gendelman, H.E. (2003). Peripheral nerve induces macrophage neurotrophic activities: Regulation of neuronal process outgrowth, intracellular signaling and synaptic function. *Journal of Neuroimmunology, 142*, 112-129.

Shimabukuro, Y. et al. (1998). Using shade fraction image segmentation to evaluate deforestation in Landsat thematic mapper images of the Amazon region. *International Journal of Remote Sensing 19*(3), 535-541.

Shukla, M. B., Kok, R., Prasher, S. O., Clark, G., & Lacroix, R. (1996). Use of artificial neural networks in transient drainage design. *Transactions of the ASAE, 39*, 119-124

Shyu, M. L., Chen, S.C., Sarinnapakorn, K., and Chang, L. (2006). Principal component-based anomaly detection scheme. In T.S. Lin, S. Ohsuga, J. Liau, & X. Hu (Eds.), *Foundations and Novel Approaches in Data Mining* (pp. 311-329) Springer-Verlag.

Silva, M. P. S. (2006). Mineração de Padrões de Mudança em Imagens de Sensoriamento Remoto [Mining patterns of change in remote sensing images] (Unpublished doctoral thesis). São José dos Campos: National Institute for Space Research (INPE).

Silva, M. P. S., Câmara, G., Souza, R. C. M., Valeriano, D. M. & Escada, M. I. S. (2005). Mining patterns of change in remote sensing image databases. J. Han & B. Wah (Eds.), In *Proceedings of the Fifth IEEE International Conference on Data Mining* (pp. 362-369).

Sinha, I. (2000, March/ April). Cost transparency: The Net's real threat to prices and brands. *Harvard Business Review*, 43-55.

Smart, J. C. (Ed.) (2005). *Higher education: Handbook of theory and research* (Vol. 20). Virginia Tech: Springer.

Smeulders, A.W.M., Worring, M., Santini, S., Gupta, A. & Jain, R. (2000). Content-based image retrieval at the end of the early years. *IEEE Transactions on Pattern Analysis and Machine Intelligence, 22*(1), 1349-1380.

Sobeih, A. (2005). *Supporting natural resource management and local development in a developing connection: Bridging the policy gap between the information society and sustainable development.* A publication of the International Institute for Sustainable Development (IISD), pp. 186-210.

Song, S. (2005). Viewpoint: Bandwidth can bring African universities up to speed. *Science in Africa,* September 2005. Retrieved April 13, 2008, from http://www.scienceinafrica.co.za/2005/september/bandwidth.htm

Sormani, A. (2005). Debt causes problems for SMEs. *European Venture Capital & Capital Equity Journal, 1*, 1.

Speth, J. G. (2004). *Red sky at morning: America and the crisis of the global environment.* Yale University Press.

Spiliopoulou, M. & Pohle, C. (2001). Data mining for measuring and improving the success of Web sites. *Data Mining and Knowledge Discover, 5*(1-2), 85-114.

Spiliopoulou, M., Pohle, C., & Faulstich, L. (1999). Improving the effectiveness of a Web site with Web usage mining. In *Proceedings of WEBKDD99* (pp. 142-162), San Diego, CA.

Srivastava, A. N., & Weigend, A. S. (1994). Computing the probability density in connectionist regression. In M. Marinara & G. Morasso (Eds.), *Proceedings ICANN, 1* (pp. 685-688). Berlin: Springer-Verlag.

Srivastava, J., Cooley, R., Deshpande, M., & Tan, P. (2000). Web usage mining: Discovery and applications of usage patterns from Web data. *SIGKDD Explorations, 1*, 12-23.

Stake, R. E. (1998). Case studies. In N. K. Denzin & Y. S. Lincoln (Eds.), *Strategies of qualitative inquiry* (pp. 86-109). Thousand Oaks, CA: Sage Publications.

Stamatakis, K, Karkaletsis, V., Paliouras, G., Horlock, J., Grover, C., Curran, J. R. & Dingare, S. (2003). Domain-specific Web site identification: The CROSSMARC focused Web crawler. In *Proceedings of the Second International Workshop on Web Document Analysis (WDA 2003)* (pp. 75-78), Edinburgh, UK.

Stefanakis, E., Vazirgiannis, M., & Sellis, T. (1999). Incorporating fuzzy set methodologies in a DBMS repository for the application domain of GIS. *International Journal of Geographic Information Science, 13*, 657-675.

Steinmueller, W. E. (2001). ICTs and the possibilities for leapfrogging by developing countries. *International Labour Review, 140*(2), 193-210.

Stolfo, S.J., Fan, W., Lee, W., Prodromidis, A., & Chan, P.K. (2000). Cost-based modeling and evaluation for data mining with application to fraud and intrusion detection: Results from the JAM project. In *Proceedings of the DARPA Information Survivability Conference*.

Stumme, G., Wille, R. & Wille, U. (1998). *Conceptual knowledge discovery in databases using formal concept analysis methods.* Berlin-Heidelberg, Germany: Springer, Verlag.

Sturges, J. & Hanrahan, K. (2004). Comparing telephone and face-to-face qualitative interviewing: a research note. *Qualitative Research, 4*(1) 107-118.

Sullivan, L. (2004). Wal-Mart's way. *Information Week*. Retrieved March 31, 2005, from http://www.informationweek.com/story/showArticle.jhtml?articleID=47902662&pgno=3

Sullivan, L. (2005). Wal-Mart assesses new uses for RFID. *Information Week*. Retrieved March 31, 2005, from http://www.informationweek.com/showArticle.jhtml?articleID=159906172

Sun, A., Lim, E. P. & Ng, W.K. (2002). Web classification using support vector machine. In *Proceedings of the Fourth ACM CIKM International Workshop on Web Information and Data Management (WIDM'02)*, McLean, Virginia.

Sutcliffe, A. (2001). Heuristic evaluation of Website attractiveness and Web usability. *Lecture notes in computer science* (Vol. 2220, pp. 183-198).

Svoboda, M., LeComte, D., Hayes, M., Heim, R., Gleason, K., Angel, J., Rippey, B., Thinker, R., Palecki, M., Stooksbury, D., Miskus, D., & Stephens, S. (2002). The drought monitor. *Bulletin of the American Meteorological Society, 83*(8), 1181-1190.

Swanson, D. R. & Smalheiser, N. R. (1997). An interactive system for finding complementary literatures: A stimulus to scientific discovery. *Artificial Intelligence, 91*(2), 183-203.

Swanson, D. R. (1986). Fish oil, Raynaud's syndrome, and undiscovered public knowledge. *Perspect Biol Med. 30*(1), 7-18.

Tadesse, T., Brown, J. F., & Hayes, M. J. (2005). A new approach for predicting drought-related vegetation stress: Integrating satellite, climate, and biophysical data over the U.S. central plains. *ISPRS Journal of Photogrammetry and Remote Sensing*, 59(4), 244-253.

Tadesse, T., Wilhite, D. A., Harms, S. K., Hayes, M. J., & Goddard, S. (2004). Drought monitoring using data mining techniques: A case study for Nebraska, USA. *Natural Hazards, 33*(1), 137-159.

Tadesse, T., Wilhite, D. A., Hayes, M. J., Harms, S. K., & Goddard, S. (2005). Discovering associations between climatic and oceanic parameters to monitor drought in Nebraska using data-mining techniques. *Journal of Climate, 18*(10), 1541-1550.

Taffler, R. & Tisshaw, H. (1977). Going, going gone - four factors which predict. *Accountancy*, March, 50-54.

Tan, C. N., & Dihardjo, H. (2001). A study on using artificial neural networks to develop an early warning predictor for credit union financial distress with comparison to the probit model. *Managerial Finance, 27*(4), 56-78.

Tan, Pang-Ning, Steinbach, M., & Kumar, V. (2005). *Introduction to data mining*. Pearson Addison Wesley.

Tan, Z. & Quektuan, C. (2007). Biological brain-inspired genetic complementary learning for stock market and bank failure prediction. *Computational Intelligence, 23*(2), 236-242.

Tango-Lowy, R. & Lewis, L. (2005). Situation management in crisis scenarios based on self-organizing neural mapping technology. In *Proceedings of the IEEE Military Communications Conference* (pp. 1-7), Atlantic City, New Jersey.

Tao, F., Murtagh, F., & Farid, M. (2003). Weighted association rule mining using weighted support and significance framework. In L. Getoor, T. E. Senator, P. Domingos & C. Faloutsos (Eds.), *Proceedings of the Ninth ACM SIGKDD International Conference on Knowledge Discovery and Data Mining* (pp. 661-666). New York: ACM Press.

Tao, V., Liang, S., Croitoru, A., Haider, Z. M., & Wang, C. (2005). GeoSwift: Open geospatial sensing services for sensor web. In S. Nittel & A. Stefanidis (Eds.), *GeoSensor Networks* (pp. 267-274). Boca Raton, FL: CRC Press.

Tapp, A. (2001). *Principles of direct marketing* (2nd ed). Prentice Hall.

Taylor, K., Walker, G., & Abel, D. (1999). A framework for model integration in spatial decision support systems. *International Journal of Geographic Information Science, 13*, 533-555.

Tepeci, M. (1999). Increasing brand loyalty in the hospitality industry. *International Journal of Contemporary Hospitality Management, 11*(5).

The New York Stock Exchange (2007). Retrieved April 13, 2008, from www.nyse.com.

The Stock Exchange of Thailand (2007). Retrieved April 13, 2008, from www.set.or.th/en/index.html.

Thearling, K. (1995). *From data mining to database marketing*. DIG White Paper 95/02. Retrieved April 13, 2008, from http://www.cs.uvm.edu/~xwu/icdm/cfp-03.shtml

Thelwall, M. (2006). Interpreting social science link analysis research: A theoretical framework. *Journal of the American Society for Information Science and Technology, 57*(1), 60-68.

Thuraisingham, B. (1999). *Data mining: Technologies, techniques, tools, and trends*. Boca Raton, FL: CRC Press LLC.

Thurston, J., Poiker, T. K., & Moore, J. P. (2003). *Integrated geospatial technologies: A guide to GPS, GIS, and data logging*. Hoboken, NJ: John Wiley & Sons.

Tigre, P. B. & Dedrick, J. (2004). E-commerce in Brazil: local adaptation of a global technology. *Electronic Markets, 14*(1) 36-40.

Ting, L. (2003). Sustainable development, the place for SDIs, and the potential of e-governance. In I. Williamson, A. Rajabifard & M. F. Feeney (Eds.), *Developing spatial data infrastructures: From concept to reality* (pp. 183-194). London, UK: Taylor & Francis.

Toivonen, H. (1996). Sampling large databases for association rules. In T. M. Vijayaraman, A. P. Buchmann, C. Mohan & N. L. Sarda (Eds.), *Proceedings of the 22nd International Conference on Very Large Data Bases* (pp. 134-145). San Francisco: Morgan Kaufmann Publishers.

TREC (Text Retrieval Conference) (2004). Retrieved April 13, 2008, from http://trec.nist.gov/.

Trippi, R. R., & DeSieno, D. (1992). Trading equity index futures with a neural-network. *Journal of Portfolio Management, 19*, 27-33.

Tsaih, R., Hsu, Y., & Lai, C. C. (1998). Forecasting S & P 500 stock index futures with a hybrid AI system. *Decision Support Systems*, *23*(2), 161-174.

Tseng, C.-C. (2004). Portfolio management using hybrid recommendation system. In *Proceedings of the 2004 IEEE International Conference on e-Technology, e-Commerce, and e-Services* (pp. 202-206). Los Alamitos, CA: IEEE Computer Society Publications.

Tug, E., Sakiroglu, M., & Arslan, A. (2006). Automatic discovery of the sequential accesses from Web log data files via a genetic algorithm. *Knowledge-Based Systems, 19*(3), 180-186.

Turner, M. G. (1989). Landscape ecology: The effect of pattern on process. *Annual Review of Ecology and Systematics, 20*, 171-197.

UNCTAD (2004). UNCTAD XI multi-stakeholder partnerships, information and communication technologies for development (ICTfD). In *Proceedings of the United Nations Conference on Trade and Development*. Retrieved April 13, 2008, http://www.unctad.org/en/docs//tdl380add1_en.pdf

UNDP (2001). *United Nations Development Program: Making new technologies work for human development.* Oxford: Oxford University Press.

Unwin, T. (2006). *Facing the challenges, dgCommunities: Open educational resources.* Retrieved April 13, 2008, from http://topics.developmentgateway.org/open-educaion

USAID (2003). *USAID Africa success stories.* Retrieved April 13, 2008, from http://africastories.usaid.gov:80/print_story.cfm?storyID=23

Utimaco (2005). *Data encryption: The foundation of enterprise security.* Foxboro, MA: Utimaco Safeware, Inc.

Van Der Zee, J. T. M. & De Jong, B. (1999). Alignment is not enough: Integrating business and information technology management with the balanced score card. *Journal of Management Information Systems, 16*(2), 137-158.

Vapnik, V.N. (2000). *The nature of statistical learning theory* (2nd ed.). New York: Springer.

Vckovski, A. & Bucher, F. (1996). Virtual data sets - Smart data for environmental applications. In *Proceedings of the Third International Conference/Workshop on Integrating GIS and Environmental Modeling*, Santa Fe, NM.

Viaene, S., Derrig, R. A., Baesens, B., & Dedene, G. (2003). A comparison of state-of-the-art classification techniques for expert automobile insurance claim fraud detection. *Journal of Risk and Insurance, 69*(3), 373-421.

Viator, J. A. & Pestorius, F. M. (2001). Investigating trends in acoustics research from 1970-1999. *Journal of the Acoustical Society of America, 109*(5), 1779-1783 Part 1.

Ville, Barry de (2001). *Microsoft data mining: Integrated business intelligence for e-commerce and knowledge management.*

Ville, Barry de (2007). *Microsoft data mining: Integrated business intelligence for e-commerce and knowledge management.* Digital Press.

Vitt, E., Luckevich, M., & Misner, S. (2002). *Business intelligence.* Microsoft Press.

Vlachos, E. (1994). GIS, DSS and the future. In *Proceedings of the 8th Annual Symposium on Geographic Information Systems in Forestry, Environmental and Natural Resources Management*, Vancouver, Canada.

Walters, D. & Lancaster, G. (1999). Value and information – Concepts and issues for management. *Management Decision, 37*(8), 643-656.

Walters, D. & Lancaster, G. (1999). Using the Internet as a channel for commerce. *Management Decision, 37*(10), 800-816.

Wang, H. & Weigend, A. S. (2004). Data mining for financial decision making. *Decision Support Systems, 37*(2004), 457-460.

Wang, J. (Ed.) (2003). *Data mining opportunities and challenges.* IRM Press.

Wang, J. T. L., Zaki, M. J., Toivonen, H. T. T., & Shasha, D. E. (2005). *Data mining in bioinformatics*. London, UK: Springer-Verlag.

Wang, J., & Wang, Z. (1997). Using neural network to determine Sugeno measures by statistics. *Neural Networks, 10*, 183-195.

Wang, J., Han, J., & Pei, J. (2003). CLOSET+: Searching for the best strategies for mining frequent closed itemsets. In L. Getoor, T. E. Senator, P. Domingos & C. Faloutsos (Eds.), *Proceedings of the Ninth ACM SIGKDD International Conference on Knowledge Discovery and Data Mining* (pp. 236-245). New York: ACM Press.

Wang, L. & Fu, X. (2005). *Data mining with computational intelligence (advanced information and knowledge processing)* (1st ed.). Springer.

Wang, L., Khan, L. & Breen, C. (2002). Object boundary detection for ontology-based image classification. In *Proceedings of the Third ACM International Workshop on Multimedia Data Mining* (pp. 51-61).

Wang, W. & Yang, J. (2005). *Mining sequential patterns from large data sets*. Secaucus, NJ: Springer-Verlag New York, Inc.

Wang, W., Yang, J., & Yu, P. (2000). Efficient mining of weighted association rules (WAR). In *Proceedings of the Sixth ACM SIGKDD International Conference on Knowledge Discovery and Data Mining* (pp. 270-274). New York: ACM Press.

Wang, X., Abraham, A., & Smith, K. (2005). Intelligent Web traffic mining and analysis. *Journal of Network and Computer Applications, 28*, 147-165.

Wang, Y. & Hu, J. (2002). A machine learning approach for table detection on the Web. In *Proceedings of the 11th International World Web Conference*, Honolulu, Hawaii.

Wasserman, S. & Faust, K. (1994). *Social network analysis: Methods and applications*. Cambridge University Press.

Watson, H., Goodhue, D., & Wixon, B. (2002). The benefits of data warehousing: Why some organizations realize exceptional payoffs. *Information & Management, 39*(6), 491-502.

Weingessel, A., Dimitriadou, E., & Hornik, K. (2003, March 20-22). An ensemble method for clustering. In *Proceedings of the 3rd International Workshop on Distributed Statistical Computing*, Vienna, Austria.

Weiss, S. M. & Indurkhya, N. (1997). *Predictive data mining: A practical guide*. Morgan Kaufmann.

Westerman, P. (2001). *Data warehousing: Using the Wal-Mart model*. San Francisco: Academic Press.

White, A. B., Kumar, P., & Tcheng, D. (2005). A data mining approach for understanding topographic control on climate-induced inter-annual vegetation variability over the United States. *Remote Sensing of Environment, 98*, 1-20.

Whiting, R. (2004). Vertical thinking. *Information Week*. Retrieved March 31, 2005, from http://www.informationweek.com/showArticle.jhtml?articleID=18201987

Wilhelmi, O. V. & Wilhite, D. A. (2002). Assessing vulnerability to agricultural drought: a Nebraska case study. *Natural Hazards, 25*(1), 37-58.

Wilhite, D. A. (2000): Drought as a natural hazard: concepts and definitions. In D. A. Wilhite (Ed.), *Drought: A global assessment* (Vol. 1, pp. 3-18). London: Routledge Publishers.

Williams, D. (2004). The strategic implications of Wal-Mart's RFID mandate. *Directions Magazine*. Retrieved October 23, 2004, from http://www.directionsmag.com/article.php?article_id=629

Williams, R. (1997). Universal solutions or local contingencies? Tensions and contradictions in the mutual shaping of technology and work organization. In I. McLoughlin & M. Harris (Eds), *Innovation, organizational change and technology*. London, UK: International Thomson Business Press.

WIPO, World Intellectual Property Organization (2002). *Interregional forum on small and medium-sized enterprises (SMEs) and intellectual property* (Tech. Rep. No. 02/01). Moscow: Document of WIPO.

Wise, S. M. & Haining, R. P. (1991). The role of spatial analysis in geographical information systems. *Westrade Fairs, 3*, 1-8.

Witten, I. & Frank, E. (1999). *Data mining, Practical machine learning tools and techniques with Java implementations*. Morgan Kaufman.

Witten, I. & Frank, E. (2005). *Data mining, practical machine learning tools and techniques* (2nd ed.). Morgan Kaufman.

Wolfgang, G. & Lars, S. (2000). Mining Web navigation path fragments. In *Proceedings of the Workshop on Web Mining for E-Commerce (KDD2000)* (pp. 105-110). Boston, MA.

WSSD (2002). *Press release for fifth partnership plenary, world summit on sustainable development*. Johannesburg, South Africa.

Wu, H., Gordon, M., DeMaagd, K., & Fan, W. (2006). Mining Web navigations for intelligence. *Decision Support Systems, 41*, 574-591.

Yan, N., Wang, Z., Shi, Y., & Chen, Z. (2005). *Classification by linear programming with signed fuzzy measures*. Working Paper, University of Nebraska at Omaha, USA.

Yang, C.-C., Prasher, S. O., & Lacroix, R. (1996). Application of artificial neural networks to land drainage engineering. *Trans. ASAE, 39*, 525-533.

Yannas, P. & Lappas, G. (2005). Web campaign in the 2002 Greek municipal elections. *Journal of Political Marketing, 4*(1), 33-50.

Yao, Y. Y., Hamilton, H. J. & Wang, X. (2002). PagePrompter: An intelligent agent for Web navigation created using data mining techniques. *Lecture notes in computer science* (Vol. 2475, pp. 506-513).

Ye, Z., Liu, X., Yao, Y., Wang, J., Zhou, X., Lu, P., & Yao, J. (2002). An intelligent system for personal and family financial service. In L. Wang, J. C. Rajapakse, K. Fukushima, S.-Y. Lee & X. Yao (Eds.), *Proceedings of the 9th International Conference on Neural Information Processing* (Vol. 5, pp. 2325-2327). Los Alamitos, CA: IEEE Computer Society Publications.

Yoon, J. P. & Kerschberg, L. (1993). A framework for knowledge discovery and evolution in databases. *IEEE Trans. On Knowledge And Data Engineering, 5*(6), 973-979.

Yu, D. L., & Gomm, J. B. (2002). Enhanced neural network modelling for a real multi-variable chemical process. *Neural Computing and Applications*, 10(4), 289-299.

Yu, H., Han, J., & Chang, K. C. (2002). PEBL: Positive example based learning for Web page classification using SVM. In *Proceedings Of The International Conference On Knowledge Discovery In Databases (KDD02)* (pp. 239-248), New York.

Yu, L., Wang, S., & Lai, K. K. (2005). Mining stock market tendency using GA-based support vector machines. In X. Deng & Y. Ye (Eds.), *Proceedings of the First International Workshop on Internet and Network Economics* (pp. 336-345). Berlin Heidelberg, Germany: Springer-Verlag.

Yu, P.L. (1985). *Multiple criteria decision making: Concepts, techniques and extensions*. New York: Plenum Press.

Zahra, S., Sisodia, R., & Matherne, B. (1999, April). Exploiting the dynamic links between competitive and technology strategies. *European Management Journal, 17*(2), 188-201.

Zaiane, O. R. (2001). Web usage mining for a better Web-based learning environment. In *Proceedings of Conference on Advanced Technology for Education* (pp. 60-64). Banff, Alberta, Canada.

Zaïane, O. R., Han, J., Li, Z.-N., & Hou, J. (1998). Mining Multimedia Data. In *Proceedings of the CASCON'98: Meeting of Minds* (pp. 83-96), Toronto, Canada.

Zaki, M. J. & Hsiao, C.-J. (2002) CHARM: An efficient algorithm for closed itemset mining. In R. L. Grossman, J. Han, V. Kumar, H. Mannila & R. Motwani (Eds.), *Proceedings of the Second SIAM International Conference on Data Mining* (Part IX No. 1). Philadelphia, PA: SIAM.

Zaki, M. J., Parthasarathy, S., Li, W., & Ogihara, W. (1997). Evaluation of sampling for data mining of association rules. In *Proceedings of the 7th International. Workshop Research Issues in Data Engineering.*

Zaki, M. J., Parthasarathy, S., Ogihara, M., & Li, W. (1997). New algorithms for fast discovery of association rules. In D. Heckerman, H. Mannila, & D. Pregibon (Eds.), *Proceedings of the Third International Conference on Knowledge Discovery and Data Mining* (pp. 283-286). Menlo Park, CA: AAAI Press.

Zaki, M. J.,Parthasarathy, S., & Li, W. (1997). A localized algorithm parallel association mining. In *Proceedings of the 9th ACM Symposium Parallel Algorithms and Architectures.*

Zavgren, C. (1985). Assessing the vulnerability to failure of American industrial firms: A logistics analysis. *Journal of Accounting Research, 22*, 59-82.

Zeiler, M. (1999). *Modeling our world: The ESRI guide to Geodatabase design.* Redlands, CA: ESRI Press.

Zenobi, G., & Cunningham, P. (2002). An approach to aggregating ensembles of lazy learners that supports explanation. *Lecture Notes in Computer Science, 2416*, 436-447.

Zhang, D. & Zhou, L. (2004). Discovery golden nuggets: Data mining in financial application. *IEEE Transactions on Systems, Man, and Cybernetics – Part C: Applications and Reviews, 34*(4), 513 –522.

Zhang, J., Shi, Y., & Zhang, P. (2005). *Several multi-criteria programming methods for classification.* Working Paper, Chinese Academy of Sciences Research Center on Data Technology & Knowledge Economy and Graduate University of Chinese Academy of Sciences, China.

Zhang, Y., Yu, J. X., & Hou, J. (2005). *Web communities: Analysis and construction.* Berlin: Springer.

Zhao, Y. & Karypis, G. (2004). Empirical and theoretical comparisons of selected criterion functions for document clustering. *Machine Learning, 55*(3), 311-331.

Zheng, J., Thylin, M., Ghorpade, A., Xiong, H., Persidsky, Y., Cotter, R., Niemann, D., Che, M., Zeng, Y., Gelbard, H. et al. (1999). Intracellular CXCR4 signaling, neuronal apoptosis and neuropathogenic mechanisms of HIV-1-associated dementia. *Journal of Neuroimmunology, 98*, 185-200.

Zheng, J., Zhuang, W., Yan, N., Kou, G., Erichsen, D., McNally, C., Peng, H., Cheloha, A., Shi, C., & Shi, Y. (2004). Classification of HIV-1-mediated neuronal dendritic and synaptic damage using multiple criteria linear programming. *Neuroinformatics, 2*, 303-326.

Zhou, C., Li, Z., Meng, Y. & Meng, Q. (2004). A data mining algorithm based on rough set theory. In *Proceedings of International Conference on Information Acquisition 2004* (pp. 413-416).

Zhou, Z., Jiang, K., & Li, M. (2005). Multi-instance learning based Web mining. *Applied Intelligence, 22*(2), 135-147.

Zhu, D. H. & Porter, A. L. (2002). Automated extraction and visualization of information for technological intelligence and forecasting. *Technological Forecasting and Social Change, 69*(5), 495-506.

Zimmermann, H.-J. (1978). Fuzzy programming and linear programming with several objective functions. *Fuzzy Sets and Systems, 1*, 45-55.

Zmijewski, M. E. (1984). Methodological issues related to the estimation of financial distress prediction models. *Journal of Accounting Research*, (Supplement), 59-82.

Zukerman, I. & Albrecht, D. (2001). Predictive statistical models for user modeling. *User Modelling and User Adapted Interaction, 11*, 5-18. tecedents of executive information system success: A path analytic approach. *Decision Support System, 22*(1), 31-43.

About the Contributors

Hakikur Rahman, PhD is the executive director and CEO of Sustainable Development Networking Foundation (SDNF), the transformed entity of the Sustainable Development Networking Programme (SDNP) in Bangladesh where he was working as the national project coordinator since December 1999. SDNP is a global initiative of UNDP and it completed its activity in Bangladesh on December 31, 2006. He is also acting as the secretary of South Asia Foundation Bangladesh Chapter. Before joining SDNP he worked as the director, Computer Division, Bangladesh Open University. He has written several books and many articles/papers on computer education for the informal sector and distance education. He is the founder-chairperson of Internet Society Bangladesh Chapter; editor, the Monthly Computer Bichitra; founder-principal and member secretary, ICMS College; head examiner (Computer), Bangladesh Technical Education Board; executive director, BAERIN (Bangladesh Advanced Education Research and Information Network) Foundation; and involved in establishment of a ICT based distance education university in Bangladesh. Graduating from the Bangladesh University of Engineering and Technology in 1981, he has done his Master's of engineering from the American University of Beirut in 1986 and completed his PhD in computer engineering from the Ansted University, BVI, UK in 2001.

* * *

Gilberto Câmara is general director of Brazil's National Institute for Space Research (INPE) for the period 2006 to 2010. INPE works in space science, space engineering, Earth observation and weather and climate studies. Previously, he was head of INPE's Image Processing Division from 1991 to 1996 and director for Earth observation from 2001 to 2005. His research interests include geographical information science and engineering, spatial databases, spatial analysis and environmental modeling. He has published more than 150 full papers on refereed journals and scientific conferences. He has also been the leader in the development of GIS technology in Brazil.

Frans Coenen has a general background in AI has been working in the field of data mining and knowledge discovery in data (KDD) for some ten years. He is a member of the IFIP WG12.2 — Machine Learning and Data Mining group and the British Computer Society's specialist group in AI. He has some 140 refereed publications on KDD and AI related research. Frans Coenen is currently a senior lecturer within the Department of Computer Science at the University of Liverpool.

Maria Isabel Sobral Escada is graduated in ecology and has her doctorate in remote sensing from National Institute for Space Research—INPE. She works in the Image Processing Division (DPI) at INPE and is vice-coordinator of GEOMA—an Amazonia modeling network composed by several Institutes of Brazilian Ministry of Science and Technology—MCT. Her research interests include Amazonia land use and land cover change, pattern analysis, models and their connection with social, economic, territorial planning, and public policy issues.

Michael J. Hayes is the director for the National Drought Mitigation Center and an associate professor in the School of Natural Resources at the University of Nebraska-Lincoln. His interests include the economic, environmental, and social impacts of drought; developing drought monitoring and impact assessment methodologies; and assisting states and Native American tribes with the development of drought plans. Dr. Hayes received a Bachelor's degree in meteorology from the University of Wisconsin-Madison, and his Master's and Doctoral degrees in atmospheric sciences from the University of Missouri-Columbia.

Ronald Neil Kostoff received a PhD in aerospace and mechanical sciences from Princeton University in 1967. He has worked for Bell Laboratories, Department of Energy, and Office of Naval Research (ONR). He has authored over 100 technical papers, served as guest editor of three journal special issues, obtained two text mining system patents, and presently manages a text mining pilot program at ONR.

Raymond George Koytcheff is a recent graduate of Columbia University, where he majored in biophysics and economics-mathematics. At the Office of Naval Research, he performed text mining of nanotechnology research. At the Naval Research Laboratory (NRL), he worked on remote sensing and tribology research.

Ali Serhan Koyuncugil, MSc, PhD is working as a statistician for Capital Markets Board of Turkey. He had his licence, MSc, and PhD degrees in statistics from Ankara University Department of Statistics. His current research interests are design and development of fraud detection, risk management, early warning, surveillance, information, decision-support and classification systems, design and development of datawarehouses and statistical databases, development of indicators, models and algorithms, conducting analysis on capital markets, finance, health, SME's, large scale statistical researchs (e.g., census), population and development, socioeconomic and demographic affairs based on data mining, statistics, quantitative decision making, operational research, optimization, mathematical programming, fuzzy set, technical demography theory and applications. He took part in a lot of international and national projects (UN, IBRD, EU, etc.). He took part in a lot of international and national conferences as an organizer, reviewer and advisor. He is member of the IASC and IASS sections of ISI, Turkish Statistical Association, Turkish Informatics Society and was former vice head of Turkish Statisticians Association.

A.V.Senthil Kumar is presently working as a senior lecturer in the Department of MCA, CMS College of Science and Commerce, Coimbatore, Tamilnadu, India. He has more than 11 years of teaching and 5 years of industrial experience. His research area includes data mining and image processing.

Georgios Lappas, PhD, is Lecturer of Informatics in the Department of Public Relations and Communication in the Technological Educational Institution (TEI) of Western Macedonia, Kastoria, Greece. He holds a BSc in physics from the University of Crete-Greece (1990), MSc in applied artificial intelligence from the University of Aberdeen-UK (1993) and he received his PhD from University of Hertfordshire-UK for his work on "Combinatorial Optimization Algorithms Applied to Pattern Classification." His research interests include: pattern classification, machine learning, neural networks, web mining, multimedia mining, the use of the Internet in politics (e-politics), in public administration (e-Government), and in campaigning (e-campaigns).

Clifford GY Lau is a research staff member with the Institute for Defense Analyses' Information Technology and Systems Division. Prior to joining IDA, he worked at the Office of Naval Research (ONR). He received a PhD in electrical engineering and computer science in 1978 from the University of California at Santa Barbara. He has published over 40 papers and served as guest editor for the IEEE Proceedings, and is a fellow of the IEEE.

Diana Luck, PhD, lectures in general and specialist marketing modules as well as in project management at the London Metropolitan University. Her interest in the interdisciplinary aspects of management research stems from her past experience in a variety of business environments. She considers business processes to be part of a gestalt rather than a set of disjointed disciplines. Her research interests revolve around CRM and corporate social responsibility. She would consider her main contribution to her field of study to be the broadening of marketing into the social arena and a focus upon accountability.

Inya Nlenanya is a program coordinator with the Iowa Resource for International Service, a nonprofit organization based in Ames, Iowa whose mission is to promote international education, development, and peace through rural initiatives. Mr. Nlenanya obtained his bachelors degree in electronic engineering from the University of Nigeria, Nsukka. He also has a Master's degree in agricultural engineering from Iowa State University. He currently resides in Ames, Iowa.

Nermin Ozgulbas, MSc, PhD is associate professor of finance at Baskent University in Turkey. She taught financial management, financial analysis and cost accounting at the Department of Health Care Management in Baskent University. She also taught financial management and cost analysis at distance education program of Administration of Health Care Organizations in Anadolu University. Her research and publication activities include finance, accounting, cost accounting and cost effectiveness in health care organizations, capital markets and SMEs. She has publications presentations and projects in many subject areas including the topics mentioned earlier. Some of the journals published her articles are: *The International Journal of Health Planning and Management, The Business Review Cambridge, Journal of Economy, Business and Finance, Journal of Productivity, The Health Care Manager, Journal of Accounting and Finance, World of the Accounting and Finance Journal, Journal of Health and Society.*

Abdul Matin Patwari is the vice chancellor of the University of Asia Pacific, Dhaka, Bangladesh. Obtaining his PhD in electrical engineering from University of Sheffield, UK in 1967 he has held the position of head of the Department of Electrical and Electronics Engineering (EEE) and dean of the faculty, Bangladesh University of Engineering & Technology (BUET). He was the vice chancellor of BUET, director general of Islamic Institute of Technology and also served many national committees as the Chairman. He served several universities as visiting professor, including Purdue University, Indiana; California State University, Pomona; The University of New Castle, Upon Tyne. Dr. Patwari has over 75 publications in the field of engineering science and visited almost all important countries of the world as the delegate head or team member.

Maira Petrini has been a professor at the *Fundação Getulio Vargas-EAESP*, in Brazil, since 2000. Her research interests include business intelligence and corporate strategic planning. Professor Petrini has also worked as an IT consultant since 2001. Her work has been published in major Brazilian journals.

Marlei Pozzebon is an associate professor at *HEC Montréal*, in Canada. She has been affiliated with this institution since 2002. Her research interests include the political and cultural aspects of information technology implementation, the use of structuration theory and critical discourse analysis in the information systems field, business intelligence and the role of information technology in local development, and corporate social responsibility. Before joining *HEC Montréal*, Professor Pozzebon had worked at three Brazilian universities. She has also been an IT consultant since 1995. Prior to this, for at least 10 years, she was a systems analyst. Her research has been published in, among others, *Journal of Management Studies, Organization Studies,* and *Journal of Strategic Information Systems and* the *Journal of Information Technology.*

Marcelino Pereira dos Santos Silva is director of the Post Graduate Department and coordinator of the Master Program in Computer Science of the Rio Grande do Norte State University (UERN). As professor of computer science, he has been on UERN since 1996. Born January 16, 1970 in São Paulo, Brazil, he earned his Bachelor's degree in computer science from the Federal University of Campina Grande in 1992 and his PhD from the National Institute for Space Research in 2006. He is a member of the Brazilian Computer Society, and his research interests include data mining, geographical information science and artificial intelligence.

Tsegaye Tadesse received the BS degree in physics from Addis Ababa University, Ethiopia (1982), his MSc from space studies from International Space University, France (1998), and his PhD in agrometeorology from the University of Nebraska-Lincoln, U.S.A (2002). Dr. Tadesse is currently a climatologist/assistant geoscientist with the National Drought Mitigation Center at the University of Nebraska-Lincoln. His current research is on the development of new drought monitoring and prediction tools that utilize remote sensing, GIS and data mining techniques. His other research includes data mining application in identifying drought characteristics and their association with satellite and oceanic indices.

R.S.D. Wahidabanu is presently head, Department of CSE, Government College of Engineering, Salem, Tamilnadu, India. She has 25 years of teaching experience. Her research areas include pattern recognition, artificial intelligence, and data mining.

Yanbo J. Wang is currently a fourth year doctoral student in the Department of Computer Science at the University of Liverpool, UK. He was awarded a Bachelor of administrative studies with honours, in information technology, by York University, Canada. Yanbo's main current research is in data mining and text mining, especially approaches for classification association rule mining, weighted association rule mining, and their applications.

Brian D. Wardlow received his BS degree in geography and geology from Northwest Missouri State University (1994), the MA degree in geography from Kansas State University (1996), and the PhD degree in geography from the University of Kansas (2005). He is currently an assistant professor with the National Drought Mitigation Center at the University of Nebraska-Lincoln. His current research is on the development of new drought monitoring and prediction tools that utilize remote sensing, GIS and data mining techniques. Dr. Wardlow's other research includes the application of remote sensing for land cover characterization/change detection, environmental monitoring, and natural resource management.

Xinwei Zheng is a fourth year PhD student in finance at Durham Business School, UK. He got his MSc accounting & finance at University of Edinburgh, UK, and Bachelor of economics at Dongbei University of Finance and Economics, China. His major research interests are market microstructure, asset pricing and investment, macroeconomics and stock market, Chinese economics, and data mining. He is also interested in the programming of PcGive and Eviews econometrics software, and high frequency data analysis of Visual FoxPro.

Index

A

additive models 150, 151
allocating pattern (ALP) 112, 113, 118, 121–131
Amazonia 55–60–69, 72, 73, 293, 294, 300
apriori algorithm 114, 121, 140
artificial neural network (ANN) 81, 112, 224
association rule (AR) 43, 45, 46, 52, 53, 81, 82, 84, 86, 111, 112, 126, 131–143, 157, 280, 283, 287, 290–298, 303–306, 311, 314, 317, 320, 322, 324
association rule mining (ARM) 53, 58, 113, 114, 115, 132, 133, 134, 298, 304, 317, 320
ata transformation 7

B

basel-II 227
Bayesian classifiers 84
bibliometrics 187, 202, 203, 219, 220, 307, 308, 313
bioinformatics 41, 111, 135, 171, 278, 295, 312, 322
biophysical 214, 282, 283, 284, 287, 288, 289, 290, 319
brain-derived neurotrophic factor (BDNF) 12
business intelligence (BI) 24, 158, 175, 177, 186, 188, 198, 241, 242, 257, 259, 260, 299, 301–321

C

causation 30
CHi-square Automatic Interaction Detector (CHAID) 225, 229, 231, 232, 240, 314
close coupling 270
closed directed graph approach 43
cluster affinity search technique (CAST) 145
cluster cleaning 145
collaborative filtering 80, 84, 85, 92, 171, 312
competitive intelligence 246, 248, 249, 256
computational intelligence iii, vi, 26
computer terminal network (CTN) 190
content-based image retrieval (CBIR) 59
customer relationship management (CRM) 32, 96–109, 137, 185, 187, 294, 297, 306, 311, 313, 316

D

data archeology 138
data cleaning 7
data pattern processing 138
decision tree 12, 14, 68, 72, 83, 140, 141, 142, 225, 229, 231, 232, 233
deep packet inspection (DPI) 177
domain knowledge 33
dynamic hashing algorithm (DHA) 49, 50, 52

E

early warning system 221–229, 234–239, 306
Earth observation (EO) systems 63, 64, 71, 73
electronic data interchange (EDI) 174, 190, 195
environmental informatics 269
extrapolation 33

F

false negatives 33
false positives 33
FLP classification 12
FLP method 6
fuzzy clustering analysis 203
fuzzy system 81

G

genetic algorithms 81, 84, 87, 158, 299
geographical information system (GIS) 61, 64, 73, 173, 177, 262–279, 283, 288, 290–303, 314, 319, 320, 321, 323
geographic data mining 274
geospatial component 268, 269, 270, 274
geovisualization 270
global positioning systems (GPS) 264
graphical user interface (GUI) 193
grid computing 263, 267

H

HIV-1 associated dementia (HAD) 1, 2, 22
Human Drug Metabolism Database (hDMdb) 179

I

image domain 65
image mining 55–75
information discovery 78, 138, 223
intelligent agents 52, 79, 84, 179
interpolation 33

J

Java Foundation Classes (JFC) 273

K

knowledge center 143, 144, 145, 154
knowledge extraction 138, 266
knowledge society 137, 163, 166, 170– 177, 183, 184, 187, 222, 223, 237, 307

L

land use change 59, 65, 66, 69, 70–73, 296
limnological 163, 178
LINDO Systems Inc 7
loose coupling 268, 270

M

machine learning 2, 17, 22, 53, 74–95, 111, 112, 134, 158, 159, 163, 171, 179, 185, 188, 222, 265, 266, 272–274, 282, 293, 299, 315–323
mazonia forest 57
multiple criteria linear programming (MCLP) 4, 12
multiple criteria programming (MCP) 1, 2, 3
multiple criteria quadratic programming (MCQP) 4

N

nanoscience 199, 200, 203, 220, 307
nanotechnology 199–212, 216–220, 307, 308
National Ecological Observatory Network (NEON) 263
National Institute for Space Research 55, 56, 60, 74, 75, 305, 318
node addition 145
node removal 145
non-linearity 30

O

online analytical processing (OLAP) 194
online transaction processing (OLTP) 194
open and distance learning xiv, 164, 172

P

Pareto principle 100
Perpetual Inventory (PI) system 195
point-of-sales (POS) data 191

Q

Query Statistics 193

R

radio frequency identification (RFID) 189, 196
recommendation systems 80, 85
Remote sensing image mining 57
return on investment (ROI) analysis 192
rough set theory (RS) 139, 159, 324

S

satellite data 56, 284, 287, 288, 290, 296
self-organizing map 81
self-organizing maps (SOM) 82
Semantic Web 79, 88, 91, 304
simple quadratic programming (SQP) 20
spatial analysis 61, 186, 268, 270, 271, 276, 279, 293, 304, 323
spatial data 57, 58, 74, 171, 177, 182, 184, 187, 266, 267, 270,–279, 290, 292, 301–320
spatial data mining 57
spatial decision support system (SDSS) 271
spatial patterns 57, 59, 60–72
spatial resolution 57
spectral resolution 57
Standardized Seasonal Greenness (SSG) 285, 286, 287
structural classifier 65, 68, 70, 72
support vector machine (SVM) 3, 20, 84, 94, 319
sustainable development 73, 172, 178, 184, 188, 262, 269, 271, 275, 279, 290, 293, 318, 323

T

Teradata Corporation 191
tight coupling 268

V

visual data mining 266, 277, 306

W

Wal-Mart 189
Web content mining 77, 78, 79, 81, 83
Web mining 76–79, 81–95, 297, 299, 300, 305, 307, 314, 317, 324
Web structure mining 79, 80, 81, 84
Web usage mining 79–95, 299, 319, 323
weighted association rule (WAR) 110, 134, 135, 311, 322
weighted association rule mining (WARM) 113, 115